Food biotechnology

Food biotechnology

S. C. Bhatia

BE (Chemical), MBA

WOODHEAD PUBLISHING INDIA PVT LTD

New Delhi

Published by Woodhead Publishing India Pvt. Ltd.
Woodhead Publishing India Pvt. Ltd.,
303, Vardaan House, 7/28, Ansari Road,
Daryaganj, New Delhi - 110002, India
www.woodheadpublishingindia.com

First published 2016, Woodhead Publishing India Pvt. Ltd.
© Woodhead Publishing India Pvt. Ltd., 2016

Woodhead Publishing India Pvt. Ltd. ISBN: 978-93-85059-18-6
Woodhead Publishing India Pvt. Ltd. e-ISBN: 978-93-85059-67-4

Typeset by Mind Box Solutions, New Delhi
Printed and bound by Replika Press Pvt. Ltd.

Contents

Food biotechnology is not new. For thousands of years people have been discovering that fruit juices ferment into wine, that milk can be used to develop products such as cheese or yogurt, or that beer can be made through the fermentation of malt and hops. These principles were used to breed hybrid corn, wheat, and many other crops in which certain traits could be selected in order to increase plant yield. Such breeding methods largely accounted for the phenomenal gains in crop productivity during the 20th century and led to modern farming practices. Today, in the arena of food, the primary goals of food biotechnology are to provide a more abundant, less expensive, and a more nutritious food supply in order to address the needs of our growing global population. Today, food biotechnology utilizes the knowledge of plant science and genetics to further this tradition. Through the use of modern biotechnology, scientists can move genes for valuable traits from one plant to another. This process results in tangible environmental and economic benefits that are passed on to the farmer and the consumer.

This book on Food Biotechnology is divided into seven sections and contains 24 chapters and a case study.

Section I discusses General Considerations. Chapter 1 is devoted to food biotechnology: A review. The chapter highlights benefits and risks of food biotechnology. Chapter 2 deals with micro-organisms associated with food. The chapter discusses the primary sources of micro-organism and factors influencing the microbial activity. Chapter 3 concentrates on fermentation biotechnology and various types of fermentation processes along with factors influencing the fermentation. Chapter 4 focuses on genetically modified foods. Genetically modified foods have the potential to solve many of the world's hunger and malnutrition problems, and to help protect and preserve the environment by increasing yield and reducing reliance upon chemical pesticides and herbicides.

Section II discusses Protein and Enzymes. Chapter 5 concentrates on cross-linking of proteins. Cross-links can be introduced to a food protein matrix by chemical, physical and enzymatic means. Chapter 6 deals with enzymes as biocatalysts. Biocatalysis may be broadly defined as the use of enzymes or whole cells as biocatalysts for industrial synthetic chemistry. They have been used for hundreds of years in the production of alcohol via fermentation, and cheese via enzymatic breakdown of milk proteins. Chapter 7 is devoted to enzymes in food industry and various applications of enzymes in food industry are discussed.

Section III discusses Milk, Dairy and Bakery Products. Chapter 8 concentrates on dairy products. Dairy products include—milk and any of the foods made from milk, including butter, cheese, ice cream, yogurt, and condensed and dried milk. Nutritional and genetic interventions to alter the milk composition for specific health and/or processing opportunities are gaining importance in dairy biotechnology. Keeping this in mind Chapter 9 focuses on designer milk. Chapter 10 deals with genetically modified cheese. Chapter 11 discusses bakery and cereal products. Baking is important heat conversion operation in the manufacture of breads, biscuits and cakes.

Section IV discusses Fruit and Vegetables. Chapter 12 deals with fruit and vegetable biotechnology. Biotechnology of fruit and vegetable production are an aid to conventional breeding and its ability to transfer genes between different organisms. Chapter 13 is devoted to genetically modified fruits. Chapter 14 concentrates on genetically modified vegetables.

Section V discusses Meat, Fish and Poultry Products. Chapter 15 focuses on biotechnology of fermented meat. The chapter discusses biotechnology of fermented meat along with raw materials preparation and fermentation sausages. Chapter 16 deals with genetically engineered fish. Chapter 17 is devoted to poultry industry. Poultry products are nutritious and add variety to the human diet. Most of the products are from chickens, but ducks and turkeys also are important sources. Chapter 18 concentrates on impact of biotechnology on poultry. The impact of biotechnology in poultry nutrition is of significant importance. Biotechnology plays a vital role in the poultry feed industry. Nutritionists are continually putting their efforts into producing better and more economical feed.

Section VI discusses Beverages. Chapter 19 deals with beverages and various types of carbonated, noncarbonated and alcoholic beverages. Chapter 20 focuses on biotechnology of brewer's yeast. Chapter 21 concentrates on genetically modified organisms in wine industry.

Section VII discusses Special Topics. Chapter 22 deals with carbon footprint of food. The carbon footprint on food is an estimate of all the emissions caused by the production (e.g. farming), manufacture and delivery to the consumer and the disposal of packaging. Chapter 23 focuses on nanotechnology in the food industry. Nanotechnology has the potential to revolutionize the agricultural and food industry with new tools for the molecular treatment of diseases, rapid disease detection, enhancing the ability of plants to absorb nutrients, etc. Chapter 24 devoted to bioethics and biotechnology. The ethical assessment of new technologies, including biotechnology, requires a different approach to ethics. Changes are necessary because new technology can have a more profound impact on the world because of the importance and difficulty of predicting consequences; and because biotechnology now manipulates humans themselves.

In the end a case study related to the American carbon footprint is discussed.

This book could not have been completed without the help of Mr Aman Bhatia (my nephew) who worked hard in locating and organising the material and spent many hours checking the manuscript. Appreciations are also extended to Mr Harinder Singh, Senior DTP operator, who drew and labelled the flow diagrams and worked long hours to bring the book on time.

I am also thankful to the editorial team of Woodhead Publishing India Pvt. for their wholehearted cooperation in bringing out the book in time. It may not be wrong to hold that this book on Food Biotechnology is essential reading for students pursuing engineering courses. Besides students, this book will prove useful to industrialists and consultants in the respective fields. It has been prepared with meticulous care, aiming at making the book error-free. Constructive suggestions are always welcome from users of this book.

S. C. Bhatia

Section I
General considerations

1

Food biotechnology: A review

1.1   Introduction

Biotechnology is the use of biological processes, organisms, or systems to manufacture products intended to improve the quality of human life. The earliest biotechnologists were farmers who developed improved species of plants and animals by cross pollenisation or crossbreeding. In recent years, biotechnology has expanded in sophistication, scope, and applicability. Biotechnology also offers the potential for new industrial processes that require less energy and are based on renewable raw materials. It is important to note that biotechnology is not just concerned with biology, but it is a truly interdisciplinary subject involving the integration of natural and engineering sciences. Biotechnology, like other advanced technologies, has the potential for misuse. Concern about this has led to efforts by some groups to enact legislation restricting or banning certain processes or programs, such as human cloning and embryonic stem-cell research. There is also concern that if biotechnological processes are used by groups with nefarious intent, the end result could be biological warfare.

Scientific advances through the years have relied on the development of new tools to improve health care, agricultural production, and environmental protection. Individuals, consumers, policymakers, and scientists must ultimately decide if the benefits of biotechnology are greater than the risks associated with this new approach.

Modern or 'new' biotechnology refers to the understanding and application of genetic information of animal and plant species. Genetic engineering modifies the functioning of genes in the same species or moves genes across *species resulting in Genetically Modified Organisms (GMOs)*.

Modern biotechnology also raises ethical issues by interfering with the genetic code of plant and animal, including human, species. As such, it may be perceived as 'unnatural' or even sacrilegious. Additionally, GM food (and feed) products and plant species can be viewed with mistrust, either because of health concerns arising from their direct consumption or because of longer-term environmental disruption arising from their uncontrolled release in nature.

In Agriculture, applications of biotechnology concentrate on the genetic modification of existing plant and animal species, by means of genetic material implantation from one species to another, where 'natural' cross-breeding does not function. In terms of commercial importance, gene modified (GM) crops, corn, soya and other oilseeds are, so far, the main applications.

1.2   Biotechnology contributes to the social, economic, and environmental sustainability of agriculture

Biotechnology contributes to the environmental sustainability of agriculture by improving the safe and effective use of pesticides, reducing the amount of insecticide used on crops, reducing greenhouse gas emissions, preserving and improving soil quality, and reducing crop losses both in the field and after harvest. Biotechnology and other precision agricultural technologies (e.g. conservation tillage, integrated pest management (IPM), and automated farming equipment systems using computerized GPS (global positioning system technology) help to increase the amount of food that can be harvested per acre of land or per animal, reducing the need to use more and more land to feed a growing population. Biotechnology and good agricultural practices improve soil quality and reduce pollution by allowing farmers to till (or mechanically work the soil) less often or not at all. Biotechnology reduces agriculture's 'carbon footprint,' with less carbon released into the air and more carbon retained in the soil. Agricultural biotechnology efforts in developing nations are being pursued with the guidance of and in cooperation with the local communities to ensure a positive social impact.

Biotechnology allows farmers to harvest more food using available farmland, vital for feeding a growing world population. Biotechnology has the potential to strengthen crops against extreme temperatures, drought, and poor soil conditions. These advancements are critical in developing nations, where crop losses can mean health and economic devastation.

1.3   Branches of biotechnology

Some of the important branches of biotechnology are discussed below.

1.3.1   Plant biotechnology

Plant biotechnology can be defined as the introduction of desirable traits into plants through genetic modification. The roots of plant biotechnology can be traced back to the time when humans started collecting seeds from their favourite wild plants and began cultivating them in tended fields. It appears that when the plants were harvested, the seeds of the most desirable plants were retained and replanted the next growing season.

1.3.2   Biotechnology: Green, red and white

Green biotechnology first wave: Biotechnology in agriculture industry, often referred to as the first wave since it was our earliest application of biotechnology. Although genetically modified crops get a great deal of press today, humans have been genetically modifying our crops for thousands of years–albeit unknowingly. Through generations of plant crossbreeding, we have created crops that are tailored to human needs.

Red biotechnology second wave: Biotechnology in medical industry usually referred to as the second wave of biotechnology. This field of biotechnology emerged in the 1970s as companies raced to produce novel vaccines, antibiotics, proteins, and hormones via genetic manipulation. Today, this industry has merged with the traditional pharmaceutical industry as the sector searches for new treatments and cures; it is characterized by very high R&D costs.

Industrial biotechnology or White biotechnology third wave: Industrial or White biotechnology uses enzymes and micro-organisms to make bio-based products in sectors such as chemicals, food and feed, detergents, paper and pulp, textiles and bioenergy (such as biofuels or biogas). In doing so, it uses renewable raw materials and is one of the most promising, innovative approaches towards lowering greenhouse gas emissions. The application of industrial biotechnology has been proven to make significant contributions towards mitigating the impacts of climate change in these and other sectors. In addition to environmental benefits, biotechnology can improve industry's performance and product value and, as the technology develops and matures, white biotechnology will yield more and more viable solutions for our environment. These innovative solutions bring added benefits for both our climate and our economy.

1.3.3   Animal biotechnology

Animal biotechnology is a branch of biotechnology in which molecular biology techniques are used to genetically engineer (i.e. modify the genome of) animals in order to improve their suitability for pharmaceutical, agricultural or industrial applications. Animal biotechnology has been used to produce genetically modified animals that synthesize therapeutic proteins, have improved growth rates or are resistant to disease.

1.3.4   Environmental biotechnology

Environmental biotechnologists combine biology and engineering to develop and use processes that remediate contaminated sites. For example, there are a variety of microbes, fungi, and bacteria capable of consuming pollutants and breaking them down into harmless components over time. Environmental biotechnologists identify, use, and develop appropriate microbes for remediating a particular area, and the pollutants unique to it. Contaminated soil may be remediated on site, or placed in containers and hauled away for treatment.

Environmental biotechnologists may also use plants to filter pollutants in soil, water, or air, convert plants to biofuels, or develop more sustainable processes to prevent pollution.

1.3.5   Molecular biotechnology

In its broadest sense, molecular biotechnology is the use of laboratory techniques to study and modify nucleic acids and proteins for applications in areas such as human and animal health, agriculture, and the environment. Molecular biotechnology results from the convergence of many areas of research, such as molecular biology, microbiology, biochemistry, immunology, genetics, and cell biology. It is an exciting field fuelled by the ability to transfer genetic information among organisms with the goal of understanding important biological processes or creating a useful product. The completion of the human genome project has opened a myriad of opportunities to create new medicines and treatments, as well as approaches to improve existing medicines. Molecular biotechnology is a rapidly changing and dynamic field. As the pace of advances accelerates, its influence will increase. The importance and impact of molecular biotechnology is being felt across the world.

The tools of molecular biotechnology can be applied to develop and improve drugs, vaccines, therapies, and diagnostic tests that will improve human and animal health. Molecular biotechnology has applications in plant and animal agriculture, aquaculture, chemical and textile manufacturing, forestry, and food processing. Every aspect of our lives in the coming decades will be affected by this dynamic field.

1.3.6   Food chemistry

Food science is an interdisciplinary study involving microbiology, biology, chemistry, engineering, and biotechnology. Food science is the application of science and engineering to the production, processing, distribution, preparation, evaluation, and utilization of food. Food chemistry encompasses the composition and properties of food components and the chemical changes they undergo during handling, processing, and storage. A food chemist must know chemistry and biochemistry and have knowledge of physiological chemistry, botany, zoology, and molecular biology to study and modify biological substances as sources of human food. Food chemists work with biological systems that are dead or dying (post-harvested plants and postmortem animal tissues) and study the changes they undergo when exposed to different environmental conditions. For example, during the marketing of fresh tomatoes, the food chemist must determine the optimal conditions to sustain the residual life in the tomatoes so the tomatoes will continue to ripen and arrive at the supermarket as a high-quality product for the consumer. Vital to understanding food science is the knowledge of the primary compounds in food. These compounds are carbohydrates, lipids, and proteins. The experiments and background information focus on the chemistry (functional properties) and structure of these compounds found in foods. Food science also includes biotechnology, which is the use of biological processes to make new foods, enzymes, supplements, drugs, and vaccines. For thousands of years, people have been using microorganisms in the fermentation of beer and in the making of cheeses, wines, and breads.

1.3.7   Food microbiology

Food microbiology focuses on a wide variety of current research on microbes that have both beneficial and deleterious effects on safety and quality of foods, and are thus concern of public health. Food microbiology is the scientific study of microorganisms, both in food and used for the production of food. This includes microorganisms that contaminate food, as well as those used in its production; for example to produce yoghurt, cheese, beer and wine.

1.3.8   Modern food biotechnology

Modern food biotechnology increases the speed and precision with which scientists can improve food traits and production practices. For centuries prior to the development of this technology, farmers have spent generations in crossbreeding plants or animals to obtain the specific beneficial traits they were looking for and avoid the traits they did not want. The process not only took a lot of time and effort, but the final outcome was far from guaranteed.

1.3.9   Food biotechnology

Today, food biotechnology utilizes the knowledge of plant science and genetics to further this tradition. Through the use of modern biotechnology, scientists can move genes for valuable traits from one plant to another. This process results in tangible environmental and economic benefits that are passed on to the farmer and the consumer.

1.4   Benefits of biotechnology

(1) *Protection of the environment:* Scientists have made some foods, such as papayas and potatoes, more resistant to disease. These crops need less chemical spray to protect them from harmful insects or viruses, which is better for water and wildlife. Other crops are protected from herbicides that are used to control weeds, thus allowing farmers to conserve soil by tilling the ground less often.

(2) *Greater crop yields:* Farmers can use biotechnology to help plants survive, warding off insects and better tolerance to herbicides. This allows a better harvest from these hardier plants.

(3) *Better tasting and fresher foods:* Sweeter peppers and tomatoes that ripen more slowly are examples of how biotechnology can produce fresher and better tasting food.

(4) *Grow more food on less land:* By the year 2050, the earth's population is estimated to be nine billion people. Using biotechnology, farmers can produce more crops on the land they already have. This way, countries do not have to devote more land to farming. In turn, developing countries can benefit most, since they will have the largest population growth.

(5) *Keep food safe to eat:* Scientists can more accurately find unwanted viruses and bacteria that may be present in food. This will cause an even lower risk of food-borne illnesses. Some types of fungus, which can be found in corn, release substances that can harm animals that eat them. These substances are already regulated in the United States, and biotechnology provides another tool that can help further reduce the amount of these substances in corn.

(6) *New food varieties:* Biotechnology can extend advances in crossbreeding, allowing for new food varieties; for example, seedless melons and mini avocadoes. Farmers can also develop food with better flavor and a better nutrient profile.

Some other benefits of biotechnology include:

• *Better tasting tomatoes year round:* Biotech tomatoes soften slower, so they can stay on the vine longer before shipping, thereby gaining added taste and nutrients.

- *Environment-friendly squash:* Biotech generated squash plants that do not require chemical sprays to combat viruses will mean not only less overall chemical use in farming, but also increased availability and lower prices for crookneck squash.
- *Healthier cooking oils:* Corn, soybeans, canola, and other plants could be modified to reduce the saturated fat content of cooking oils derived from these crops.
- *Herbicide-tolerant crops:* Breeding herbicide, pesticide, and fungicide tolerant crops allow more selective application of agricultural chemicals.

1.4.1   Health and medical benefits of biotechnology

- Modern food biotechnology may help promote public health, providing fruits, vegetables and grains with more nutritional benefits. These include more proteins, vitamins and minerals, or less fat and saturated fat. Already some oils have a better fatty acid profile, less saturated fat and trans fat, and more monounsaturated fat. This can promote heart health.
- For those with food allergies, biotechnology is seeking ways to reduce allergens in peanuts, wheat and other crops.
- Non-food applications of biotechnology may result someday in new vaccines and medications to treat heart disease, cancer and diabetes. For example, some fruits and vegetables will contain more antioxidants, such as vitamins C and E. Scientists have already developed a type of rice containing vitamin A and iron, thus reducing the risk of blindness and anemia where this is a main staple in their diet.

1.5   Facts about biotechnology

- It is safe to consume foods produced through biotechnology. Numerous studies conducted over the past three decades have supported the safety of foods produced through biotechnology, and consumers have been eating biotech foods safely since 1996, with no evidence of harm demonstrated anywhere in the world.
- Agricultural technologies, including biotechnology, are currently providing benefits to consumers, farmers, and the environment worldwide. Hardier, disease-free crops keep prices stable for consumers and ensure a reliable supply of nutritious, wholesome foods.

In developing nations, where a failed crop means the farmer cannot buy food and other essentials for his or her family, biotechnology has helped improve crop quality and consistency. In addition, herbicide-tolerant crops allow for better weed management, which gives farmers choice and flexibility. It also allows them to reduce soil tillage, protecting soil quality, reducing water pollution, and reducing agriculture's carbon footprint for generations to come. Thanks in part to biotechnology, farmers are able to use less insecticide.

- The regulation of foods produced through biotechnology is coordinated by the FDA, EPA, and USDA to ensure the safety of the U.S. food supply. In 1993, FDA determined that currently available food and animal feed derived from biotechnology are safe. These foods are held to the same rigorous safety standards as all other foods.

- Biotechnology has prevented entire food crops from being destroyed by pests or disease. When there was simply no other solution to the plant diseases destroying them, biotechnology was used to develop plums and Hawaiian papaya protected from viruses that threaten these crops. Scientists are now working to leverage biotechnology against extreme climate conditions such as drought, which is of increasing concern with climate change.

- Consumers are informed through labeling requirements for all foods, including those produced through biotechnology. The FDA requires labeling based on the nutrition and safety of the food, rather than how it was produced. Special labeling of foods is required if–a major food allergen is introduced; the nutritional content of the food has changed; or there are any other substantial changes to the food's composition.

- Foods produced through plant biotechnology are widely grown and consumed both in the United States and worldwide. In 2012, 17.3 million farmers in 28 countries grew biotech crops on 420.8 million acres. Notably, more than 15 million of those farmers were small, resource-poor farmers in developing countries.

- The use of biotechnology itself does not cause food allergies or increase the potential for a food to cause an allergic reaction or a new food allergy. During FDA's extensive review of a new biotech food product, the presence of any of the major food allergens (milk, eggs, wheat, fish, shellfish, tree nuts, soya, or peanuts) would trigger extensive testing. If the product were ever permitted in the food supply, it would require special allergen labeling to alert allergic consumers.

- Foods from biotechnology are just as nutritious as conventional foods, and some are higher in certain nutrients. Independent, peer-reviewed research, as well as regulatory review, has confirmed that current foods developed using biotechnology provide the same nutritional value as conventional foods, except where nutritional improvements have been made, such as cooking oils that deliver more healthful fats.

- Animal biotechnology, such as genetic engineering and cloning, is a safe way to produce fish, meat, milk, or eggs. Animal biotechnology includes a number of advanced breeding practices, as well as products such as the protein hormone given to dairy cows, recombinant bovine somatotropin (rbST). The safety of milk and other dairy products from cows given rbST has been established and reinforced through decades of research.

- With so much discussion of antibiotics in animal agriculture, it is important to note that there is no association between foods produced through biotechnology and resistance to antibiotics. FDA-approved antibiotics are available to farmers through livestock veterinarians to help prevent and treat disease in farm animals. Antibiotic use on the farm is closely regulated to ensure safety for the animals and for people consuming meat, milk, and eggs. In addition, a waiting period is enforced to ensure that food animals are clear of any antibiotics before entering the food supply.

- Biotechnology does not increase the prevalence of 'super weeds.' Insects and weeds can become tolerant to any pest control technique, whether used in biotechnology, conventional, or organic agriculture.

- Biotechnology increases the amount of food that can be produced on the same amount of land. It is estimated that the world population will reach 9 billion people by the year 2050, which would increase food needs by 70%. Biotechnology will need to be a part of the solution, as it encourages sustainable farm practices to protect precious non-renewable resources. In addition, herbicide-tolerant and insect- and disease- protected crops are allowed to thrive through better weed and insect control, allowing farmers to harvest a greater quantity of healthy, damage free crops. Also in development are crops that can grow even in regions where water is scarce, or where soil and water contain high levels of salt.

1.6   Risks of biotech foods

There are two issues of primary concern to food consumers: (i) the potential introduction of food allergens and (ii) marker genes that would increase human resistance to antibiotics. The potential for food allergens in biotechnology products is monitored by the FDA. Each food is evaluated for its allergenic potential as part of the regulatory process and labeling is required if a known allergen is transferred to a food source not normally associated with that allergen. Presently, no food products are on the US market with this designation. In fact, some products have been pulled from the review process precisely because of this concern. There is no current scientific evidence of increased antibiotic resistance as a result of genetically modified foods. (This would be more likely to result from overuse of prescription antibiotics). However, because of public concern, crops are now being developed without such antibiotic-resistant genes.

Additionally, people are concerned about the environment and the introduction of 'super' weeds or plants that are herbicide resistant or harmful to insects. Scientists are collecting data about both of these issues as part of their work to carefully assess the risks associated with the use of biotechnology. The EPA monitors the environmental impact of biotechnology, including its use for food production.

1.6.1   Concerns about biotechnology

As biotechnology has become widely used, questions and concerns have also been raised. The most vocal opposition has come from European countries. One of the main areas of concern is the safety of genetically engineered food. In assessing the benefits and risks involved in the use of modern biotechnology, there are a series of issues to be addressed so that informed decisions can be made. In making value judgments about risks and benefits in the use of biotechnology, it is important to distinguish between technology-inherent risks and technology-transcending risks. The former includes assessing any risks associated with food safety and the behavior of a biotechnology-based product in the environment. The latter involves the political and social context in which the technology is used, including how these uses may benefit or harm the interests of different groups in society. The health effects of foods grown from genetically engineered crop depend on the composition of the food itself. Any new product may have either beneficial or occasional harmful effects on human health. For example, a biotech-derived food with a higher content of digestible iron is likely to have a positive effect if consumed by iron-deficient individuals. Alternatively, the transfer of genes from one species to another may also transfer the risk for exposure to allergens. These risks are systematically evaluated by FDA and identified prior to commercialisation.

Individuals allergic to certain nuts, for example, need to know if genes conveying this trait are transferred to other foods such as soyabeans. Labeling would be required if such crops were available to consumers.

Among the potential ecological risks identified is increased weediness, due to cross-pollination from genetically modified crops spreads to other plants in nearby fields. This may allow the spread of traits such as herbicide-resistance to non-target plants that could potentially develop into weeds. This ecological risk is assessed when deciding if a plant with a given trait should be released into a particular environment, and if so, under what conditions. Other potential ecological risks stem from the use of genetically modified corn and cotton with insecticidal genes from *Bacillus thuringiensis* (*Bt* genes). This may lead to the development of resistance to *Bacillus thuringiensis* in insect populations exposed to the biotech-derived crop.

There also may be risks to non-target species, such as birds and butterflies, from the plants with *Bacillus thuringiensis* genes. The monitoring of these effects of new crops in the environment and implementation of effective risk management approaches is an essential component of further research. It is also important to keep all risks in perspective by comparing the products of biotechnology and conventional agriculture.

The reduction of biodiversity would represent a technology-transcending risk. Reduced biological diversity due to destruction of tropical forests, conversion of land to agriculture, overfishing, and the other practices to feed a growing world population is a significant loss far more than any potential loss of biodiversity due to biotech-derived crop varieties.

Improved governance and international support are necessary to limit loss of biodiversity.

There is no evidence that genetic transfers between unrelated organisms pose human health concerns that are different from those encountered with any new plant or animal variety. The risks associated with biotechnology are the same as those associated with plants and microbes developed by conventional methods.

1.6.2   Consumer and food industry perspectives

Survey research over the past decade shows that biotechnology is not likely to become an important issue for most American consumers. Consumers find biotechnology acceptable when they believe it offers benefits and it is safe. Surveys have consistently found that a majority of American consumers are willing to buy insect-protected food crops developed through biotechnology that use fewer chemical pesticides, as well as more nutritious foods. American consumers also appreciate the role that biotechnology can play in feeding the world. Research shows that European consumers are much less supportive of all biotechnology applications.

In contrast, European consumers express the most trust in those groups that oppose biotechnology. They have much less confidence in government, industry, or even scientists. American culture is more supportive and rewarding of new technology. Europeans tend to view food differently from US consumers. In fact, some Europeans reject all American food products. Europeans also want to protect their small farms to maintain open space and rural employment.

Most of the industry leaders interviewed are quite enthusiastic about the benefits of biotechnology—especially in terms of increased food availability, enhanced nutrition, and environmental protection. Most feel that biotechnology has already provided benefits to consumers. Almost all recognize that foods developed through biotechnology have already been part of consumers' everyday diet. They clearly do not agree with most of the opponents' claims and tend to have almost no trust in such groups.

Their main concerns involve lack of consumer acceptance—not the safety of the foods. They express high levels of confidence in the science and the regulatory process. In fact, almost none feel that biotechnology should not be used because of uncertain, potential risks. Most food industry leaders do not feel it is necessary to have special labels on biotech-derived foods. They express concerns that such labels would be perceived as a warning by consumers. They are also worried that the need to segregate commodities would pose financial and logistical burdens on everyone in the system including consumers. Food industry leaders recognize a major need to educate the public about biotechnology. They look to third parties, such as university and government scientists to provide such leadership.

Research shows that consumers will accept biotech foods if they see a benefit to themselves or society and if the price is right. Their responses to foods developed through biotechnology are basically the same as for any other food-taste, nutrition, price, safety and convenience are the major factors that influence our decisions about which foods to eat. How seeds and food ingredients are developed will only be relevant for a relatively small group of concerned.

1.6.3   National and international biotechnology policy

National governments and international policy making bodies rely on food scientists and others to develop innovations that will create marketable food products and increase food supplies. Governments also rely on scientific research because they are responsible for setting health and safety standards regarding new developments. International organizations can suggest policy approaches and help develop international treaties that are ratified by national governments.

Economic success in the competitive international market demands that food production become more efficient and profitable. National governments and international organizations support food biotechnology as a means to avoid global food shortages. Many policy-making bodies are also trying to balance support of the food biotechnology industry with public calls for their regulation. Such regulations are necessary to protect public health and safety, to promote international trade, conserve natural resources, and account for ethical issues.

The majority of processed foods on the market contain soya or corn ingredients that come from GM plants. To date, none have posed a food safety risk. The chief safety concerns are the potential to alter nutrient content or introduce allergens. Federal agencies involved in biotechnology regulation include the US Department of Agriculture (USDA) which evaluates agricultural production processes for all foods; the Food and Drug Administration (FDA), which evaluates whole non-animal foods (seafood), food ingredients, and food additives and the Environmental Protection Agency (EPA), which evaluates plants with insecticidal properties.

Developers of GM plants and biotech-derived foods are required to consult with FDA prior to the commercialisation of the product. This consultation procedure entails a science-based safety assessment of the product that focuses on the protection of the consumer, development and the environment. Thus developers, have a strong incentive to cooperate fully with FDA and the other agencies prior to marketing their products.

1.6.4   Labelling issues

Labelling food derived from GM plants and animals is an important, but complex issue. Some consumers and consumer groups believe they have a right to know whether biotechnology was used to produce food. Others believe labelling is not necessary if foods are essentially equivalent in composition. Food labels are regulated by FDA and, in some cases, by USDA. Regulatory agencies are concerned with ensuring that food labels are both true and not misleading. In accordance with their statutory mandate, the FDA has determined that a food product should be labeled as a product of biotechnology only if it has been changed in some significant way. The US Court of Appeals recently reaffirmed that the FDA's labeling policy is correct.

1.7   Food biotechnology safety and regulations

Food biotechnology is not new. For thousands of years people have been discovering that fruit juices ferment into wine, that milk can be used to develop products such as cheese or yogurt, or that beer can be made through the fermentation of malt and hops. These principles were used to breed hybrid corn, wheat, and many other crops in which certain traits could be selected in order to increase plant yield. Such breeding methods largely accounted for the phenomenal gains in crop productivity during the 20th century and led to modern farming practices.

Today, in the arena of food, the primary goals of food biotechnology are to provide a more abundant, less expensive, and a more nutritious food supply in order to address the needs of our growing global population.

1.8   Different techniques associated with food biotechnology

Old: Older food biotechnology techniques include conventional crossbreeding, which refers to the random recombination of genes through sexual reproduction leading to a new organism with improved traits. Crossbred plants, for instance, may require several generations to achieve a particular trait due to the randomness of gene transfer. Examples of such traits are improved crop yield, aesthetic qualities, increased tolerance to physical stress such as cold temperatures, and increased resistance to disease and insects.

New: Modern food biotechnology techniques include the joining of two pieces of DNA from different organisms leading to a single piece of DNA. Individual 'specific' genes are transferred from one organism to another in order to improve the nutrient levels of a food, for example, such as fortifying a fruit or vegetable. Modern techniques are much faster and more precise. It is possible to quickly transfer a specific gene of interest rather than waiting on the random shuffling of genes over several generations.

1.8.1   Food products have been or are being developed with biotechnology

Examples of products developed through food biotechnology include corn varieties containing a bacterial gene that kills insects and soybeans inserted with gene that renders them resistant to weed killers such as Roundup. Cotton, squash, and papaya are other examples of commodities in which biotechnology was used to reduce pesticide use, increase profitability through greater yield, and ultimately reduce the cost of commodities at the consumer level.

Examples of foods developed through biotechnology to increase the levels of nutrients or to address a health concern include oils, such as canola, in which the levels of nutritionally essential fatty acids are increased, varieties of wheat that do not contain gluten, and potatoes (protein), kiwi (resveratrol), and lettuce (iron).

1.8.2   Safety concerns associated with food biotechnology

The current connotation of food biotechnology—also known as Genetically Modified Organism (GMO), or Genetically Engineered (GE), among others— is a food product developed through the genetic modification of a plant, animal, or micro-organism in a laboratory by scientists.

Food biotechnology is an umbrella term covering a vast variety of processes for using living organisms—such as plants, animals, microbes, or any part of these organisms—to develop new or improved food products. It includes the newer forms of food biotechnology that offer a faster and more precise means to develop food products.

1.8.3   Biotechnology for 21st century

Experts in United States anticipate the world's human population in 2050 to be approximately 9 billion. The world's population is growing, but its surface area is not. Compounding the effects of population growth is the fact that most of the earth's ideal farming land is already being utilized. To avoid damaging environmentally sensitive areas, such as rain forests, we need to increase crop yields for land currently in use. By increasing crop yields through the use of biotechnology, the constant need to clear more land for growing food is reduced.

Countries in Asia, Africa, and elsewhere are grappling with how to continue feeding a growing population. They are also trying to benefit more from their existing resources. Biotechnology holds the key to increasing the yield of staple crops by allowing farmers to reap bigger harvests from currently cultivated land, while preserving the land's ability to support continued farming.

Increased crop yield, greater flexibility in growing environments, less use of chemical pesticides and improved nutritional content make agricultural biotechnology, quite literally, the future of the world's food supply.

Microorganisms associated with food

2.1 Introduction

Food serves as an interacting medium between various living species because it is a source of nutrients for humans, animals as well as microorganisms. This is a natural consequence of cohabitation. Human and animal food is basically derived from plant and animal sources. Food fit for human consumption is also a medium for the growth and activity of microorganisms. Hence human food is always associated with a variety of microorganisms. Since the primary function of microorganisms is self perpetuation, they use the human or animal food as a source of nutrients for their own growth and activity. Microbial activity in a food can be beneficial in certain cases but in most cases it leads to deterioration of the food and renders it unfit for human consumption. Microorganisms can be used as processing aids in the production of fermented foods. A variety of food chemicals and additives may be produced by fermentation involving select species of microorganisms.

Pathogenic microorganisms grow in the food utilising the nutrients in the food and produce toxins which are detrimental to the health of the consumer when such food is consumed. Food also serves as a vector or medium for certain pathogens that cause food infections and diseases.

The metabolic activity of various microorganisms not only utilizes the nutrients in food but also causes the spoilage of food through undesirable enzymatic changes affecting the quality of the food.

2.2 Bacteria, yeast and molds

Thousands of genera and species of microorganisms have been identified and classified. Several hundreds of these are associated in one way or other with food products. Many of them are of industrial importance as they find use in the production of new foods and food chemicals by fermentation and also in the preservation of food products. Micro-organisms are capable of spoiling food and causing diseases. The microorganisms which are of importance in food microbiology include bacteria, yeasts and molds.

2.2.1 Bacteria

Bacteria are microscopic single-celled organisms that thrive in diverse environments. They can live within soil, in the ocean and inside the human gut. Humans' relationship with bacteria is complex. Sometimes they lend a helping hand, by curdling milk into yogurt, or helping with our digestion. At other times they are destructive, causing diseases like pneumonia and Methicillin-resistant *Staphylococcus Aureus* (MRSA).

Based on the relative complexity of their cells, all living organisms are broadly classified as either prokaryotes or eukaryotes. Bacteria are prokaryotes. The entire organism consists of a single cell with a simple internal structure. Unlike eukaryotic DNA, which is neatly packed into a cellular compartment called the nucleus, bacterial DNA floats free, in a twisted thread-like mass called the nucleoid.

Bacterial cells also contain separate, circular pieces of DNA called plasmids. Bacteria lack membrane-bound organelles, specialized cellular structures that are designed to execute a range of cellular functions from energy production to the transport of proteins. However, both bacterial and eukaryotic cells contain ribosomes. These spherical units are where proteins are assembled from individual amino acids, using the information encoded in a strand of messenger RNA.

2.2.2   Yeasts

Yeasts consist of one cell, and belong to the taxonomic group called fungi, which also contains molds. There are many species of yeasts. The most common yeast known is *Saccharomyces cerevisiae*, which is used in the baking- and brewing industry. Yeasts also play an important role in the production of wine, kefir and some other products. Most yeasts used in the food industry are round and divide themselves through budding. This budding is a characteristic used to recognize them through a microscope. During budding the cells appear in 8-shaped forms.

Yeasts need sugar to grow. They produce alcohol and carbon dioxide from sugar. This reaction makes yeast so important for the food industry. Yeasts also produce pleasant aroma components. These aroma compounds play a very important role for the flavor of the end product. In beer the yeast is needed to produce the alcohol and the carbon dioxide for the brim. In the bread industry, both alcohol and carbon dioxide are formed; the alcohol evaporates during baking.

Yeasts can be found everywhere in nature, especially on plants and fruits. After fruits fall off the tree, fruits become rotten through the activity of molds, which form alcohol and carbon dioxide from the sugars in it.

Yeasts are grown in the industry in big tanks with sugary water in the presence of oxygen. When the desired amount of yeast is reached the liquid is pumped out, and the yeast is then dried. Nothing else is added in the production of yeast.

2.2.3   Molds

Molds are microscopic fungi that live on plant or animal matter. No one knows how many species of fungi exist, but estimates range from tens of thousands to perhaps 300,000 or more. Most are filamentous (threadlike) organisms and the production of spores is characteristic of fungi in general. These spores can be transported by air, water, or insects.

Unlike bacteria that are one-celled, molds are made of many cells and can sometimes be seen with the naked eye. Under a microscope, they look like skinny mushrooms. In many molds, the body consists of:

- Root threads that invade the food it lives on
- A stalk rising above the food
- Spores that form at the ends of the stalks

The spores give mold the color you see. When airborne, the spores spread the mold from place to place like dandelion seeds blowing across a meadow. Molds have branches and roots that are like very thin threads. The roots may be difficult to see when the mold is growing on food and may be very deep in the food. Foods that are moldy may also have invisible bacteria growing along with the mold. Bacteria, yeasts and molds attack virtually all the constituents of foods. Depending on their nature and availability of enzymes some ferment sugars and hydrolyze starches and cellulose while other hydrolyze fats and produce rancidity.

2.3   Primary sources of microorganisms commonly associated with food

The primary source of microorganisms associated with food is the environment with which the species are associated. The genera and species that cause deterioration of foods are found normally in food products. Each genus has its own particular requirements, and each is affected in predictable ways by the parameters of its environment.

Eight environmental sources of micro-organism found in food are given below:

1. Soil and water
2. Plants and plant products
3. Food utensils

4. Intestinal tract of human and animals

5. Food handlers

6. Animal feeds

7. Animal hides

8. Air and dust

Soil and water: These two environments are place together because many of the bacteria and fungi that inhibit both share a lot in common. Soil organism may enter the atmosphere by the action of wind and latter enter the water bodies when it rains. They also enter water when rainwater flows over soils into bodies of water. Aquatic organisms can be deposited onto soils through the actions of cloud formation and subsequent rainfall. Alteromonus spp. is aquatic forms that requires seawater salinity for growth and would not be expected to persist in soils.

Plant and plant products: Many or most soil and water organisms contaminate plants. Relatively, only a small numbers of organisms find the plant environment suitable to their overall well-being. Among others that are commonly associated with plants are bacterial plant pathogens in the genera *Corenebacterium, Curtobacterium, Pseudomonas* and *Xanthomonas* and *fungal pathogens* among several genera of molds.

Food utensils: When vegetables are harvested in containers and utensils, one would expect to find some or all of the surface organisms on the products, contact surfaces. In a similar way, the cutting block in a meat market along with cutting knives and grinders are contaminated from initial samples and this process leads to build up of organisms.

Intestinal track of human and animals: This flora becomes a water source when polluted water is used to wash raw food products. The intestinal flora consist of many organisms that do not persist as long in water as do others and notable among these are pathogens such as *Salmonellae*.

Food handlers: The micro flora on the hands and other garments of handlers generally reflects the environment and habits of individuals, and the organisms in question may be those from soils, water, dust, and other environmental sources. Other important sources are those that are common in nasal cavities and the mouth and the skin and those from gastrointestinal track that may enter foods through poor personal hygienic practices.

Animal feeds: This is an important source of salmonellae to poultry and other farm animals. Listeria monocytogenes is found in dairy and meat animals. The organisms in dry animal feed are spread throughout the animal environment and may be expected to occur on animal hides.

Animal hides: The types of organisms found in raw milk can be reflection of the flora of the udder when proper procedures are not followed in milking. From both the udder and hide, organisms can contaminate the general environment.

Air and dust: Many organisms are found in air and dust in a food processing operation. Among fungi, a number of molds may be expected to occur in air and dust along with some yeast. In general the types of organisms in air and dust would be those that are constantly reseeded to the environment.

2.4   Factors influencing microbial activity

The growth and activity of the microorganisms depend on the nature of the food and its composition. The factors that govern the microbial activity in a food include both intrinsic and extrinsic factors. The intrinsic factors include pH, water activity, oxidation-reduction potential, nutrients content and presence of inhibitors in the food. The extrinsic factors include temperature, relative humidity and the atmosphere surrounding the food. These factors operate individually as well as in combination and affect the growth and activity of microorganisms in foods.

Some intrinsic factors are interlinked with some extrinsic factors. For example, water activity rises with increasing temperature, there is an increase in water activity of 0.03 with a 10°C rise in temperature.

2.4.1   Intrinsic factors

pH

The intracellular pH of any organism must be maintained above the pH limit that is critical for that organism. The control of intracellular pH is required in order to prevent the denaturation of intracellular proteins. Each organism has a specific requirement and pH tolerance range; some are capable of growth in more acid conditions than others. Most microorganisms grow best at neutral pH (7.0). Yeasts and molds are typically tolerant of more acidic conditions than bacteria but several species of bacteria will grow down to pH 3.0. These species are typically those that produce acid during their metabolism such as the acetic or lactic acid bacteria.

Bacterial pathogens are usually unable to grow below pH 4.0. The type of microbial growth typically seen in a particular food is partly related to the pH of that product. Fruits are naturally acidic, which inhibits the growth of many bacteria, therefore spoilage of these products is usually with yeasts and molds. Meat and fish however have a natural pH much nearer neutral and they are therefore susceptible to the growth of pathogenic bacteria. Individual strains of a particular species can acquire acid resistance or acid tolerance compared to the normal pH range for that organism. For example acid-adapted *Salmonella* have been reported that are capable of growth at pH 3.8. There is a broad distinction between high and low acid foods; with low acid foods being those with a pH above 4.6, and high acid, below this. This is because pH 4.6 is the lower limit for the growth of mesophilic *Clostridium botulinum*. Foods with a pH greater than 4.6 must either be chilled, or if ambient stored, undergo a thermal process to destroy C. *botulinum* spores, or have a sufficiently low water activity to prevent its growth.

Different foods tend to spoil in different ways. For example, carbohydrate rich foods often undergo acid hydrolysis when they spoil; this usually reduces the pH, and tends to reduce the risk of pathogen growth. This principle is used in the fermentation of dairy and lactic meat fermentations. In contrast, protein-rich foods tend to increase in pH when they spoil, making them possibly less safe, as the pH rise to the zone where more pathogens can grow. Micro-organisms are able to grow in an environment with a specific pH, as shown in Table 2.1.

Table 2.1 Growth of microorganisms in an environment.

Micro-organisms	Minimum pH value	Optimum pH value	Maximum pH value
Gram +ve bacteria	4.0	7.0	8.5
Gram –ve bacteria	4.5	7.0	9.0
Yeasts	2.0	4.0–6.0	8.5–9.0
Molds	1.5	7.0	11.0

Some bacteria are:

- Acidophilic bacteria, e.g. Lactic acid bacteria (pH 3.3–7.2) and acetic acid bacteria (pH 2.8–4.3).
- Basophilic bacteria e.g. *Vibrio parahaemolyticus* (pH 4.8–11.0) and *Enterococcus* spp. (pH 4.8– 10.6).
  - Increasing the acidity of foods either through fermentation or the addition of weak acids could be used as a preservative method.

Moisture content and water activity

The water requirements of microorganisms are generally expressed in terms of water activity of the substrate in which they grow.

Water activity (a$_w$)

- Water activity is a measure of the water available for microorganisms to grow or reactions to take place, i.e. measure of the amount of water disposable for the microorganisms.
- It is a ratio of water vapour pressure of the food substance to the vapour pressure of pure water at the same temperature.
- Water activity is expressed as:
- Water activity $(a_w) = P/P_w$ where P = water vapour pressure of the food substance and P_w = water vapour pressure of pure water $(P_w = 1.00)$.
- The growth of microorganisms is limited due to minimum water activity values (Table 2.2).

Table 2.2 Growth of microorganisms due to minimum water activity values.

Micro-organisms	Minimum water activity (a$_w$) values
Gram +ve bacteria	9.5
Gram –ve bacteria	0.91
Yeasts	0.88
Molds	0.80

Oxidation reduction

Oxidation-reduction potential (O–R potential, E$_h$): Micro-organisms are sensitive to the oxidation-reduction potential of the substrate. The O–R potential of a substrate is defined as the ease with which the substrate loses or gains electrons and is expressed by the symbol E_h. A substance that loses electrons readily is a good reducing agent while a substance which gains electrons is a good oxidant.

When electrons are transferred from one compound to another, a potential difference exists between the two compounds, which is expressed in mill volts (mV). A highly oxidized substance has more positive potential while the more reduced substance has a negative potential.

When the concentrations of oxidant and reductant are equal the potential difference is zero. Fruit juices are highly oxidized substrates with E_h values in the range of +400 mV. Solid meat is a reducing medium with E_h values in the range of –200 mV while minced meat is an oxidizing medium with E_h values in the range of +200 mV.

Aerobic microorganisms such as aerobic bacteria require positive E_h values (oxidized) for their growth while anaerobes such as *Clostridium* species require negative E_h values (reduced). Some aerobic bacteria grow better in slightly reduced conditions and these organisms are referred to as *microaerophiles*. Examples of microaerophilic bacteria include lactobacilli and streptococci. Facultative organisms are capable of growing under aerobic as well as in anaerobic conditions.

The O–R potential of a food is determined by: (i) the characteristic O–R potential of the original food, (ii) the poising capacity, i.e. the resistance to change the potential of the food, (iii) the oxygen tension of the atmosphere surrounding the food and (iv) the access of atmosphere to the food. Most fresh plant and animal foods have a low and well-poised O–R potential in their interior. The presence of reducing sugars and ascorbic acid in plants and fruits and the presence of –SH groups in meat are responsible for the low O–R potential. In living plant and animal cells these reducing substances tend to poise the O–R potential at a low level resisting the effect of oxygen due to respiration and diffusion from the outside. Thus a cut of fresh meat or a whole fruit would have aerobic conditions only close to the surface that supports the growth of aerobic microorganisms while the interior of the meat or fruit supports the growth of anaerobic bacteria. Processing of foods alters the poising power of food by destroying or altering the reducing or oxidizing substances and also may allow more diffusion of oxygen. Thus clear fruit juices do not contain the natural reducing substances which are lost during extraction and filtration and become substrates with more positive E_h values. Micro-organisms affect the E_h values of foods during their growth. Thus aerobes consume oxygen and reduce the E_h values while anaerobes cannot affect the E_h values. In the presence of limited amounts of oxygen the same aerobic and facultative organisms may produce incompletely oxidized products such as organic acids from carbohydrates while in the presence of large amounts of oxygen, complete oxidation of carbohydrates to carbon dioxide and water occur. Similarly, protein decomposition under anaerobic conditions results in putrefaction while under aerobic conditions the products may be less obnoxious.

Nutrient content

The kinds and proportions of nutrients in the food are all-important in determining what organism is most likely to grow. Consideration must be given to (i) foods for energy, (ii) foods for growth and (iii) accessory food substances, or vitamins, which may be necessary for energy or growth.

Foods for energy: The carbohydrates, especially the sugars, are most commonly used as an energy source, but other carbon compounds may serve, e.g. esters, alcohols, peptides, amino acids, and organic acids and their salts.

Complex carbohydrates, e.g. cellulose, can be utilized by comparatively few organisms, and starch can be hydrolyzed by only a limited number of organisms. Micro-organisms differ even in their ability to use some of the simpler soluble sugars. Many organisms cannot use the disaccharide lactose (milk sugar) and therefore do not grow well in milk. Maltose is not attacked by some yeasts. Bacteria often are identified and classified on the basis of their ability or inability to utilize various sugars and alcohols. Most organisms, if they utilize sugars at all, can use glucose.

Foods for growth: Micro-organisms differ in their ability to use various nitrogenous compounds as a source of nitrogen for growth. Many organisms are unable to hydrolyze proteins and hence cannot get nitrogen from them without help from a proteolytic organism. One protein may be a better source of nitrogenous food than another because of different products formed during hydrolysis, especially peptides and amino acids. Peptides, amino acids, urea, ammonia, and other simpler nitrogenous compounds may be available to some organisms but not to others or may be usable under some environmental conditions but not under others.

Accessory food substances or vitamins: Some microorganisms are unable to manufacture some or all of the vitamins needed and must have them furnished. Most natural plant and animal foodstuffs contain an array of these vitamins, but some may be low in amount, or lacking. Thus meats are high in B vitamins and fruits are low, but fruits are high in ascorbic acid. Egg white contains biotin but also contains avidin, which ties up biotin, making it unavailable to microorganisms and eliminating as possible spoilage organisms those which must have biotin supplied. The processing of foods often reduces the vitamin content. Thiamine, pantothenic acid, the folic acid group, and ascorbic acid (in air) are heat-labile, and drying causes a loss in vitamins such as thiamine and ascorbic acid. Even storage of foods for long periods, especially if the storage temperature is elevated, may result in a decrease in the level of some of the accessory growth factors.

Presence of inhibitory substances

Inhibitory substances, originally present in the food, added purposely or accidentally, or developed there by growth of microorganisms or by processing methods, may prevent growth of all microorganisms or, more often, may deter certain specific kinds. Examples of inhibitors naturally present are the lactenins and anticoliform factor in freshly drawn milk, lysozyme in egg white, and benzoic acid in cranberries.

A micro-organism growing in a food may produce one or more substances inhibitory to other organisms, products such as acids, alcohols, peroxides, and even antibiotics. Propionic acid produced by the propionibacteria in Swiss cheese is inhibitory to molds; alcohol formed in quantity by wine yeasts inhibits competitors; and nisin produced by certain strains of *Streptococcus lactis* may be useful in inhibiting lactate-fermenting, gas-forming clostridia in curing cheese and undesirable in slowing down some of the essential lactic acid *streptococci* during the manufacturing process. There also is the possibility of the destruction of inhibitory compounds in foods by microorganisms. Certain molds and bacteria are able to destroy some of the phenol compounds that are added to meat or fish by smoking or benzoic acid added to foods; sulphur dioxide is destroyed by yeasts resistant to it; and *lactobacilli* can inactivate nisin. Heating foods may result in the formation of inhibitory substances. Heating lipids may hasten autoxidation and make them inhibitory, and browning concentrated sugar sirups may result in the production of furfural and hydroxymethyl furfural, which are inhibitory to fermenting organisms. Long storage at warm temperatures may produce similar results.

2.4.2   Extrinsic factors

For the growth of the microorganisms certain external factors should be favorable.

Temperature

Micro-organisms grow over a wide range of temperatures. The lowest temperature at which a micro-organism grows is about −34°C and the highest temperature is about 90°C. The various microorganisms are usually grouped on the basis of their temperature requirements for growth. Thus organisms that grow well below 20°C and have optimum temperature in the range of 20–30°C are referred to as psychrophiles or psychrotrophs. Organisms that grow well between 20°C and 45°C with optimum temperature in the range of 30–40°C are called mesophiles. Thermophiles are those organisms that grow well at temperatures above 45°C with optimum temperature in the range of 55–65°C.

At temperatures higher than an organism's optimum growth range, cells die rapidly. Lower temperatures still result in cell death but at a slower rate. Temperature can therefore be used to eliminate or control the growth of microorganisms. Heat treatments (pasteurisation or sterilisation) eliminate contaminating microorganisms via the application of heat for a specific time period (time and temperatures used being dependent upon the target organism). Refrigeration of a food can prevent spoilage by controlling the growth of thermophilic or mesophilic organisms. Most pathogens are capable of growth at refrigeration temperatures and therefore cannot be controlled via refrigeration alone. Some, for example Listeria, can grow at very low temperatures.

Relative humidity of environment

The relative humidity of the storage environment must be such that excessive drying of the food or absorption of moisture by the food does not occur. If the growth of microorganisms in a food is controlled by the water activity of a product then it is very important that the food be stored under relative humidity conditions which will not allow the uptake of moisture from the air, and therefore an increase in water activity. Packaging can be used to limit the migration of moisture into the product.

Atmosphere

It is possible to inhibit surface spoilage of foods by a controlled atmosphere surrounding the food.

As all microorganisms have specific requirements for oxygen and carbon dioxide, by altering the atmosphere within a food package the growth of microorganisms can be controlled. Vacuum packing food removes available oxygen and thereby prevents the growth of aerobic organisms; it does however still allow the growth of anaerobes such as *C. botulinum*. Modified atmosphere packaging (MAP) allows the food producer to select the atmosphere within the package using varying combinations of oxygen, carbon dioxide and nitrogen, depending upon the product type and target microorganisms. The majority of MAP foods have varying combinations of carbon dioxide and nitrogen.

2.5   Micro-organisms important in food industry

In addition to natural microflora determined by type of plant or animal and environmental conditions, every food may be contaminated from outside sources on the way from the field to the processing plant, or during storage, transport and distribution. There are thousands of different types of microorganisms everywhere in air, soil and water, and consequently on foods, and in the digestive tract of animals and human. Fortunately, the majority of microorganisms perform useful functions in the environment and also in some branches of food industry, such as production of wine, beer, bakery products, dairy products etc. On the other hand unwanted spoilage of foods is generally caused by microorganisms and contamination of food with pathogens causes food safety problems.

The microorganisms occurring on and/or in foods are from a practical point of view divided into three groups: molds, yeast and bacteria. Molds are generally concerned in the spoilage of foods; their use in the food industry is limited (e.g. mold ripened cheese). Yeasts are the most widely used microorganisms in the food industry due to their ability to ferment sugars to ethanol and carbon-dioxide. Some types of yeast, such as baker's yeasts are grown industrially, and some may be used as protein sources, mainly in animal feed.

Bacteria important in food microbiology may be divided into groups according to the product of fermentation, e.g. lactic acid bacteria, acetic acid bacteria, propionic acid bacteria. Bearing in mind the food constituent attacked (used as food for microorganisms), proteolytic, saccharolytic and lipolytic bacteria may be distinguished. Their systematic classification is based primarily on morphological and physiological properties (e.g. aerobic and anaerobic bacteria, gas forming bacteria, etc.). Lactic acid bacteria are widely used in the dairy industry, and acetic acid bacteria in vinegar production. Many bacteria are known as microorganisms that cause spoilage and some are pathogens (e.g. *salmonellae, staphylococci,* etc.).

2.5.1   Importance of molds in foods

Main groups

1. *Deuteromycota (fungi imperfecti):* This is a class of septate fungi in which a sexual stage (perfect state) of the life cycle has not been observed. The Fungi imperfecti are essentially a provisional taxonomic grouping; if a sexual stage is observed in a fungus assigned to this group, the organism will then be assigned a new name that reflects its perfect state.

    (a) *Aspergillus:* Very important genus that includes species used in industrial fermentations and the production of enzymes and organic acids. It also includes several spoilage microorganisms and two species, *A. flavus* and *A. parasiticus,* which produce carcinogenic aflatoxins.

    (b) *Botrytis:* Cause grey mold rot in fruits.

    (c) *Penicillium:* Like *Aspergillus,* this genus includes industrially important species as well as others that are involved in food spoilage and the production of mycotoxins.

2. *Ascomycota:* Septate mycelium, sexual reproduction characterized by the formation of a sac-like structure called an ascus. The ascus eventually ruptures and the enclosed spores are liberated. This is a very large class of fungi that includes the genera *Byssochlamys,* and the perfect states of some fungi formerly classified under *Aspergillus.*

(a) *Byssochlamys:* Form heat resistant spores that can survive heat treatment and result in the spoilage of high-acid canned foods, especially fruits.

3. *Zygomycota:* A diverse group of lower, filamentous fungi which have aseptate mycelium and are capable of rapid growth. This group has sexual and asexual modes of reproduction. Members include the genus Rhizopus, which holds several species that cause bread molds as well as *R. oligosporus*, a species used in oriental soyabean fermentations (tempeh).

(a) *Yeasts:* Classified into the same group as molds, most do not live in soil but are instead found in environments with high sugar content such as the nectar of flowers or the surface of fruits.

Ascomycota

Debaromyces: One of the most prevalent genera in dairy products. *D. hansenii* is an important food spoilage species, it can grow in 24% NaCl and at a_w as low as 0.65.

Saccharomyces: Includes the important industrial species, *S. cerevisiae*, used in bread and alcoholic fermentations. *S. bailli* is another important species because it is involved in the spoilage of several types of foods and is resistant to the antifungal preservatives benzoate and sorbate.

Deuteromycota

Candida: Includes the pathogen *C. albicans*, this genus also includes the most common species of yeasts in fresh ground beef and poultry, a few species are involved in industrial fermentations.

2.5.2   Moisture and air loving molds

Probably the best known microorganisms, molds are widely distributed in nature and grow under a variety of conditions in which air and moisture are present. They are also plants and a part of the fungi family.

Nearly everyone has seen mold growth on damp clothing and old shoes. So, many may find it hard to believe that mold is a micro-organism. However, the mold we see with the naked eye is actually a colony of millions of mold cells growing together. Molds vary in appearance. Some are fluffy and filament-like; others are moist and glossy; still others are slimy.

Growth and importance in food industry

Unlike bacteria, molds are made up of more than one cell. Vegetative cells sustain the organism by taking in food substances for energy and the production of new cell material. Reproductive cells produce small 'seed' cells called spores. Unlike bacterial spores, mold spores are the source of new mold organisms. Bacterial spores generally form only when environmental conditions are unfavorable.

Molds produce a stem consisting of several cells. Together, these cells form a 'fruiting body'. The fruiting body produces the spores, which detach and are carried by air currents and deposited to start new mold colonies whenever conditions are favorable.

Mold spores are quite abundant in the air. So, any food allowed to stand in the open soon becomes contaminated with mold if adequate moisture is present. Some types of molds are also psychrophiles and can cause spoilage of refrigerated foods.

Molds are important to the food industry. Among their many contributions are the flavor and color they add to cheeses and the making of soya sauce. They also play a role in the making of such chemicals as citric and lactic acid and many enzymes.

Molds can also cause problems in foods. Certain kinds can produce poisons called mycotoxins.

Probably the best known use of molds is in the drug industry, where they help produce such antibiotics as penicillin.

2.6   Importance of bacteria in the food industry

Bacteria that play significant roles in foods are often grouped on the basis of their activity in foods without regards to their systematic classification. Bacteria make up the largest group of microorganisms. People often think of them only as germs and the harm they do. Actually, only a small number of bacteria types are *pathogenic* (disease causing). Most are harmless and many are helpful.

Micro-organisms, including bacteria, can also be grouped according to their requirement for oxygen. Some grow only in the presence of oxygen (*aerobes*). Others grow only in the absence of oxygen (*an-aerobes*). Some are able to grow with or without oxygen (*facultative anaerobes*).

Bacterial eating habits: Bacteria and other microorganisms need food in order to grow and multiply. They vary in their food needs, but nearly everything we consider as food can also be used as food by some type of bacteria.

To be used by bacteria, a food substance must pass into the cell where it can be processed into energy and new cell material. Because most foods are too complex to move into a bacterial cell, they must be broken down into simpler substances. Enzymes do this by increasing the rate of biochemical reactions. Produced within the bacterial cell, enzymes move through the cell wall to break down the food on the outside into a form bacteria can use.

Growth and importance in food industry

Bacteria reproduce by a process called binary fission–one cell divides and becomes two. Some can reproduce at a very rapid rate under proper conditions. If food and moisture are adequate and the temperature is right, certain bacteria can reproduce in as little as 20 minutes. Within 20 minutes, one cell becomes two; in 40 minutes, there will be four, and so on. In only eight hours, the original cell will have multiplied to nearly 17 million new bacteria. Of course, conditions don't remain favorable for such a rate of reproduction for long. If they did, we could be buried in bacterial cells.

Lactic acid bacteria and its uses in food

Lactic acid bacteria refers to a large group of beneficial bacteria that have similar properties and all produce lactic acid as an end product of the fermentation process. They are widespread in nature and are also found in our digestive systems. Although they are best known for their role in the preparation of fermented dairy products, they are also used for pickling of vegetables, baking, wine making, curing fish, meats and sausages.

Without understanding the scientific basis, people thousands of years ago used lactic acid bacteria to produce cultured foods with improved preservation properties and with characteristic flavors and textures different from the original food.

Similarly today, a wide variety of fermented milk products including liquid drinks such as kefir and semi-solid or firm products like yoghurt and cheese respectively, make good use of these illustrious microbial allies.

The manufacture involves a microbial process by which the milk sugar, lactose is converted to lactic acid. As the acid accumulates, the structure of the milk protein changes (curdling) and thus the texture of the product. Other variables such as temperature and the composition of the milk, also contribute to the particular features of different products.

2.7   Importance of yeasts in foods

Saccharomyces species are the most widely used yeasts. The leading species *S. cerevisiae* is used in the manufacture of many foods, with special strains used for the leavening of bread and for the production of ale, wine, alcohol, glycerol and invertase. Top yeasts are active fermenters and grow rapidly at

20°C. The clumping of the cells and the rapid evolution of carbon dioxide sweep the cells to the surface and hence the name top yeasts. Bottom yeasts do not clump and settle to the bottom, hence the name bottom yeasts. They grow more slowly and ferment at lower temperatures of 10–15°C.

S. cerevisiae var ellipsoideus is a high alcohol yielding variety used in the production of industrial alcohol, wine and distilled liquors. *S. uvarum* is a bottom yeast used in beet manufacture. *S. Fragilis* and *S. lactis* have the ability to ferment lactose and hence are important in milk products.

Zygosaccharomyces species are osmophilic and are involved in the spoilage of honey, syrups and molasses. They find use in the fermentation of soya sauce and some wines.

Candida species spoil foods high in acid salt content. *C. utilis* is grown for food and *C. krusei* is used with dairy starter cultures to maintain the activity and longevity of lactic acid bacteria. *C. lipolytica* spoils butter and oleomargarine.

Brettanomyces species are involved in the fermentation of Belgian and English beers and in French wines, while *Kloeckera* species are found commonly on fruits and flowers and in soil. *Trichosporon* species (*T. pullulans*) grow at low temperatures and are found in breweries and on chilled beef.

Rhodotorula species cause discolouration of foods by forming coloured spots on meat and sauerkraut.

2.7.1   Sugar-loving yeasts

Yeasts are small, single-celled plants. They are members of the family *fungi* (singular, *fungus*), which also includes mushrooms. Fungi differ from other plants in that they have no chlorophyll.

Bacteria thrive on many different types of food. But most yeasts can live only on sugars and starches. From these, they produce carbon dioxide gas and alcohol. Thus, they have been useful to man for centuries in the production of certain foods and beverages. They are responsible for the rising of bread dough and the fermentation of wines, whiskey, brandy and beer. They also play the initial role in the production of vinegar.

2.7.2   Growth and importance in food industry

Some yeasts are *psychrophilic* and so they can grow at relatively low temperatures. In fact, the fermentation of wines and beer is often carried out at temperatures near 40°F. Because some kinds are psychrophiles, they can create a spoilage problem in meat and other refrigerated storage areas.

Unlike bacteria, which multiply by binary fission, yeasts reproduce by a method called *budding*. A small knob or bud forms on the parent cell, grows and finally separates to become a new yeast dell. Although this is the most common method of reproduction, yeasts also multiply by the formation of spores.

Because they can grow under conditions of high salt or sugar content, they can cause the spoilage of certain foods in which bacteria would not grow. Examples are honey, jellies, maple syrup and sweetened condensed milk. Foods produced by the bacterial fermentation process, such as pickles and sauerkraut, can also be spoiled by yeasts which interfere with the normal fermentative process.

Certain yeasts are pathogenic. However, yeast infections are much less common than are bacterial infections. Today, the impact of yeasts on food and beverage production extends beyond the original and popular notions of bread, beer and wine fermentations by *Saccharomyces cerevisiae*.

The commercial significance of yeasts in food and beverage production shown in Table 2.3.

Table 2.3 The commercial and community significance of yeasts in food and beverage production.

Production of fermented foods and beverages
Production of ingredients and additives for food processing
Spoilage of foods and beverages
Biocontrol of spoilage microorganisms
Probiotic and biotherapeutic agents
Source of food allergens
Source of opportunistic, pathogenic yeasts

In a positive context, they contribute to the fermentation of a broad range of other commodities, where various yeast species may work in concert with bacteria and filamentous fungi. Many valuable food ingredients and processing aids are now derived from yeasts. Some yeasts exhibit strong antifungal activity, enabling them to be exploited as novel agents in the biocontrol of food spoilage. The probiotic activity of some yeasts is another novel property that is attracting increasing interest. Unfortunately, there is also a darker side to yeast activity. Their ability to cause spoilage of many commodities, with major economic loss, is well known in many sectors of the food and beverage industries, while the public health significance of yeasts in foods and beverages is a topic of emerging concern.

Fermentation biotechnology

3.1   Introduction

Fermentation is traditionally a process which enables to preserve food and as such has been used for centuries until present. However nowadays, the main purpose of food fermentation is not to preserve, since other preservation techniques are known, but to produce a wide variety of fermentation products with specific taste, flavor, aroma and texture. Using various microbial strains, fermentation conditions (microorganisms, substrates, temperature, time of fermentation, etc.) and chemical engineering achievements, enable us to manufacture hundreds of types of dairy (cheeses, fermented milk products), vegetable (sauerkraut, pickles, olives), meat (fermented sausages) products, breads, alcoholic beverages (wine, beer, cider), vinegar and other food acids, as well as oils. In such a wide variety of products, tastes and textures, surprising is that in the majority of cases, only two types of fermentations are used: lactic acid and ethanolic fermentation.

The function of both is to change conditions, so unwanted spoiling or pathogenic microorganisms would not grow and alter the food. Historically, fermentation products were mainly food products, but in recent years an increased interest has been observed in the production of bulk chemicals (ethanol and other solvents), specialty chemicals (pharmaceuticals, industrial enzymes), biofuels and food additives (flavour modifiers). Fermentation processes are also used in agriculture.

Fermentation processes can be classified as spontaneous and induced (e.g. making bread dough by the addition of baking yeast to flour). Fermentation products contain chemical energy, which means that are not fully oxidized and their complete mineralization requires oxygen. Fermentation is less energetically efficient than oxidative phosphorylation (ATP is produced only by substrate-level phosphorylation). While fermentation of 1 molecule of glucose yields 2 molecules of ATP, in aerobic respiration 36 ATP molecules are formed. The final step of fermentation, transformation of pyruvate into end products, does not generate the energy, but produces NAD^+ that is required for glycolysis.

Fermentation processes utilize microorganisms to convert solid or liquid substrates into various products. The substrates used vary widely, any material that supports microbial growth being a potential substrate. Similarly, fermentation derived products show tremendous variety. Commonly consumed fermented products include bread, cheese, sausage, pickled vegetables, cocoa, beer, wine, citric acid, glutamic acid and soya sauce.

3.2   Types of fermentation

Most commercially useful fermentations may be classified as either solid-state or submerged cultures. In solid-state fermentations, the microorganisms grow on a moist solid with little or no 'free' water, although capillary water may be present. Examples of this type of fermentation are seen in mushroom cultivation, bread-making and the processing of cocoa, and in the manufacture of some traditional foods, e.g. miso (soya paste), sake, soya sauce, tempeh (soyabean cake) and gari (cassava), which are now produced in large industrial operations. Submerged fermentations may use a dissolved substrate, e.g. sugar solution, or a solid substrate, suspended in a large amount of water to form a slurry. Submerged fermentations are used for pickling vegetables, producing yoghurt, brewing beer and producing wine and soya sauce.

Solid-state and submerged fermentations may each be subdivided— into oxygen-requiring aerobic processes, and anaerobic processes that must be conducted in the absence of oxygen. Examples of aerobic fermentations include submerged-culture citric acid production by *Aspergillus niger* and solid-state koji fermentations (used in the production of soya sauce). Fermented meat products such as bologna sausage (polony), dry sausage, pepperoni and salami are produced by solid-state anaerobic fermentations utilizing acid-forming bacteria, particularly *Lactobacillus*, *Pediococcus* and *Micrococcus* species. A submerged-culture anaerobic fermentation occurs in yoghurt making.

Fermentations may require only a single species of micro-organism to effect the desired chemical change. In this case the substrate may be sterilized, to kill unwanted species prior to inoculation with the desired micro-organism. However, most food fermentations are non-sterile. Typically fermentations used in food processing require the participation of several microbial species, acting simultaneously and/or sequentially, to give a product with the desired properties, including appearance, aroma, texture and taste. In non-sterile fermentations, the culture environment may be tailored specifically to favor the desired microorganisms. For example, the salt content may be high, the pH may be low, or the water activity may be reduced by additives such as salt or sugar.

Factors influencing fermentations: A fermentation is influenced by numerous factors, including temperature, pH, nature and composition of the medium, dissolved O_2, dissolved CO_2, operational system (e.g. batch, fedbatch, continuous), feeding with precursors, mixing (cycling through varying environments), and shear rates in the fermenter. Variations in these factors may affect: the rate of fermentation; the product spectrum and yield; the organoleptic properties of the product (appearance, taste, smell and texture); the generation of toxins; nutritional quality; and other physico-chemical properties.

The formulation of the fermentation medium affects the yield, rate and product profile. The medium must provide the necessary amounts of carbon, nitrogen, trace elements and micronutrients (e.g. vitamins). Specific types of carbon and nitrogen sources may be required, and the carbon: nitrogen ratio may have to be controlled. An understanding of fermentation biotechnology is essential for developing a medium with an appropriate formulation. Concentrations of certain nutrients may have to be varied in a specific way during a fermentation to achieve the desired result. Some trace elements may have to be avoided—for example, minute amounts of iron reduce yields in citric acid production by *Aspergillus niger*. Additional factors, such as cost, availability, and batch-to-batch variability also affect the choice of medium.

3.2.1   Submerged fermentations

Fermentation systems

Industrial fermentations may be carried out either batch wise, as fedbatch operations, or as continuous cultures (Fig. 3.1). Batch and fedbatch operations are quite common, continuous fermentations being relatively rare. For example, continuous brewing is used commercially, but most beer breweries use batch processes.

In batch processing, a batch of culture medium in a fermenter is inoculated with a micro-organism (the 'starter culture'). The fermentation proceeds for a certain duration (the 'fermentation time' or 'batch time'), and the product is harvested. Batch fermentations typically extend over 4–5 days, but some traditional food fermentations may last months. In fedbatch fermentations, sterile culture medium is added either continuously or periodically to the inoculated fermentation batch. The volume of the fermenting broth increases with each addition of the medium, and the fermenter is harvested after the batch time. In continuous fermentations, sterile medium is fed continuously into a fermenter and the fermented product is continuously withdrawn, so the fermentation volume remains unchanged. Typically, continuous fermentations are started as batch cultures and feeding begins after the microbial population has reached a certain concentration. In some continuous fermentations, a small part of the harvested culture may be recycled, to continuously inoculate the sterile feed medium entering the fermenter (Fig. 3.1d). Whether continuous inoculation is necessary depends on the type of mixing in the fermenter. 'Plug flow' fermentation devices (Fig. 3.1d), such as long tubes that do not allow back mixing, must be inoculated continuously. Elements of fluid moving along in a plug flow device behave like tiny batch fermenters. Hence, true batch fermentation processes are relatively easily transformed into continuous operations in plug flow fermenters, especially if pH control and aeration are not required. Continuous cultures are particularly susceptible to microbial contamination, but in some cases the fermentation conditions may be selected (e.g. low pH, high alcohol or salt content) to favor the desired microorganisms compared to potential contaminants.

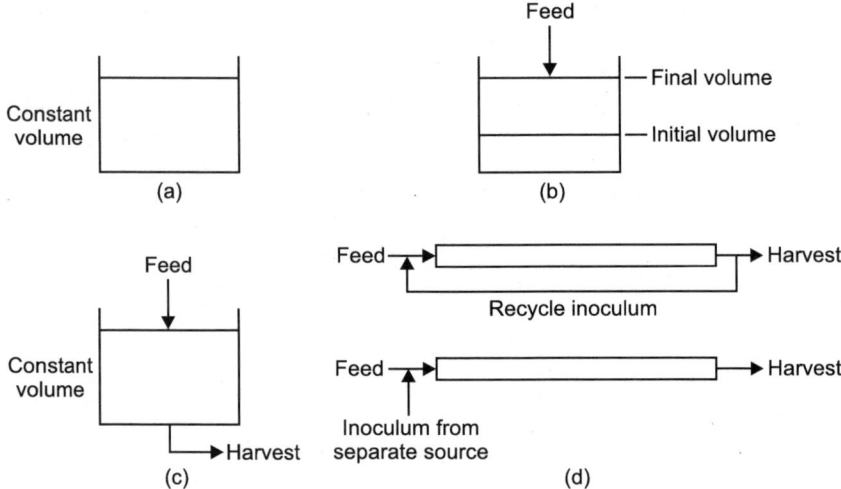

Figure 3.1 Fermentation methodologies: (a) Batch fermentation, (b) fedbatch culture, (c) continuous-flow well-mixed fermentation and (d) continuous plug flow fermentation, with and without recycling of inoculum

In a 'well-mixed' continuous fermenter (Fig. 3.1c), the feed rate of the medium should be such that the dilution rate, i.e. the ratio of the volumetric feed rate to the constant culture volume, remains less than the maximum specific growth rate of the micro-organism in the particular medium and at the particular fermentation conditions. If the dilution rate exceeds the maximum specific growth rate, the micro-organism will be washed out of the fermenter. Industrial fermentations are mostly batch operations. Typically, a pure starter culture (or seed), maintained under carefully controlled conditions, is used to inoculate sterile Petri dishes or liquid medium in the shake flasks. After sufficient growth, the pre-culture is used to inoculate the 'seed' fermenter. Because industrial fermentations tend to be large (typically 150–250 m³), the inoculum is built up through several successively larger stages, to 5–10% of the working volume of the production fermenter. A culture in rapid exponential growth is normally used for inoculation. Slower-growing microorganisms require larger inocula, to reduce the total duration of the fermentation. An excessively long fermentation time (or batch time) reduces productivity (amount of product produced per unit time per unit volume of fermenter), and increases costs. Sometimes inoculation spores, produced as seeds, are blown directly into large fermentation vessels with the in going air.

Microbial growth

Microbial growth in a newly inoculated batch fermenter typically follows the pattern shown in Fig. 3.2. Initially, in the lag phase, the cell concentration does not increase very much. The length of the lag phase depends on the growth history of the inoculum, the composition of the medium, and the amount of culture used for inoculation. An excessively long lag phase ties up the fermenter unproductively–hence the duration of the lag phase should be minimized. Short lag phases occur when: the composition of the medium and the environmental conditions in the seed culture and the production vessel are identical (hence less time is needed for adaptation); the dilution shock is small (i.e. a large amount of inoculum is used); and the cells in the inoculum are in the late exponential phase of growth. The lag phase is essentially an

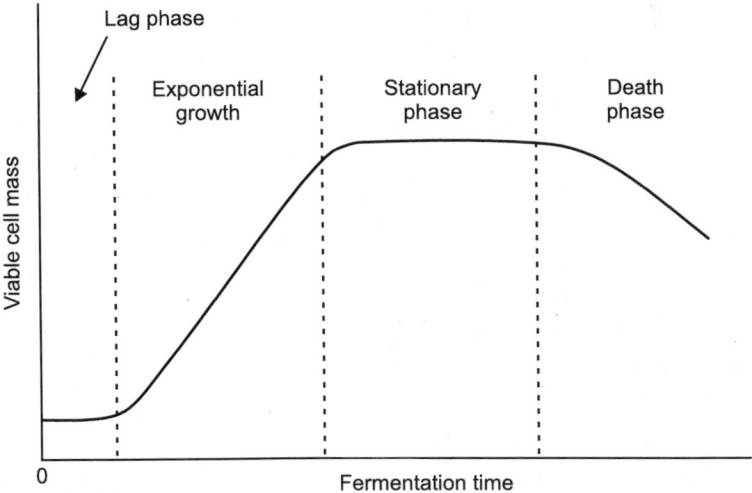

Figure 3.2 Typical growth profile of microorganisms in a submerged culture

adaptation period in a new environment. The lag phase is followed by exponential growth, during which the cell mass increases exponentially. Eventually, as the nutrients are exhausted and inhibitory products of metabolism build up, the culture enters a stationary phase. Ultimately, starvation causes cell death and lysis, and hence the biomass concentration declines.

Exponential growth can be described by Equation 3.1:

$$\frac{dX}{dt} = \mu X - k_d X \qquad\qquad ...(3.1)$$

where, X is the biomass concentration at time t; is the specific growth rate (i.e. growth rate per unit cell mass); and k_d is the specific death rate. During exponential growth, the specific death rate is negligible and Equation 3.1 reduces to Equation 3.2:

$$\frac{dX}{dt} = \mu X \qquad\qquad ...(3.2)$$

For a cell mass concentration X_0 at the beginning of the exponential growth (X_0 usually equaling the concentration of inoculum in the fermenter), and taking the time at which exponential growth commences as zero, Equation 3.2 can be integrated to produce Equation 3.3:

$$\text{In} \qquad\qquad \frac{X}{X_0} = \mu t \qquad\qquad ...(3.3)$$

Using Equation 3.3, the biomass doubling time, t_d, can be derived (Equation 3.4):

$$t_d = \frac{\ln 2}{\mu} \qquad\qquad ...(3.4)$$

Doubling times typically range over 45–160 min. Bacteria generally grow faster than yeasts, and yeasts multiply faster than molds. The maximum biomass concentration in submerged microbial fermentations is typically 40–50 kg m^{-3}.

The specific growth rate μ depends on the concentration S of the growth-limiting substrate, until the concentration is increased to a non-limiting level and μ attains its maximum value μ_{max}. The dependence of the growth rate on substrate concentration typically follows Monod kinetics. Thus the specific growth rate is given as Equation 3.5.

$$\mu = \mu_{max} \frac{S}{k_s + S} \qquad\qquad ...(3.5)$$

where, k_s is the saturation constant. Numerically, k, is the concentration of the growth-limiting substrate when the specific growth rate is half its maximum value.

An excessively high substrate concentration may also limit growth, for instance by lowering water activity. Moreover, certain substrates inhibit product formation, and in yet other cases, a fermentation product may inhibit biomass growth. For example, ethanol produced in the fermentation of sugar by yeast can be inhibitory to cells. Multiple lag phases (or diauxic growth) are sometimes seen when two or more growth-supporting substrates are available. As the preferentially utilized substrate is exhausted, the cells enter a lag phase while the biochemical machinery needed for metabolizing the second substrate is developed. Growth then resumes.

3.2.2   Aeration and oxygen demand

Submerged cultures are most commonly aerated by bubbling with sterile air. Typically, in small fermenters, the maximum aeration rate does not exceed 1 volume of air per unit volume of culture broth. In large bubble columns and stirred vessels, the maximum superficial aeration velocity tends to be < 0.1 m s^{-1}. Superficial aeration velocity is the volume flow rate of air divided by the cross-sectional area of fermenter. Significantly higher aeration rates are achievable in airlift fermenters. In these, aeration gas is forced through perforated plates, perforated pipes or single-hole spargers located near the bottom of the fermenter. Because O_2 is only slightly soluble in aqueous culture broths, even a short interruption of aeration results in the available O_2 becoming quickly exhausted, causing irreversible damage to the culture. Thus uninterrupted aeration is necessary. Prior to use for aeration, any suspended particles, microorganisms and spores in the gas are removed by filtering through microporous membrane filters.

The O_2 requirements of a fermentation depend on the microbial species, the concentration of cells, and the type of substrate. O_2 supply must at least equal O_2 demand, or the fermentation will be O_2-limited. O_2 demand is especially difficult to meet in viscous fermentation broths and in broths containing a large concentration of O_2-consuming cells.

As a general guide, the capability of a fermenter in terms of O_2 supply depends on the aeration rate, the agitation intensity and the properties of the culture broth. In large fermenters, supplying O_2 becomes difficult when demand exceeds 4–5 kg m^{-3} h^{-1}.

At concentrations of dissolved O_2 below a critical level, the amount of O_2 limits microbial growth. The critical dissolved O_2 level depends on the micro-organism, the culture temperature and the substrate being oxidized. The higher the critical dissolved O_2 value, the greater the likelihood that O_2 transfer will become limiting. Under typical culture conditions, fungi such as *Penicillium chrysogenum* and *Aspergillus oryzae* have a critical dissolved O_2 value of about 3.2×10^{-4} kg m^{-3}. For bakers' yeast and *Escherichia coli*, the critical dissolved O_2 values are 6.4×10^{-5} kg m^{-3} and 12.8×10^{-5} kg m^{-3} respectively.

The aeration of fermentation broths generates foam. Typically, 20–30% of the fermenter volume must be left empty to accommodate the foam and allow for gas disengagement. In addition, mechanical 'foam breakers' and chemical antifoaming agents are commonly used. Typical antifoams are silicone oils, vegetable oils and substances based on low-molecular-weight polypropylene glycol or polyethylene glycol. Emulsified antifoams are more effective, because they disperse better in the fermenter. Antifoams are added in response to signals from a foam sensor. The excessive use of antifoams may interfere with some downstream separations, such as membrane filtrations– hydrophobic silicone antifoams are particularly troublesome.

3.2.3   Heat generation and removal

All fermentations generate heat. In submerged cultures, typically 3–15 kW m^{-4} comes from microbial activity. In addition, mechanical agitation of the broth produces up to 15 kW m^{-3}. Consequently, a fermenter must be cooled to prevent a rise in temperature and damage to the culture. Heat removal tends to be difficult, because typically the temperature of the cooling water is only a few degrees lower than that of the fermentation broth. Therefore industrial fermentations are commonly limited by their heat-transfer capability. The ability to remove heat depends on: the surface area available for heat exchange; the temperature difference between the broth and the cooling water; the properties of the broth and the coolant; and the turbulence in these fluids. The geometry of the fermenter determines the surface area that can be provided for heat exchange. Heat generation due to metabolism depends on the rate of O_2 consumption, and heat removal in large vessels becomes difficult as the rate of O_2 consumption approaches 5 kg m^{-3}.

A fermenter must provide for heat transfer during sterilization and subsequent cooling, as well as removing metabolic heat. Liquid medium, or a slurry, for a batch fermentation may be sterilized using batch or continuous processes. In batch processes, the medium or some of its components and the fermenter itself are commonly sterilized together in a single step, by heating the medium inside the fermenter. Steam may be injected directly into the medium, or heating may take place through the fermenter wall.

Heating to high temperatures (typically 121°C) during sterilization often leads to undesirable reactions between components of the medium. Such reactions reduce the yield, by destroying nutrients or by generating compounds which inhibit growth. This thermal damage can be prevented or reduced by sterilizing only certain components of the medium in the fermenter and adding other, separately sterilized components, later. Sugars and nitrogen-containing components are often sterilized separately.

Dissolved nutrients that are especially susceptible to thermal degradation may be sterilized by passage through hydrophilic polymer filters, which retain particles of 0.45 μm or more. Even finer filters (e.g. retaining particles of 0.2 μm) are also available.

The heating and cooling of a large fermentation batch takes time, and ties up a fermenter unproductively. In addition, the longer a medium remains at a high temperature, the greater the thermal degradation or loss of nutrients. Therefore, continuous sterilization of the culture medium en route to a pre-sterilized fermenter is preferable, even for batch fermentations.

Continuous sterilization is rapid and it limits nutrient loss—however, the initial capital expense is greater, because a separate sterilizer is necessary.

Photosynthetic microorganisms

Photosynthetic cultures of microalgae and cyanobacteria require light and CO_2 as nutrients. Microalgae such as *Chlorella* and the cyanobacterium *Spirulina* are produced commercially as health foods in Asia. Algae are also cultivated as aquaculture feeds for shellfish. Typically, open ponds or shallow channels are used for the outdoor photosynthetic culture of micro-algae. Culture may be limited by the availability of light, but under intense sunlight, photoinhibition limits productivity. Temperature variations also affect performance. More controlled production is achieved in outdoor tubular photobioreactors, bubble columns and airlift systems. Tubular bioreactors use a 'solar receiver', consisting of either a continuous tube looped into several U-shapes to fit a compact area, or several parallel tubes connected to common headers at either end. The continuous looped-tube arrangement is less adaptable, because the length of the tube cannot exceed a certain value: photosynthetically produced O_2 builds up along the tube, and high levels of dissolved O_2 inhibit photosynthesis. The parallel-tube arrangement can be readily scaled up by increasing the number of tubes. Typically, the tubes are 0.05–0.08 m in diameter and the continuous-run length of any tube does not exceed 50 m. However, greater lengths may be feasible, depending on the flow velocity in the tube. The tubular solar receivers may be mounted horizontally, or horizontal tubes may be stacked in a ladder configuration, forming the rungs of the ladder. The latter arrangement reduces the area of land required.

The culture is circulated through the tubes by an airlift pump or other suitable low-shear mechanism. The maximum flow rate is limited by the tolerance of the algae to hydrodynamic stress. The flow velocity is usually $0.3–0.5$ m s^{-1}. The tube diameter is limited by the need to achieve adequate penetration of light. This declines as the cell concentration increases, due to self-shading. Closed, temperature-controlled outdoor tubular systems attain significantly higher productivity than open channels. The protein content of the algal biomass, and the adequacy of the development of color (chlorophyll) affect the acceptability of the product.

Among other types of culture system, airlift devices tend to perform better than bubble columns because only part of the airlift system is aerated and hence the penetration of light is less affected by air bubbles.

Conventional external-loop airlift devices may not be suitable because of the relatively high hydrodynamic shear rates they generate. However, concentric-tube airlift devices, with gas forced into the draft tube (zone of poor light penetration), are likely to perform well. Also, split-cylinder types of airlift system may be suitable. However, the volume of the aerated zone in any airlift device for microalgal culture should not exceed approximately 40% of the total volume of the circulating zones. This way the light blocking effect of bubbles remains confined to a small zone.

Submerged-culture fermenters

The major types of submerged-culture bio-reactor are:

- Stirred-tank fermenter
- Bubble column
- Airlift fermenter
- Fluidized-bed fermenter
- Trickle-bed fermenter

These are shown in Fig. 3.3.

Stirred-tank fermenter (see Fig. 3.3a). This is a cylindrical vessel with a working height-to-diameter ratio (aspect ratio) of 3–4. A central shaft supports three to four impellers, placed about 1 impeller-diameter apart. Various types of impeller, that direct the flow axially (parallel to the shaft) or radially (outwards from the shaft) may be used. Sometimes axial- and radial-flow impellers are used on the same shaft. The vessel is provided with four equally spaced vertical baffles that extend from near the walls into the vessel. Typically, the baffle width is 8–10% of the vessel diameter.

Bubble column (see Fig. 3.3b). This is a cylindrical vessel with a working aspect ratio of 4–6. It is sparged at the bottom, and the compressed gas provides agitation. Although simple, it is not widely used because of its poor performance relative to other systems. It is not suitable for very viscous broths or those containing large amounts of solids.

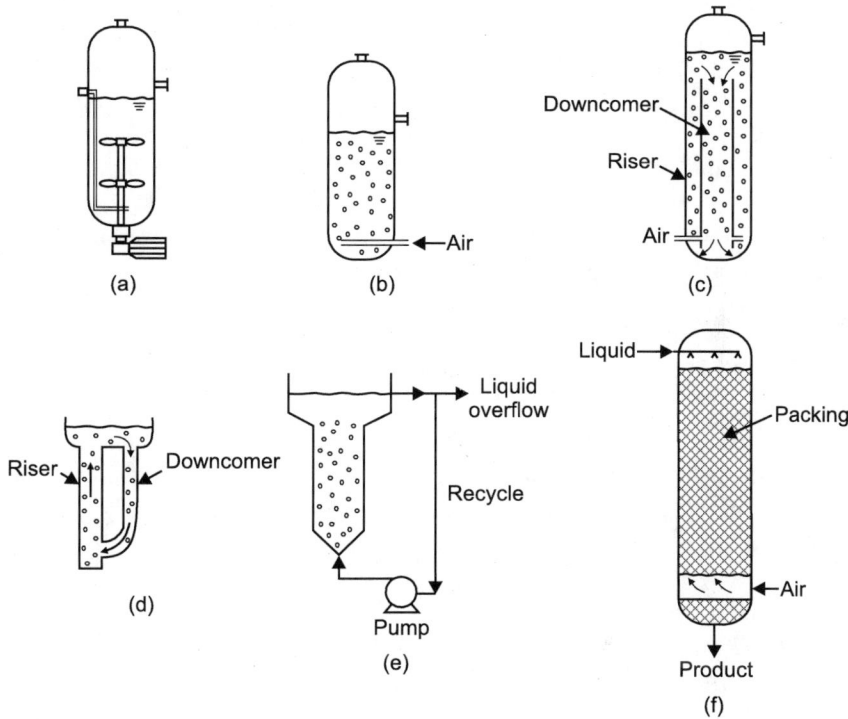

Figure 3.3 Types of submerged-culture fermenter: (a) Stirred-tank fermenter, (b) bubble column, (c) internal-loop airlift fermenter, (d) external-loop airlift fermenter, (e) fluidized-bed fermenter and (f) trickle-bed fermenter.

Airlift fermenters (see Figs. 3.3c and 3.3d). These come in internal-loop and external-loop designs. In the internal-loop design, the aerated riser and the unaerated downcomer are contained in the same shell. In the external-loop configuration, the riser and the downcomer are separate tubes that are linked near the top and the bottom. Liquid circulates between the riser (upward flow) and the downcomer (downward flow). Generally, these are very capable fermenters, except for handling the most viscous broths. Their ability to suspend solids and transfer O_2 and heat is good. The hydrodynamic shear is low. The external-loop design is relatively little used in industry.

Fluidized-bed fermenter (see Fig. 3.3e). These are similar to bubble columns with an expanded cross section near the top. Fresh or recirculated liquid is continuously pumped into the bottom of the vessel, at a velocity that is sufficient to fluidize the solids or maintain them in suspension. These fermenters need an external pump. The expanded top section slows the local velocity of the upward flow, such that the solids are not washed out of the bioreactor.

Trickle-bed fermenter (see Fig. 3.3f). These consist of a cylindrical vessel packed with support material (e.g. woodchips, rocks, plastic structures). The support has large open spaces, for the flow of liquid and gas and the growth of microorganisms on the solid support. A liquid nutrient broth is sprayed onto the top of the support material, and trickles down the bed. Air may flow up the bed, countercurrent to the liquid flow. These fermenters are used in vinegar production, as well as in other processes. They are suitable for liquids with low viscosity and few suspended solids.

3.2.4   Solid-state fermentations

Solid-state (substrate) fermentation (SSF) has been defined as the fermentation process occurring in the absence or near-absence of free water. SSF processes generally employ a natural raw material as carbon and energy source. SSF can also employ an inert material as solid matrix, which requires supplementing a nutrient solution containing necessary nutrients as well as a carbon source. Solid substrate (matrix), however, must contain enough moisture. Depending upon the nature of the substrate, the amount of water absorbed could be one or several times more than its dry weight, which leads relatively high water activity (a_w) on the solid/gas interface in order to allow higher rate of biochemical process. Low diffusion of nutrients and metabolites takes place in lower water activity conditions whereas compaction of substrate occurs at higher water activity. Hence, maintenance of adequate moisture level in the solid matrix along with suitable water activity are essential elements for SSF processes. Solid substrates should have generally large surface area per unit volume (say in the range of 10^3–10^6 m^2/cm^3 for the ready growth on the solid/gas interface). Smaller substrate particles provide larger surface area for microbial attack but pose difficulty in aeration/respiration due to limitation in inter-particle space availability. Larger particles provide better aeration/respiration opportunities but provide lesser surface area. In bioprocess optimization, sometimes it may be necessary to use a compromised size of particles (usually a mixed range) for the reason of cost effectiveness. For example, wheat bran, which is the most commonly used substrate in SSF, is obtained in two forms, fine and coarse. Former contains particles of smaller size (mostly smaller than 500–600 μ) and the latter mostly larger than these.

Most of SSF processes use a mix of these two forms at different ratios for optimal production.

Solid substrates generally provide a good dwelling environment to the microbial flora comprising bacteria, yeast and fungi. Among these, filamentous fungi are the best studied for SSF due to their hyphal growth, which have the capability to not only grow on the surface of the substrate particles but also penetrate through them. Several agro crops such as cassava, barley, etc. and agro-industrial residues such as wheat bran, rice bran, sugarcane bagasse, cassava bagasse, various oil cakes (e.g. coconut oil cake, palm kernel cake, soyabean cake, ground nut oil cake, etc.), fruit pulps (e.g. apple pomace), corn cobs, saw dust, seeds (e.g. tamarind, jack fruit), coffee husk and coffee pulp, tea waste, spent brewing grains, etc. are the most often and commonly used substrates for SSF processes. During the growth on such substrates hydrolytic exo-enzymes are synthesized by the microorganisms and excreted outside the cells, which create and help in accessing simple products (carbon source and nutrients) by the cells. This in turn promotes biosynthesis and microbial activities. Apart from these, there are several other important factors, which must be considered for development of SSF processes. These include physico-chemical and biological factors such as pH of the medium, temperature and period of incubation, age, size and type of inoculum, nature of substrate, type of micro-organism employed, etc.

Significance of SSF

SSF has been considered superior in several aspects to submerged fermentation (SMF) due to various advantages it renders. It is cost effective due to the use of simple growth and production media comprising agro-industrial residues, uses little amount of water, which consequently releases negligible or considerably less quantity of effluent, thus reducing pollution concerns. SSF processes are simple, use low volume equipment (lower cost), and are yet effective by providing high product titres (concentrated products). Further, aeration process (availability of atmospheric oxygen to the substrate) is easier since oxygen limitation does not occur as there is a increased diffusion rate of oxygen into moistened solid substrate, supporting the growth of aerial mycelium. These could be effectively used at smaller levels also, which makes them suitable for rural areas also.

General aspects of SSF

There are several important aspects, which should be considered in general for the development of any bioprocess in SSF. These include selection of suitable micro-organism and substrate, optimization of process parameters and isolation and purification of the product. Going by theoretical classification based on water activity, only fungi and yeast were termed as suitable microorganisms for SSF. It was thought that due to high water activity requirement, bacterial cultures might not be suitable for SSF. However, experience has shown that bacterial cultures can be well managed and manipulated for SSF processes. It has been generally claimed that product yields are mostly higher in SSF in comparison to SMF. However, so far there is not any established scale or method to compare product yields in SSF and SMF in true terms. The exact reasoning for higher product titres in SSF is not well known currently. The logical reasoning given is that in SSF microbial cultures are closer to their natural habitat and probably hence their activity is increased.

Selection of a proper substrate is another key aspect of SSF. In SSF, solid material is non-soluble that acts both as physical support and source of nutrients. Solid material could be a naturally occurring solid substrate such as agricultural crops, agro-industrial residues or inert support. However, it is not necessary to combine the role of support and substrate but rather reproduce the conditions of low water activity and high oxygen transference by using a nutritionally inert material soaked with a nutrient solution. In relation to selection of substrate, there could be two major considerations; one that there is a specific substrate, which requires suitable value-addition and/or disposal. The second could be related with the goal of producing a specific product from a suitable substrate. In the latter case, it would be necessary to screen various substrates and select the most suitable one. Similarly it would be important to screen suitable microorganisms and select the most suitable one. If inert materials such as polyurethane foam are used, product isolation could be relatively simpler and cheaper than using naturally occurring raw materials such as wheat bran because while extracting the product after fermentation, along with the product, several other water-soluble components from the substrate also leach out and may pose difficulties in purification process. Inert materials have been often used for studying modelling or other fundamental aspects of SSF.

Other relevant issues here could be the selection of process parameters and their optimization. These include physicochemical and biochemical parameters such as particle size, initial moisture, pH and pre-treatment of the substrate, relative humidity, temperature of incubation, agitation and aeration, age and size of the inoculum, supplementation of nutrients such as N, P and trace elements, supplementation of additional carbon source and inducers, extraction of product and its purification, etc. Depending upon the kind, level and application of experimentation, single and/or multiple variable parameters optimization method could be used for these.

Biochemical engineering aspects of SSF

In recent years, several excellent reports have appeared providing a great deal of knowledge and understanding of the fundamental aspects of SSF and today we know much better information about the heat and mass transfer effects in SSF processes, which have been considered as the main difficulties in handling SSF systems. However, there still remains much to be done in this regard. During SSF, a large amount of heat is generated, which is directly proportional to the metabolic activities of the micro-organism. The solid materials/matrices used for SSF have low thermal conductivities; hence heat removal from the process could be very slow. Sometimes accumulation of heat is high, which denatures the product formed and accumulated in the bed. Temperature in some locations of the bed could be 20°C higher than the incubation temperature. In the early phases of SSF, temperature and concentration of oxygen remain uniform throughout the substrate but as the fermentation progresses, oxygen transfer takes place resulting in the generation of heat. The transfer of heat into or out of SSF system is closely related with the aeration of fermentation system. The temperature of the substrate is also very critical in SSF as it ultimately affects the growth of the micro-organism, spore formation and germination, and product formation. High moistures results in decreased substrate porosity, which in turn prevents oxygen penetration. This may help bacterial contamination. On the other hand, low moisture content may lead to poor accessibility of nutrients resulting in poor microbial growth.

Water relations in SSF must be critically evaluated. Water activity (a_w) of the substrate has determinant influence on microbial activity. In general, the type of micro-organism that can grow in SSF systems are determined by a_w. The a_w of the medium has been attributed as a fundamental parameter for mass transfer of the water and solutes across the microbial cells. The control of this parameter could be used to modify the metabolic production or excretion of a micro-organism.

Substrate characteristics

Water activity: Typically, solid-state fermentations are carried out with little or no free water. Excessive moisture tends to aggregate the substrate particles, and hence aeration is made difficult. For example steamed rice, a common substrate, becomes sticky when the moisture level exceeds 30–35% w/w. Percentage moisture by itself is unreliable for predicting growth: for a given micro-organism growing on different substrates, the optimum moisture level may differ widely.

This water activity correlates with microbial growth. The water activity of the substrate is the ratio of the vapor pressure of water in the substrate to the saturated vapor pressure of pure water at the temperature of the substrate. Water activity equals 1/100th of the relative humidity (RH%) of the air in equilibrium with the substrate. Typically, water activities of < 0.9 do not support bacterial growth, but yeasts and fungi can grow at water activities as low as 0.7. Thus the low-moisture environment of many solid-state fermentations favors yeasts and fungi.

The water activity depends on the concentrations of dissolved solutes, and so sometimes salts, sugars or other solutes are added to alter the water activity. Different additives may influence the fermentation differently, even though the change in water activity produced may be the same. Furthermore, the fermentation process itself leads to changes in the water activity, as products are formed and the substrate is hydrolyzed, e.g. the oxidation of carbohydrates produces water. During fermentation, the water activity is controlled by aeration with humidified air and, sometimes, by intermittent spraying with water.

Particle size: The size of substrate particles affects the extent and the rate of microbial colonization, air penetration and CO_2 removal, as well as the downstream extraction and handling characteristics. Small particles, with high surface-to-volume ratios, are preferred because they present a relatively large surface for microbial action. However, particles that are too small and shapes that pack together tightly (e.g. flat flakes, cubes) are undesirable because close packing reduces the inter particle voids that are essential for aeration. Similarly, too many fine particles in a batch of larger particles will fill up the voids.

Substrate pH: The pH is not normally controlled in solid-state fermentations, but initial adjustments may be made during the preparation of the substrate. The buffering capacity of many substrates effectively checks large changes in pH during fermentation. This is particularly true of protein-rich substrates, especially if deamination of the protein is minimal. Some pH stability can be obtained by using a combination of urea and ammonium sulphate as the nitrogen source in the substrate. In the absence of other contributing nitrogen sources, an equimolar combination of ammonium sulphate and urea is expected to yield the greatest pH stability.

Aeration and agitation

Aeration plays an important role in removing CO_2 and controlling temperature and moisture. In some cases, an increased concentration of CO_2 may be severely inhibitory, while an increase in the partial pressure of O_2 may improve productivity. Deep layers and heaps of substrate may require forced aeration and agitation. Forced aeration rates vary widely, a typical range being $(0.05–0.2) \times 10^{-3}$ m^3 kg^{-1} min^{-3}. Occasional turning and mixing improve O_2 transfer and reduce compaction and mycelial binding of the substrate particles.

However, excessive agitation is undesirable because continuous agitation damages the surface hyphae—although mixing suppresses sporulation, which is often unwanted. The frequency of agitation may be purely experience-based, as in the occasional turning of a fermenting heap of cocoa beans, or it may be adjusted in response to a temperature controller.

Heat transfer

The biomass levels in solid-state fermentations, at 10–30 kg in^{-3}, are lower than those in submerged cultures. However, because there is little water, the heat generated per unit of fermenting mass tends to be much greater in solid-state fermentations than in submerged cultures, and again because there is little water to absorb the heat, the temperature can rise rapidly. The cumulative metabolic heat generation in fermentations producing koji, for the manufacture of a variety of products, has been noted at 419–2387 kJ per kilogram solids. Higher values, up to 13 398 kJ kg^{-1}, have been observed during composting. Peak heat generation rates in koji processes lie in the range 71–159 kJ kg^{-1} h^{-1} but average rates are more moderate, at 25–67 kJ kg^{-1} h^{-1}. The peak rate of production of metabolic heat during the fermentation of readily oxidized substrates, such as starch, can be much greater than that associated with typical koji processes.

The substrate temperature is controlled mostly through evaporative cooling hence drier air provides a better cooling effect. The intermittent spraying of cool water is sometimes necessary to prevent dehydration of the substrate. The air temperature and humidity are also controlled. Occasionally, the substrate-containing metal trays may also be cooled (by circulating a coolant), even though most substrates are relatively dry and porous, and hence are poor conductors.

The intermittent agitation of substrate heaps further aids heat removal. However, despite much effort, temperature gradients in the substrate do occur, particularly during peak microbial growth.

3.2.5   Koji fermentations

Koji fermentations are widely practiced. Koji comprises soyabeans or grain on which mold is growing, and has been used in oriental food preparation for thousands of years. Koji is a source of fungal enzymes, which digest proteins, carbohydrates and lipids into nutrients which are used by other microorganisms in subsequent fermentations. Koji is available in many varieties, which differ in terms of the mold, the substrate, the method of preparation and the stage of harvest. The production of soya sauce, miso and sake involves koji fermentation. Koji technology is also employed in the production of citric acid in Japan. The production of soya sauce (shoyu in Japanese) koji is detailed below, as an example of a typical industrial solid-substrate fermentation.

The koji for soya sauce is made from soyabeans and wheat. Soyabeans, or defatted soyabean flakes or grits are moistened and cooked (e.g. for 0.25 min or less, at about 170°C) in continuous pressure cookers. The cooked beans are mixed with roasted, cracked wheat, the ratio of wheat to beans varying with the variety of shoyu.

The mixed substrate is inoculated with a pure culture of *Aspergillus oryzae* (or A. *sojae*), the fungal spore density at inoculation being about 2.5×10^8 spores per kilogram of wet solids. After a 3-day fermentation, the substrate mass becomes green-yellow because of sporulation. The koji is then harvested, for use in a second submerged fermentation step. Koji production is highly automated and continuous—processes producing up to 4150 kg h^{-1} of koji have been described. Similar large-scale operations are used to produce koji for miso and sake in Japan.

Solid-state fermenters

Solid-state fermentation devices vary in technical sophistication, from very primitive banana-leaf wrappings, bamboo baskets and substrate heaps to the highly automated machines used mainly in Japan. Some 'less sophisticated' fermentation systems, e.g. the fermentation of cocoa beans in heaps, are quite effective at large-scale processing. Also, some of the continuous highly mechanized processes for the fermentation of soya sauce that are successful in Japan, are not suitable for less highly developed locations in Asia. Thus, fermentation practice must be tailored to local conditions.

The use of pressure vessels is not the norm for solid-state fermentation. The commonly used devices are:

- Tray fermenter
- Static-bed fermenter
- Tunnel fermenter
- Rotary disc fermenter
- Rotary drum fermenter
- Agitated-tank fermenter
- Continuous screw fermenter

These are described below:

Large concrete or brick fermentation chambers, or koji rooms, may be lined with steel, typically Type 304 stainless steel and are corrosion-resistant.

Tray fermenter: This is a simple type of fermenter, widely used in small- and medium-scale Koji operations (see Fig. 3.4). The trays are made of wood, metal or plastic, and often have a perforated or wire-mesh base to achieve improved aeration. The substrate is fermented in shallow 0.15 m deep layers. The trays may be covered with cheese cloth to reduce contamination, but processing is non-sterile. Single or stacked trays may be located in chambers in which the temperature and humidity are controlled, or simply in ventilated areas. Inoculation and occasional mixing are done manually, although the handling, filling, emptying and washing of trays may be automated. Despite some automation, tray fermenters are labor-intensive, and require a large production area. Hence the potential for scaling up production is limited.

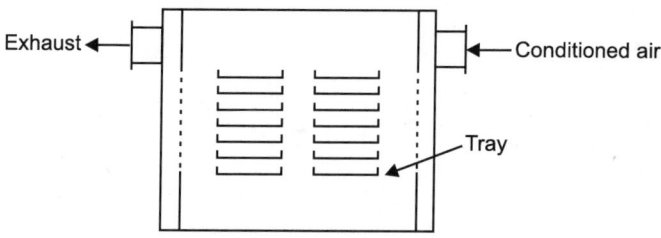

Figure 3.4 Tray fermenter

Static-bed fermenter: This is an adaptation of the tray fermenter (Fig. 3.5). It employs a single, larger and deeper, static bed of substrate located in an insulated chamber. O_2 is supplied by forced aeration through the bed of substrate.

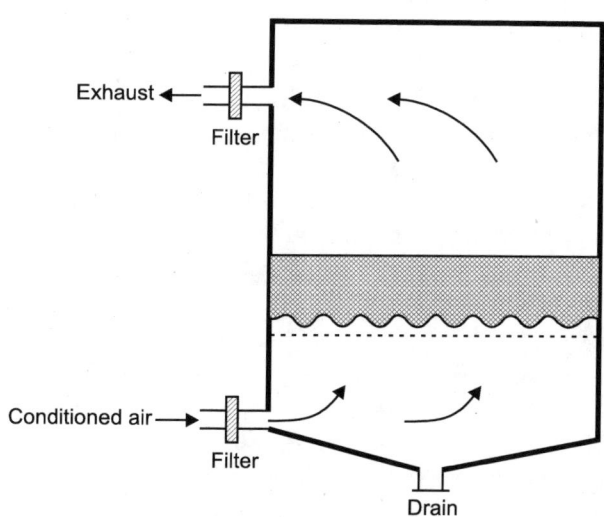

Figure 3.5 Static-bed fermenter

Tunnel fermenter: This is an adaptation of the static-bed device (Fig. 3.6). Typically, the bed of solids is quite long but no deeper than 0.5 m. Fermentation using this equipment may be highly automated, by way of mechanisms for mixing, inoculation, continuous feeding and harvest of the substrate.

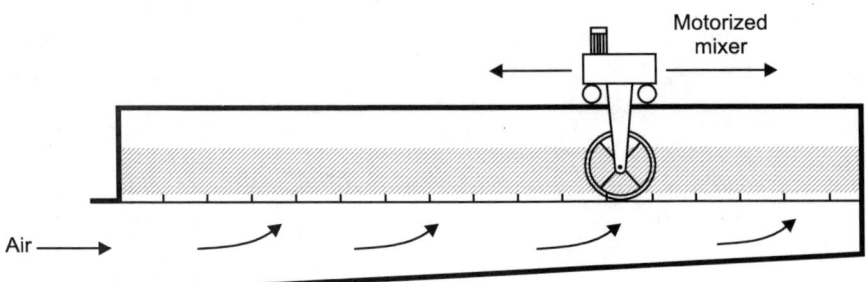

Figure 3.6 Tunnel fermenter

Rotary disc fermenter: The rotary disc fermenter consists of upper and lower chambers, each with a circular perforated disc to support the bed of substrate A common central shaft rotates the discs. Inoculated substrate is introduced into the upper chamber, and slowly moved to the transfer screw. The upper screw transfers the partly fermented solids through a mixer to the lower chamber, where further fermentation occurs. The fermented substrate is harvested using the lower transfer screw. Both chambers are aerated with humidified, temperature-controlled air. Rotary disc fermenters are used in large-scale koji production in Japan.

Rotary drum fermenter: The cylindrical drum of the rotary drum fermenter is supported on rollers, and rotated at 1–5 rpm around the long axis (Fig. 3.7). Rotation may be intermittent, and the speed may vary with the fermentation stage. Straight or curved baffles inside the drum aid in the tumbling of the substrate, hence improving aeration and temperature control. Sometimes the drum can be inclined, causing the substrate to move from the higher inlet end to the lower outlet during rotation. Aeration occurs through coaxial inlet and exhaust nozzles.

Agitated-tank fermenter: In this type of fermenter, either one or more helical-screw agitators are mounted in cylindrical or rectangular tanks, to agitate the fermented substrate (Fig. 3.8). Sometimes, the screws extend into the tank from mobile trolleys that ride on horizontal rails located above the tank. Another stirred-tank configuration is the paddle fermenter. This is similar to the rotary drum device, but the drum is stationary and periodic mixing is achieved by motor-driven paddles supported on a concentric shaft.

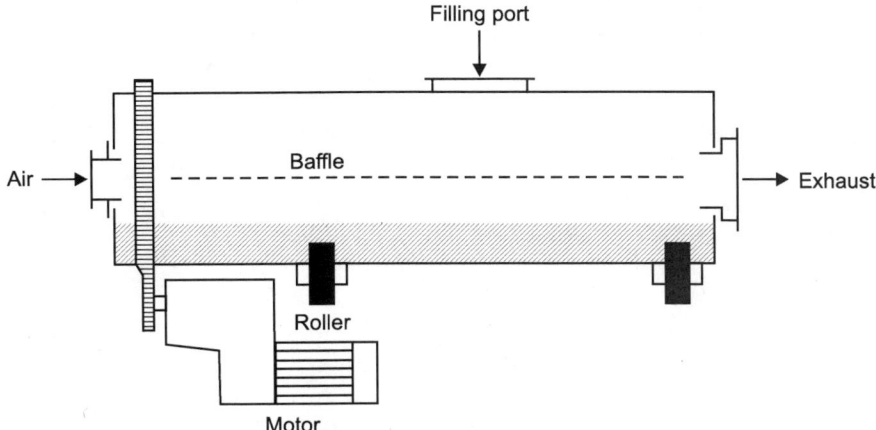

Figure 3.7 Rotary drum fermenter

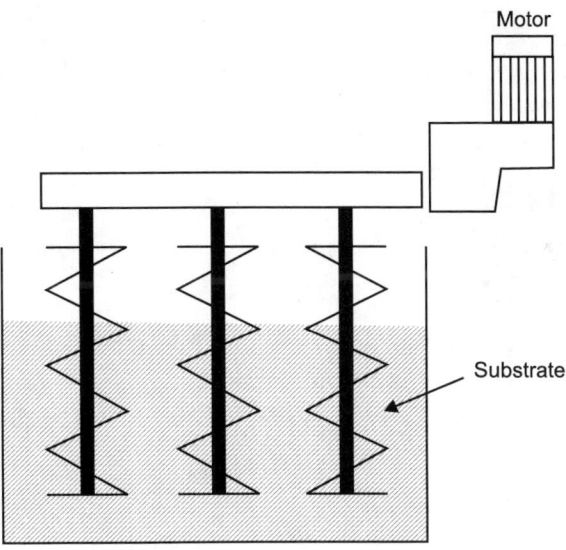

Figure 3.8 Agitated-tank fermenter

Continuous screw fermenter: In this type of fermenter, sterilized, cooled and inoculated substrate is fed in through the inlet of the non-aerated chamber (Fig. 3.9). The solids are moved towards the harvest port by the screw and the speed of rotation and the length of the screw control the fermentation time. This type of fermenter is suitable for continuous anaerobic or micro-aerophilic fermentations.

Sterile substrate
and inoculum

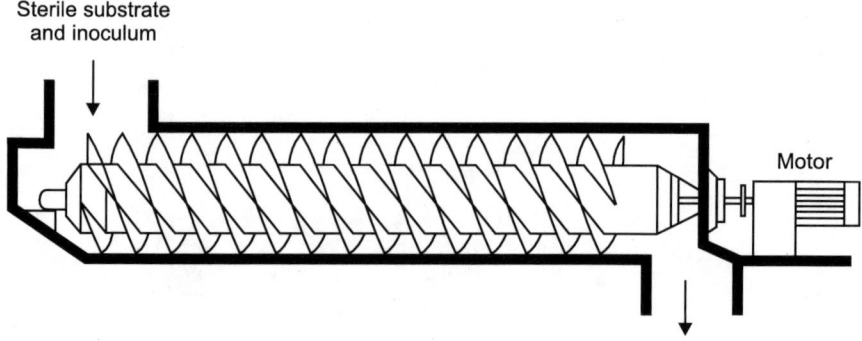

Motor

Figure 3.9 Continuous screw fermenter

3.2.6   Safe fermentation practice

The microorganisms used in certain industrial fermentations are potentially harmful. Certain strains have caused fatal infections in immuno compromised individuals, and rare cases of fatal disease in previously healthy adults have also been reported. Microbial spores and fermentation products, as well as microbes, have been implicated in occupational diseases. Most physiologically active fermentation products are potentially disruptive to health, and certain products are highly toxic.

The product spectrum of a given micro-organism often depends on the fermentation conditions. Under certain environmental conditions, some organisms, e.g. *Aspergillus flavus* and *A. oryzae*, are known to produce lethal toxins, and specific strains of the blue-veined cheese mold *Penicillium* roqueforti also produce mycotoxins under narrowly defined environmental conditions. Poor operational practice and failings in process and plant design can increase the risks.

4
Genetically modified foods

4.1   Introduction

Genetically modified foods have the potential to solve many of the world's hunger and malnutrition problems, and to help protect and preserve the environment by increasing yield and reducing reliance upon chemical pesticides and herbicides. Yet there are many challenges ahead for governments, especially in the areas of safety testing, regulation, international policy and food labelling. Many people feel that genetic engineering is the inevitable wave of the future and that we cannot afford to ignore a technology that has such enormous potential benefits. However, we must proceed with caution to avoid causing unintended harm to human health and the environment as a result of our enthusiasm for this powerful technology.

The term GM foods or GMOs (genetically modified organisms) is most commonly used to refer to crop plants created for human or animal consumption using the latest molecular biological techniques. These plants have been modified in the laboratory to enhance desired traits such as increased resistance to herbicides or improved nutritional content. The enhancement of desired traits has traditionally been undertaken through breeding, but conventional plant breeding methods can be very time consuming and are often not very accurate. Genetic engineering, on the other hand, can create plants with the exact desired trait very rapidly and with great accuracy. For example, plant geneticists can isolate a gene responsible for drought tolerance and insert that gene into a different plant. The new genetically modified plant will gain drought tolerance as well. Not only can genes be transferred from one plant to another, but genes from non-plant organisms also can be used. The best-known example of this is the use of *Bacillus thuringiensis* genes in corn and other crops. *Bacillus thuringiensis* or *Bt* is a naturally occurring bacterium that produces crystal proteins that are lethal to insect larvae. *Bacillus thuringiensis* crystal protein genes have been transferred into corn, enabling the corn to produce its own pesticides against insects such as the European corn borer.

4.2   Advantages of GM foods

The world population has topped 7.3 billion people and is predicted to double in the next 50 years. Ensuring an adequate food supply for this booming population is going to be a major challenge in the years to come. GM foods promise to meet this need in a number of ways:

- *Pest resistance:* Crop losses from insect pests can be staggering, resulting in devastating financial loss for farmers and starvation in developing countries. Farmers typically use many tons of chemical pesticides annually. Consumers do not wish to eat food that has been treated with pesticides because of potential health hazards, and run-off of agricultural wastes from excessive use of pesticides and fertilizers can poison the water supply and cause harm to the environment. Growing GM foods such as *Bacillus thuringiensis* corn can help eliminate the application of chemical pesticides and reduce the cost of bringing a crop to market.

- *Herbicide tolerance:* For some crops, it is not cost-effective to remove weeds by physical means such as tilling, so farmers will often spray large quantities of different herbicides (weed killer) to destroy weeds, a time-consuming and expensive process that requires care so that the herbicide doesn't harm the crop plant or the environment. Crop plants genetically engineered to be resistant to one very powerful herbicide could help prevent environmental damage by reducing the amount of herbicides needed. For example, Monsanto has created a strain of soyabeans genetically modified to be not affected by their herbicide product Roundup ready soyabeans. A farmer grows these soyabeans which then only require one application of weed killer instead of multiple applications, reducing production cost and limiting the dangers of agricultural waste run-off.

- *Disease resistance:* There are many viruses, fungi and bacteria that cause plant diseases. Plant biologists are working to create plants with genetically engineered resistance to these diseases.

- *Cold tolerance:* Unexpected frost can destroy sensitive seedlings. An antifreeze gene from cold water fish has been introduced into plants such as tobacco and potato. With this antifreeze gene, these plants are able to tolerate cold temperatures that normally would kill unmodified seedlings.

- *Drought tolerance/salinity tolerance:* As the world population grows and more land is utilized for housing instead of food production, farmers will need to grow crops in locations previously unsuited for plant cultivation. Creating plants that can withstand long periods of drought or high salt content in soil and groundwater will help people to grow crops in formerly inhospitable places.

- *Nutrition:* Malnutrition is common in third world countries where impoverished peoples rely on a single crop such as rice for the main staple of their diet. However, rice does not contain adequate amounts of all necessary nutrients to prevent malnutrition. If rice could be genetically engineered to contain additional vitamins and minerals, nutrient deficiencies could be alleviated. For example, blindness due to vitamin A deficiency is a common problem in third world countries. Researchers at the Swiss Federal Institute of Technology for Plant Sciences have created a strain of 'golden' rice containing an unusually high content of beta-carotene (vitamin A).

- *Pharmaceuticals:* Medicines and vaccines often are costly to produce and sometimes require special storage conditions not readily available in third world countries. Researchers have developed edible vaccines in tomatoes and potatoes. These vaccines are much easier to ship, store and administer than traditional injectable vaccines.

- *Phytoremediation:* Not all GM plants are grown as crops. Soil and groundwater pollution continues to be a problem in all parts of the world. Plants such as poplar trees have been genetically engineered to clean up heavy metal pollution from contaminated soil.

4.2.1   Effectiveness of GM crops

According to the FDA and the United States Department of Agriculture (USDA), there are over 40 plant varieties that have completed all of the Federal requirements for commercialization. Some examples of these plants include tomatoes and cantaloupes that have modified ripening characteristics, soyabeans and sugarbeets that are resistant to herbicides, and corn and cotton plants with increased resistance to insect pests. While there are very few genetically modified whole fruits and vegetables available on produce stands, highly processed foods, such as vegetable oils or breakfast cereals, most likely contain some tiny percentage of genetically modified ingredients because the raw ingredients have been pooled into one processing stream from many different sources. Also, the ubiquity of soyabean derivatives as food additives in the modern American diet virtually ensures that all US consumers have been exposed to GM food products.

4.3   Criticisms against GM foods

Environmental activists, religious organizations, public interest groups, professional associations and other scientists and government officials have all raised concerns about GM foods, and criticized agribusiness for pursuing profit without concern for potential hazards, and the government for failing to exercise adequate regulatory oversight. It seems that everyone has a strong opinion about GM foods. Most concerns about GM foods fall into three categories: environmental hazards, human health risks, and economic concerns.

4.3.1   Environmental hazards

- *Unintended harm to other organisms:* Recently a research study was published Transgenic pollen harms monarch larvae (*Nature*, Vol 399, No 6733, p. 214, May 1999) showing that pollen from *Bacillus thuringiensis* corn caused high mortality rates in monarch butterfly caterpillars. Monarch caterpillars consume milkweed plants, not corn, but the fear is that if pollen from *Bacillus thuringiensis* corn is blown by the wind onto milkweed plants in neighbouring fields, the caterpillars could eat the pollen and perish. The results seemed to support this viewpoint. Unfortunately, *Bacillus thuringiensis* toxins kill many species of insect larvae indiscriminately; it is not possible to design a *Bacillus thuringiensis* toxin that would only kill crop damaging pests and remain harmless to all other insects.

- *Reduced effectiveness of pesticides:* Just as some populations of mosquitoes developed resistance to the now-banned pesticide DDT, many people are concerned that insects will become resistant to *Bacillus thuringiensis* or other crops that have been genetically modified to produce their own pesticides.

- *Gene transfer to non-target species:* Another concern is that crop plants engineered for herbicide tolerance and weeds will cross-breed, resulting in the transfer of the herbicide resistance genes from the crops into the weeds. These 'super weeds' would then be herbicide tolerant as well. Other introduced genes may cross over into non-modified crops planted next to GM crops.

There are several possible solutions to the three problems mentioned above. Genes are exchanged between plants via pollen. Two ways to ensure that non-target species will not receive introduced genes from GM plants are to create GM plants that are male sterile (do not produce pollen) or to modify the GM plant so that the pollen does not contain the introduced gene. Cross-pollination would not occur, and if harmless insects such as monarch caterpillars were to eat pollen from GM plants, the caterpillars would survive.

Another possible solution is to create buffer zones around fields of GM crops. For example, non-GM corn would be planted to surround a field of *Bacillus thuringiensis* GM corn, and the non-GM corn would not be harvested. Beneficial or harmless insects would have a refuge in the non-GM corn, and insect pests could be allowed to destroy the non-GM corn and would not develop resistance to *Bacillus thuringiensis* pesticides. Gene transfer to weeds and other crops would not occur because the wind-blown pollen would not travel beyond the buffer zone. Estimates of the necessary width of buffer zones range from 6 meters to 30 meters or more. This planting method may not be feasible if too much acreage is required for the buffer zones.

4.3.2   Human health risks

- *Allergenicity:* Many children in the US and Europe have developed life-threatening allergies to peanuts and other foods. There is a possibility that introducing a gene into a plant may create a new allergen or cause an allergic reaction in susceptible individuals.
- *Unknown effects on human health:* There is a growing concern that introducing foreign genes into food plants may have an unexpected and negative impact on human health.

On the whole, with the exception of possible allergenicity, scientists believe that GM foods do not present a risk to human health.

4.3.3   Economic concerns

Bringing a GM food to market is a lengthy and costly process, and of course agri-biotech companies wish to ensure a profitable return on their investment. Many new plant genetic engineering technologies and GM plants have been patented, and patent infringement is a big concern of agribusiness. Yet consumer advocates are worried that patenting these new plant varieties will raise the price of seeds so high that small farmers and Third World countries will not be able to afford seeds for GM crops, thus widening the gap between the wealthy and the poor.

Patent enforcement may also be difficult, as the contention of the farmers that they involuntarily grew Monsanto-engineered strains when their crops were cross-pollinated shows one way to combat possible patent infringement is to introduce a 'suicide gene' into GM plants. These plants would be viable for only one growing season and would produce sterile seeds that do not germinate. Farmers would need to buy a fresh supply of seeds each year. However, this would be financially disastrous for farmers in third world countries who cannot afford to buy seed each year and traditionally set aside a portion of their harvest to plant in the next growing season. In an open letter to the public, Monsanto has pledged to abandon all research using this suicide gene technology.

4.4   Regulation of GM foods and role of government

Governments around the world are hard at work to establish a regulatory process to monitor the effects of and approve new varieties of GM plants. Yet depending on the political, social and economic climate within a region or country, different governments are responding in different ways.

In Japan, the Ministry of Health and Welfare has announced that health testing of GM foods will be mandatory as of April 2001. Currently, testing of GM foods is voluntary. Japanese supermarkets are offering both GM foods and unmodified foods, and customers are beginning to show a strong preference for unmodified fruits and vegetables.

India's government has not yet announced a policy on GM foods because no GM crops are grown in India and no products are commercially available in supermarkets yet. India is, however, very supportive of transgenic plant research. It is highly likely that India will decide that the benefits of GM foods outweigh the risks because Indian agriculture will need to adopt drastic new measures to counteract the country's endemic poverty and feed its exploding population. Some states in Brazil have banned GM crops entirely, and the Brazilian Institute for the Defense of Consumers, in collaboration with Greenpeace, has filed suit to prevent the importation of GM crops. Brazilian farmers, however, have resorted to smuggling GM soyabean seeds into the country because they fear economic harm if they are unable to compete in the global marketplace with other grain-exporting countries.

In Europe, anti-GM food protestors have been especially active. In the last few years Europe has experienced two major foods scares: bovine spongiform encephalopathy (mad cow disease) in Great Britain and dioxin-tainted foods originating from Belgium. These food scares have undermined consumer confidence about the European food supply, and citizens are disinclined to trust government information about GM foods. In response to the public outcry, Europe now requires mandatory food labelling of GM foods in stores, and the European Commission (EC) has established a 1% threshold for contamination of unmodified foods with GM food products.

In the United States, the regulatory process is confused because there are three different government agencies that have jurisdiction over GM foods. To put it very simply, the EPA evaluates GM plants for environmental safety, the USDA evaluates whether the plant is safe to grow, and the FDA evaluates whether the plant is safe to eat. The EPA is responsible for regulating substances such as pesticides or toxins that may cause harm to the environment. GM crops such as *Bacillus thuringiensis* pesticide-laced corn or herbicide-tolerant crops but not foods modified for their nutritional value fall under the purview of the EPA. The USDA is responsible for GM crops that do not fall under the umbrella of the EPA such as drought-tolerant or disease-tolerant crops, crops grown for animal feeds, or whole fruits, vegetables and grains for human consumption. The FDA historically has been concerned with pharmaceuticals, cosmetics and food products and additives, not whole foods. Under current guidelines, a genetically modified ear of corn sold at a produce stand is not regulated by the FDA because it is a whole food, but a box of cornflakes is regulated because it is a food product. The FDA's stance is that GM foods are substantially equivalent to unmodified, 'natural' foods, and therefore not subject to FDA regulation.

The EPA conducts risk assessment studies on pesticides that could potentially cause harm to human health and the environment, and establishes tolerance and residue levels for pesticides. There are strict limits on the amount of pesticides that may be applied to crops during growth and production, as well as the amount that remains in the food after processing. Growers using pesticides must have a license for each pesticide and must follow the directions on the label to accord with the EPA's safety standards.

Government inspectors may periodically visit farms and conduct investigations to ensure compliance. Violation of government regulations may result in steep fines, loss of license and even jail sentences.

The USDA has many internal divisions that share responsibility for assessing GM foods. Among these divisions are the Animal Health and Plant Inspection Service (APHIS) which conducts field tests and issues permits to grow GM crops, the Agricultural Research Service which performs in-house GM food research, and the Cooperative State Research, Education and Extension Service which oversees the USDA risk assessment program. The USDA is concerned with potential hazards of the plant itself. The USDA has the power to impose quarantines on problem regions to prevent movement of suspected plants, restrict import or export of suspected plants, and can even destroy plants cultivated in violation of USDA regulations. Many GM plants do not require USDA permits from APHIS. A GM plant does not require a permit if it meets these 6 criteria: (i) the plant is not a noxious weed, (ii) the genetic material introduced into the GM plant is stably integrated into the plant's own genome, (iii) the function of the introduced gene is known and does not cause plant disease, (iv) the GM plant is not toxic to non-target organisms, (v) the introduced gene will not cause the creation of new plant viruses and (vi) the

GM plant cannot contain genetic material from animal or human pathogens.

4.5   Labelling of GM foods

Labelling of GM foods and food products is also a contentious issue. On the whole, agribusiness industries believe that labelling should be voluntary and influenced by the demands of the free market. If consumers show preference for labelled foods over nonlabelled foods, then industry will have the incentive to regulate itself or risk alienating the customer. Consumer interest groups, on the other hand, are demanding mandatory labelling. People have the right to know what they are eating, argue the interest groups, and historically industry has proven itself to be unreliable at self-compliance with existing safety regulations. The FDA's current position on food labelling is governed by the Food, Drug and Cosmetic Act which is only concerned with food additives, not whole foods or food products that are considered Generally Recognized As Safe (GRAS).

There are many questions that must be answered if labelling of GM foods becomes mandatory. First, are consumers willing to absorb the cost of such an initiative? If the food production industry is required to label GM foods, factories will need to construct two separate processing streams and monitor the production lines accordingly. Farmers must be able to keep GM crops and non-GM crops from mixing during planting, harvesting and shipping. It is almost assured that industry will pass along these additional costs to consumers in the form of higher prices.

Secondly, what are the acceptable limits of GM contamination in non-GM products? The EC has determined that 1% is an acceptable limit of cross-contamination, yet many consumer interest groups argue that only 0% is acceptable. Some companies such as Gerber baby foods and Frito-Lay have pledged to avoid use of GM foods in any of their products. But who is going to monitor these companies for compliance and what is the penalty if they fail? Once again, the FDA does not have the resources to carry out testing to ensure compliance.

What is the level of detectability of GM food cross-contamination? Scientists agree that current technology is unable to detect minute quantities of contamination, so ensuring 0% contamination using existing methodologies is not guaranteed. Yet researchers disagree on what level of contamination really is detectable, especially in highly processed food products such as vegetable oils or breakfast cereals where the vegetables used to make these products have been pooled from many different sources. A 1% threshold may already be below current levels of detectability.

Finally, who is to be responsible for educating the public about GM food labels and how costly will that education be? Food labels must be designed to clearly convey accurate information about the product in simple language that everyone can understand. This may be the greatest challenge faced by a new food labelling policy: how to educate and inform the public without damaging the public trust and causing alarm or fear of GM food products. In January 2000, an international trade agreement for labelling GM foods was established. More than 130 countries, including the United States, the world's largest producer of GM foods, signed the agreement. The policy states that exporters must be required to label all GM foods and that importing countries have the right to judge for themselves the potential risks and reject GM foods, if they so choose. This new agreement may spur the U.S. government to resolve the domestic food-labelling dilemma more rapidly.

4.5.1   Labelling of GMO products: Freedom of choice for consumers

Exactly what must be labelled and how it is to be labelled, and why – is explained in the following.

Labelling guide

A basic principle applies to most food products: if genetically modified plants or microorganisms have been used in production, this must be clearly indicated.

Labelling: Yes

However, under certain conditions, numerous products are exempt from labelling obligations. These exemptions primarily concern additives and processing aids, but also apply to meat, milk and eggs.

Labelling: No

The status of flavours, additives and enzymes in regard to labelling obligations is complex. The use of genetic engineering is common, but there is no general labelling practice.

Labelling: Flavours, additives and enzymes

Labelling is also required for foods which are offered by restaurants, canteens and take-aways although there are exceptions.

Organic products without genetic engineering

By law, the use of genetical engineering is prohibited for products defined as 'organic'. Nevertheless, these products are permitted in certain cases to contain slight traces of genetically modified organisms.

Labelling: Organic products without genetic engineering

GMO labelling in the European Union: Basic principles

All food, and any ingredients, directly produced from a GMO must be labelled, even if this GMO is undetectable in the final product.

Labelling requirements: For neutral information only not for warning

Labelling empowers the buyer: In order to choose between products with or without genetically modified organisms, consumers need transparent, controllable and straightforward labelling regulations. However, the extent and breadth of these regulations are decided politically.

4.6   Most common genetically modified goods

Genetically modified material sounds a little bit like science fiction territory, but in reality, much of what we eat on a daily basis is a genetically modified organism (GMO). Whether or not these modified foods are actually healthy is still up for debate—and many times, you don't even know that you are buying something genetically modified. It is not required to label GMOs in the U.S. and Canada, but there are substantial restrictions, and even outright bans, on GMOs in many other countries. However, by 2018, Whole Foods Market will start labelling GMOs in the United States. This grocery chains' locations in Britain already provide GMO labelled products, as required by the European Union. According to the EU, GMO refers to plants and animals 'in which the genetic material has been altered in a way that does not occur naturally by mating and/or natural recombination.'

Some of the most common genetically modified foods are briefly discussed below:

Corn: Almost 85 per cent of corn grown in the U.S. is genetically modified. Even Whole Foods' brand of corn flakes was found to contain genetically modified corn. Many producers modify corn and soya so that they are resistant to the herbicide glyphosate, which is used to kill weeds.

Soya: Soya is the most heavily genetically modified food in the United States. The largest US producer of hybrid seeds for agriculture, Pioneer Hi-Bred International, created a genetically engineered soyabean, which was approved in 2010. It is modified to have a high level of oleic acid, which is naturally found in olive oil. Oleic acid is a monounsaturated omega-9 fatty acid that may lower LDL cholesterol (traditionally thought of as 'bad' cholesterol) when used to replace other fats.

Alfalfa: Cultivation of genetically engineered alfalfa was approved in 2011, and consists of a gene that makes it resistant to the herbicide Roundup, allowing farmers to spray the chemical without damaging the alfalfa.

Canola: Canola is genetically engineered form was approved in 1996, and as of 2006, around 90 percent of U.S. canola crops are genetically modified.

Sugar beets: A very controversial vegetable, sugar beets were approved in 2005, banned in 2010, then officially deregulated in 2012. Genetically modified sugar beets make up half of the U.S. sugar production, and 95 percent of the country's sugar beet market.

Milk: To increase the quantity of milk produced, cows are often given rBGH (recombinant bovine growth hormone), which is also banned in the European Union, as well as in Japan, Canada, New Zealand and Australia.

Zucchini: Genetically modified zucchini contains a toxic protein that helps make it more resistant to insects. This introduced insecticide, has recently been found in human blood, including that of pregnant women and fetuses. This indicates that some of the insecticide is making its way into our bodies rather than being broken down and excreted.

Yellow squash: Yellow squash has also been modified with the toxic proteins to make it insect resistant. This plant is very similar to zucchini, and both have also been modified to resist viruses.

Papaya: Genetically modified papaya trees have been grown in Hawaii since 1999. These Papayas are sold in the United States and Canada for human consumption. These papayas have been modified to be naturally resistant to Papaya ringspot virus, and also to delay the maturity of the fruit. Delaying maturity gives suppliers more time to ship the fruit to supermarkets.

4.7   Pros and cons of genetically modified foods

Many people today take for granted exactly where the foods they eat come from. In fact, genetically modified foods have become a common place thing in America, even though few people understand just what 'genetically modified' means. While there are some benefits that genetically modified foods may offer, there are also some risks and negative effects that these foods can cause as well.

When the term 'genetically modified' is used to describe a food, it means that the genetic makeup of one of the ingredients in that food has been altered. This is achieved by a very special set of technologies that combine the genes from different organisms, with the resulting organism being called a genetically modified food. In most cases, the specific genes that are combined have been hand-picked for the specific traits that they have. Those traits could include everything from the resistance to insects to specific nutritional value. These genetically modified foods can be in anything from corn to canola oil, which are quite common ingredients in many foods found on the market today such as snacks, cereals and sodas.

Pros of genetically modified foods

There are several benefits that have been linked to genetically modified foods, including:

- *Resistance to disease:* Genes can be modified to make crops more resilient when it comes to disease, especially those spread through insects. This can lead to higher crop yields, which many experts argue can help to feed people in developing countries.

- *Cost:* Because foods can be more resistant to disease, it reduces the cost necessary for pesticides and herbicides. And although genetically modified seeds are a more costly investment initially, this reduction in cost along with fewer lost crops leads to more profits. In many cases, that lower cost is passed onto the consumer through lower food prices.

- *Quality:* Some genetically modified foods, particularly fruits and veggies, have a longer shelf life than natural products.

- *Taste:* Some people claim that genetically modified foods have a better taste. In some cases, the genes can be altered in order to improve taste, although this is still one factor that varies from person to person.

- *Nutritional content:* Foods are often genetically modified in order to increase their nutritional content. This is especially helpful for certain populations where a specific nutrient is lacking in the local diet.

4.7.1   Cons of genetically modified foods

Although there are some benefits to genetically modified foods, there are some risks that have been associated with these foods. Some of these risks include:

- *Allergens and toxins:* Some genetically modified foods may contain higher levels of allergens and toxins, which can have negative effects on the personal health of those who eat them. This may be especially dangerous for people with serious food allergies.

- *Antibiotic resistance:* Because genetically modified foods are often developed to fight off certain pesticides and herbicides, there may be an increased risk that people who eat those foods may be more resistant to antibiotics.
- *New diseases:* Viruses and bacteria are used in the process of modifying foods, which means that there is a possibility that they could cause the development of a new disease.
- *Nutritional content:* Not all genetically modified foods are changed to increase their nutritional content. Instead, these foods may actually lose nutritional content in the process of altering their genetic makeup.
- *Loss of biodiversity:* Genetically modified foods could potentially cause damage to other organisms in the ecosystems where they are grown. If these organisms are killed off, it leads to a loss of biodiversity in the environment while also putting other organisms at risk by creating an unstable ecosystem.

Section II
Proteins and enzymes

Cross-linking of proteins

5.1   Introduction

Foods are multicomponent matrices with complex structures. The macro-molecular structure of food affects its mechanical and physical properties, chemical and microbiological stability, diffusion properties, product engineering, sensory properties and nutrition.

The textural and water-holding properties of food play a major role in food product quality, as they are recognized by consumers and are factors behind food choice. Proteins along with carbohydrates and fats are the main components affecting the textural and water-holding properties of foods. Cross-linking and aggregation of protein molecules into three-dimensional networks is an essential mechanism for developing food structures with desirable mechanical properties. In most cases protein cross-linking takes place in a food manufacturing process in which proteins are denatured, i.e. transformed from biologically active molecules to disordered structures.

The functional properties of a protein network are defined by the type and number of chemical bonds and interactions that hold individual protein molecules together. Hydrophobic and electrostatic interactions, van der Waals forces as well as hydrogen, ionic, and covalent bonds all contribute to the formation of protein gel networks, covalent bonds having clearly the highest binding energy. In proteins covalent cross-links can be formed either within a protein (intramolecular cross-linking) or between proteins (inter-molecular cross-linking).

Cross-links can be introduced to a food protein matrix by chemical, physical and enzymatic means. Enzymatic cross-linking is an attractive approach due to the high specificity of the enzyme catalysis. Moreover, enzymatic cross-linking can be controlled to a certain degree by changing the temperature or pH. The best-known enzymatically created cross-links are the isopeptide bonds catalysed by transglutaminases (TGs), the bonds in which tyrosine acts as a counterpart catalysed by tyrosinases and peroxidases as well as disulphide bonds catalysed by sulphydryl oxidases. Cross-linking can be a result of a direct enzymatic catalysis as in the case of TGs and sulphydryl oxidases, or it can occur indirectly by enzymatic production of reactive cross-linking agents, such as quinones, radicals or hydrogen peroxide, which in turn are able to react non-enzymatically with certain chemical groups in proteins with a subsequent cross-link formation. Tyrosinases, laccases, peroxidases, as well as glucose and hexose oxidases are examples of enzymes performing this type of catalysis. Enzymatically catalyzed cross-links in proteins are formed between reactive groups of amino acid residues, such as the primary amino groups of glutamine and lysine, the phenolic group of tyrosine and the thiol group of cysteine. The efficiency of the reactions is dependent on the accessibility of the target amino acids in a protein, the treatment or process conditions used and the type of enzyme. Cross-linking changes the molecular size and/or conformation of a protein, often leading to changes in its functional properties.

The effects of cross-linking enzymes are many and they are highly dependent on the applications. In the cereal sector TG has been used to improve the strength of weak and low-quality flour, texture and volume of bread and texture of pasta after cooking.

Cross-linking can also be exploited in milk products such as yoghurts in order to prevent syneresis or to make the soft texture firmer. Casein micelles in milk have been reported to be stabilized by TG against e.g. high pressure and heating in the presence of ethanol. The effects of cross-linking enzymes in meat systems, emphasizing the role of TG, are reviewed in detail in the following sections.

The high costs of meat production require ways to use the meat raw material in an efficient way. One approach is to restructure or reshape a value-added meat product from lower-value fresh meat parts by binding them together. Improvement of binding between meat pieces with the help of targeted and controlled formation of covalent cross-links by enzymes is an approach already adopted in the meat industry. TG has shown to be an efficient binding agent. Moreover, many processed meat products such as sausages and hams are today often associated with a negative health image due to their high salt content. As salt contributes to many properties of the meat products, both technological and sensory, reduction of its concentration is not straightforward.

The ability of TG to cross-link different meat proteins has been generally known for over two decades. In early studies the TG used was purified from mammalian sources such as human plasma, bovine plasma or guinea pig liver, all of which are Ca^{2+}-dependant enzymes. Industrial application of TG in meat binding started about a decade ago when microbial TGs were shown to polymerize various food proteins. Microbial TGs are far easier to produce than the mammalian enzymes and are more feasible due to their Ca^{2+} independence.

Potential cross-linking enzymes for food structure engineering are summarised in Table 5.1.

Table 5.1 Potential cross-linking enzymes for food structure engineering.

Enzyme	Reaction	Target amino acid(s) in proteins	Remarks
Transglutaminase (EC 2.3.2.13)	Formation of an isopeptide bond	Protein-bound glutamine, lysine	
Tyrosinase (EC 1.12.18.1)	Oxidation of the phenolic moiety of tyrosine	Tyrosine	
Laccase (EC 1.10.3.2)	Radical-generating oxidation of aromatic compounds	Tyrosine	Acts with carbohydrates containing ferulic acid residues
Peroxidase (EC 1.11.1.7)	Radical-generating oxidation of aromatic compounds	Tyrosine	Acts with carbohydrates containing ferulic acid residues
Glucose andhexose oxidases (EC 1.1.3.4 and EC 1.1.3.5)	Formation of H_2O_2 in conjunction with glucose or hexose oxidation		Indirect reaction with proteins due to H_2O_2 formation
Sulphydryl oxidases (EC 1.8.3)	Oxidation of cysteine residues to form disulphide cross-links	Cysteine	

Some of the TG-induced effects of non-meat food proteins are summarized in Table 5.2.

5.2   Maillard Reaction

The Maillard Reaction (also known as browning) is a type of non-enzymatic browning which involves the reaction of simple sugars (carbonyl groups) and amino acids (free amino groups). They begin to occur at lower temperatures and at higher dilutions than caramelization (Fig. 5.1). Browning, or the Maillard reaction, creates flavour and changes the colour of food, the taste and colour to baked bread and even the turning of beer brown. Maillard reactions generally only begin to occur above 285°F (140°C). Until the Maillard reaction occurs meat will have less flavour.

Table 5.2 Effects of TGs on non-meat food proteins.

Protein	Origin of TG	Effect
Soya proteins	Guinea pig liver	Protein polymerization
	Bovine plasma	Protein polymerization
	Human placenta	Gel formation
	Streptomyces sp.	Gel formation
	Bacterial	Protein polymerization
Milk proteins	*Streptomyces* sp.	Formation of as1-casein gels
		Improved strength of Na-caseinate gels
		Improved acid-induced gelation of micellar casein
		Altered Ca-sensitivity, hydrophobicity and viscosity of casein micelles
	Guinea pig liver	Polymerization of α-lactalbumin and β-lactoglobulin
	Streptomyces lydicus	Polymerization of α-lactalbumin and β-lactoglobulin
	Streptomyces mobaraense	Increased surface shear viscosity of Nacaseinate and casein protein films
		Improvement of water vapour permeability of a casein-gelatin film
		Improved gel formation rate, rheological properties and gel microstructure of acidified caseinate gels
		Improved physical and sensory properties of non-fat yogurt
		Cross-linking of micellar caseins in the presence of glutathione and without heat pre-treatment
Egg white proteins	Bacterial	Protein polymerization, improved gelation
Cereal proteins	*Streptomyces* sp.	Polymerization of glutenin subunits, reinforcement of gluten network
		Polymerization of oat globulin, altered textural properties, improved water and fat binding
		Improved thermal stability of gluten from damaged flour
	Bacterial	Wheat and barley protein polymerization

The Maillard reaction is a chemical reaction between an amino acid and a reducing sugar, usually requiring the addition of heat. The reactive sugar interacts with the amino group of the amino acid, and an interesting but poorly characterized odour and flavour molecules result. This process accelerates in

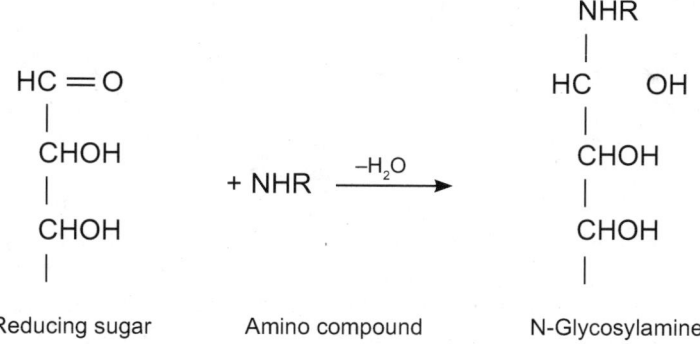

Figure 5.1 Maillard reaction

an alkaline environment because the amino groups do not neutralize. This reaction is the basis of the flavouring industry, since the type of amino acid determines the resulting flavour. In simple terms, certain foods contain carbohydrates in the form of sugars, while others contain amino acids in the form of proteins. These sugars and amino acids often exist side-by-side, as in the case of raw meats. They may also be blended together, as in the case of bread dough. As long as there is no outside catalyst, or cause for change, the meat remains red and the bread dough remains white. The Maillard reaction is the catalyst for change, primarily by the addition of heat. When bread dough or meat is introduced to a hot oven, a complex chemical reaction occurs on the surface. The carbon molecules contained in the sugars, or carbohydrates, combine with the amino acids of the proteins. This combination cannot occur without the additional heat source. The end result of this chemical recombination is the Maillard reaction. The surface of the heated bread dough is now brown, as is the outer layer of the roasted meat. Among food scientists Maillard Reaction (MR) is important because of colour and flavour formation in an enormous variety of processed foods. Such an interest devoted to a single chemical reaction might appear weird at a first glance; but actually the reaction between carbohydrate and proteins is the corner stone of many of the most important reactions taking place in food during processing because it is the origin of the colour, flavour, texture and taste of many heat-treated products. This was exactly what Louis Camille Maillard anticipated when he discovered the reaction envisaging that it could play a fundamental role in many different research fields, food chemistry, food technology, fundamental biology, diabetics, eye health and nutritional science.

The application of heat during industrial processing and household cooking of foodstuffs encompasses a variety of processes, such as boiling, frying, steaming, baking, stewing and roasting, in traditional and microwave and steam ovens. Industrial thermal treatment of foodstuffs includes many of the processes also listed for household cooking. In addition, heat has been used in traditional transformation processes other than cooking, such as toasting, kilning coffee roasting, drying processes, canning, pasteurization and related technology (UHT treatment) smoking and extrusion cooking. It is important to note that these processes can be controlled much better on industrial scale than on household level. The quality of food, from the nutritional, microbial safety point of view and sensory aspects depends on a range of variables from farm to fork, including the quality of the raw material, processing techniques, packaging and cooking. The main purpose of industrial food processing is to provide safe and high quality food as demanded by the consumer. The conduction of thermal processing in an appropriate way is the key to obtain safe food and in many cases also with enhanced nutritional functionality respect to the starting raw material.

5.2.1   Relevance in different foods

Thermal processes are frequently used in food manufacturing to obtain safe products with a prolonged shelf-life and may have a strong impact on the final quality of foods. Baking, toasting, frying, roasting, sterilization, etc. result in desired and undesired effects which all stem from the chemical reactions, namely Maillard reaction (MR), caramelization and to a minor extent, lipid oxidation, occurring while the foods are heated. One of the purposes of thermal processes is to alter the sensory properties of foods, to improve palatability and to extend the range of colours, tastes, aromas and textures in foods produced from similar raw materials. Heating also destroys enzymes and microorganisms and lowers the water activity of the food to some extent thereby preserving the foods.

In milk and dairy products MR is almost always undesired with the exception of some of the Norwegian brown cheese. In dry pasta the accurate control on MR during drying allowed to obtain dried pasta with better technological features (less starch leaching during cooking), better visual appearance (golden yellow colour) and improved storage performance. In all bakery products MR products are the main determinants of sensory quality influencing colour, flavour and texture. Moving on towards the seeds thermal treatments because of the reduced amount of water the MR rapidly produce a dark colour and great amount of melanoidins. In particular coffee represent one of the most important contributors for the dietary intake of Maillard reaction products and it has been estimated that up to two grams per day of melanoidins can come from coffee brews.

5.2.2   Importance of Maillard Reaction

The Maillard reaction is, together with fat oxidation, the chemical phenomenon that contributed to the quality of processed foods. Therefore the efforts for its understanding and control are of utmost importance to support the development of food industries. Beside the relevance in the applicative fields of food science a deep understanding of the reaction mechanisms are essential for fundamental biological issue such as those related to the ingestion of these products. The issue of the dietary AGEs (which is basically a different way to name the MR products) comes up in the last years as a possible cause of dietary related disease. Some studies demonstrating the interaction of these AGEs with endogenous receptor (named AGE receptors – RAGE) which is in turn able to trigger biochemical pathways underpinning inflammation and oxidative stress. There is a lot of debate about the relevance of dietary AGEs in this process which is also occurring with endogenously formed AGEs. *In vivo* human trials measuring a number of physiological parameters after consumption of severely heated foods gave contradictory results; however, they concur in showing some risks associated to dietary AGEs consumption for specific categories of patients such as those suffering of renal diseases, uremic and diabetics.

Potentially hazardous compounds the case of acrylamide

Beside the positive effects, some detrimental consequences of thermal processes must be carefully evaluated. The loss of thermo labile compounds such as vitamins as well as essential amino acids (lysine, triptophane) and/ or the formation of undesired tastes and off-flavours are well-established phenomena bringing about a loss in the nutritional value and sensorial quality of heated foods. All the same, the major concern arising from heating processes come from the formation of hazardous compounds, the so-called food-borne toxicants, i.e. compounds that are not naturally present in foods, but that may be developed during heating or preservation and that reveal harmful effects such as mutagenic, carcinogenic and citotoxic effects. Well-known examples of these food-borne toxicants are heterocyclic ammines, nitrosamines and polycyclic aromatic hydrocarbons. Recently two food-borne toxicants have gained much interest because of their high toxicological potential and their wide occurrence in foods: Acrylamide and hydroxymethylfurfural (HMF). Acrylamide has been added to the list of food-borne toxicants since in 2002 Swedish National Food Administration found out relevant amount of acrylamide in several heat-treated, carbohydrate- rich foods such as potato chips and crisps, coffee and bread.

Shortly after its discovery in foods, it has been clearly established that the major pathway for acrylamide formation in foods is Maillard reaction with free asparagine as main precursor. Asparagine can thermally decomposes by deamination and decarboxylation but when a carbonyl source is present the yield of acrylamide from asparagine is much higher explaining the high concentration of acrylamide detected in foods rich in reducing sugars and free asparagine such as fried potatoes and bakery products. Other minor reaction routes for acrylamide formation in foods have been postulated from acrolein, acrylic acid and from wheat gluten. Finally, acrylamide can be generated by deamination of 3-amino-propionamide (3-APA). 3-APA is an intermediate in MR, which can also form by enzymatic decarboxylation of free asparagine and can yield acrylamide upon heating even in absence of a carbonyl source.

5.3   Cross-linking of proteins

Protein–protein interactions comprise the underlying molecular mechanism of a multitude of complex biological processes. Interaction between intra-cellular proteins can be transient, wherein two or more proteins associate in the catalysis of a step in a biosynthetic or signal-transduction pathway; alternatively, stable multiprotein complexes are required in the performance of several biological functions.

Molecular chaperone complexes represent a very good example of multi-protein assemblies where the cooperative action of several components is required for folding/unfolding and transmembrane trafficking of proteins. Signal transduction pathways are illustrative of highly complex protein interaction networks displaying transient and semi-stable association of a series of proteins. Chemical cross-linking offers a direct method of identifying both transient and stable interactions. This technique involves the formation of covalent bonds between two proteins by using bifunctional reagents containing reactive end groups that react with functional groups–such as primary amines and sulphydryls–of amino acid residues. In the cell, a single protein may often engage in transient interaction with a variety of other partners in a given pathway. Where purified proteins are available, chemical cross-linking is the ideal strategy for an unambiguous demonstration of protein-protein interactions, *in vitro*. If two proteins physically interact with each other, they can be covalently cross-linked. The formation of cross-links between two distinct proteins is a direct and convincing evidence of their close proximity. In addition to information on the identity of the interacting proteins, cross-linking experiments can reveal the regions of contact between them.

Cross-linking agents: Cross-linkers (CL) are either homo- or hetero-bifunctional reagents with identical or non-identical reactive groups, respectively, permitting the establishment of inter- as well as intra-molecular cross-linkages. Inter-subunit cross-links have been used for determination of the quaternary structure and arrangement of subunits within homo-oligomeric proteins and intra-subunit cross-links for maintenance of stable tertiary structure. Ligand-induced conformational changes in proteins can be analyzed by a comparison of the rate/extent of cross-linking in the ligand-bound versus the unliganded states. In studies of hetero-oligomeric enzymes or multiprotein complexes–containing several different polypeptides–a cross-linking offers a reliable tool for unravelling spatial relationships of the components.

Homo-bifunctional reagents, specifically reacting with primary amine groups (i.e. ε-amino groups of lysine residues), have been used extensively as they are soluble in aqueous solvents and can form stable inter- and intra-subunit covalent bonds. Glutaraldehyde, a popular reagent, has been used in a variety of applications where maintenance of structural rigidity of protein is important. Homo-bifunctional imidoesters, with varying lengths of the spacer arm between their reactive end groups, are particularly useful in determination of the distances between linked residues on surfaces of neighbouring subunits in oligomeric proteins. Some examples of imidoesters are dimethyl adipimidate (DMA), dimethyl suberimidate (DMS) and dimethyl pimelimidate (DMP) with spacer arms of 8.6Å, 11Å and 9.2Å, respectively. DMA and DMS have been used to verify the quaternary structures of many oligomeric enzymes. For identification of unknown interacting partners of a given target protein, reversible homo-bifunctional cross-linkers such as N-hydroxysuccinimide (NHS) esters are advantageous in that the interacting proteins can be recovered and identified.

Examples of reversible NHS-esters are [dithiobis(succinimidyl propionate)] (DSP), dithiobis (succinimidylpropionate), which is water insoluble and the soluble derivative dithiobis (sulphosuccinimidylpropionate), [3,3′-dithiobis (sulphosuccinimidyl propionate)] (DTSSP). These reagents can be cleaved by treatment with thiols, such as β–mercaptoethanol or dithiothreitol. A crude cellular extract is treated with the cross-linking reagent and immune-precipitation with antibodies specific for the target protein is used to recover the assorted complexes containing it. The immunoprecipitate is then treated with a reducing agent thereby separating the target from its partner protein(s).

This can be done simply by dissolving the immunoprecipitate in the denaturing sample buffer, prior to resolution by Sodium dodecyl sulphate-Polyacrylamide gel electrophoresis (SDS-PAGE). The stained, polypeptide bands, corresponding to the partner proteins, can be excised and identified by Matrix-assisted laser desorption ionization-time-of-flight mass spectrometry (MALDI-TOFMS), following in-gel trypsin hydrolysis.

Protein–protein interactions can also be analysed by employing hetero-bifunctional reagents that can form stable thioester bonds between two interacting proteins. For instance, cross-linkers with one amine-reactive end and a sulphydryl-reactive moiety at the other end are selected in situations where the catalytic site of one of the protein contains an amine. To avoid damaging the active site, it is preferable to target other side chains for modification. Hetero-bifunctional cross-linkers with a NHS ester at one end and an SH-reactive groups–maleimides or pyridyl disulphides–can be used. Another approach is to modify the lysine residues of one protein to thiols and the second protein is modified by addition of maleimide groups followed by formation of stable thioester bonds between the proteins. If one of the proteins has native thiols, these groups can be reacted directly with maleimide attached to the other protein.

Alternatively, hetero-bifunctional reagents containing a photo-reactive group, such as Bis[2-(4-azidosalicylamido)ethyl)] disulphide, (BASED), can also be used. In the next step, the reaction between the two proteins is performed giving rise to stable thioester bonds. With all hetero-bifunctional reagents the cross-linking experiment is conducted in two steps. One protein is modified first with one reactive group of the hetero-bifunctional reagent; the remaining free reagent is removed. In the next step the modified protein is mixed with the second protein, which is then allowed to react with modifier group at the other end of the reagent. Some examples of such reagents, among others, are [Succinimidyl 3-(2-pyridyldithio) propionate] (SPDP) and a related compound, [Succinimidyl *trans*-4-(maleimidylmethyl) cyclohexane-1-carboxylate] (SMCC).

5.3.1   Procedures

Before starting cross-linking experiments, it is important to select the requisite cross-linker for optimal reaction with the proteins. Features such as solubility, membrane permeability, spacer arm length and reversibility of cross-links should be taken into consideration. Moreover, the experimental procedure should be such that it can be conducted under mild conditions of pH and temperature to preserve the native structure of the proteins. Homo-bifunctional reagents such as DMS, DMA, and glutaraldehyde work very well with *Neurospora* proteins.

Glutaraldehyde: Treatment with cross-linkers should be conducted in buffers free from amines. Phosphate buffers at pH 7.5–8.0 and HEPES buffers are suitable, whereas Tris-HCl should be avoided. For glutar-aldehyde treatment, reaction mixtures with 50–100 µg of interacting proteins in

20 mM HEPES buffer (pH 7.5) in a total volume of 100 μl are treated with 5 μl of 2.3% freshly prepared solution of glutaraldehyde for 2–5 minutes at 37°C. The reaction is terminated by addition of 10 μl of 1 M Tris-HCl, pH 8.0. Cross-linked proteins are solubilized by addition of an equal volume of Laemmli sample buffer to which a few μl of 0.1% bromophenol blue is added and electrophoresis is conducted in 5–20% SDS-polyacrylamide gels.

Dimethyl adipimidate (DMA) and dimethyl suberimidate (DMS): Cross-linking reactions with DMA and DMS are performed in 20 mM buffer (pH 7.5). The cross-linkers are dissolved in the buffer to a concentration of 6 mg/ml and pH is adjusted to 8.5 by addition of NaOH. A solution of 1 mg/ml of protein containing a concentration of the diimidoester at 1 or 2 mg/ml is allowed to cross-link for 3 hours at room temperature. The reaction is terminated by addition of the denaturing sample buffer, bromophenol blue and the mixture is resolved by SDS-PAGE.

Stained gels of oligomeric proteins, for instance tetrameric proteins, on treatment with the above homo-bifunctional reagents will show cross-linked dimers, trimers and tetramers, where inter-subunit cross-linking has occurred, as well as cross-linked monomers with intra-subunit cross-links, along with the unmodified monomers. Computer-assisted densito-metric analysis can be used to estimate the relative quantities of monomeric, dimeric, trimeric and tetrameric species, resolved by SDS-PAGE. If the arrangement of subunits in the native tetramer is symmetrical, the contacts between all of the individual protomers in the quaternary structure should be identical. On statistical, theoretical grounds, therefore, the frequency of single cross-links per tetramer will be the higher than that for two cross-links and that for 3 and 4/tetramer will be correspondingly lower (uncross-linked monomer > dimer > trimer > tetramer). If the arrangement of protomers in the tetramer is asymmetrical, the distribution of cross-linked species will not conform to this pattern. A preponderance of cross-linked dimeric species, relative to monomers and cross-linked trimers and tetramers is indicative of a 'dimer-of-dimer' structural arrangement.

Reversible, thio-cleavable bifunctional reagents: The advantage of using reversible reagents is the ability to liberate the participating proteins from the cross-linked product. As a result, proteins interacting with a given protein (X) can be identified. Cross-linking reactions are conducted by mixing crude cellular extracts or partially purified samples, with protein X. Cross-linking of *Neurospora* proteins, in a crude extract, works well with DSP. The reagent is dissolved in DMSO at 25 mM. The reaction buffer for cross-linking is 20 mM sodium phosphate, 0.15 M NaCl, pH 7.5 (PBS), or HEPES may be used instead. To the protein solution (0.25 to 1 mg/ml containing protein X and the crude extract) is added DSP stock solution to a final concentration of 2 mM. After incubation for 2 hours on ice, 1 M Tris-HCl (pH 7.5) is added to a concentration of 20 mM to stop the reaction.

The mixture is incubated for 15 min. The complexes are recovered by overnight treatment with an antibody-matrix (immobilized anti-X IgG) and the immune precipitate is collected by centrifugation. The precipitate is washed with PBS to remove the unbound material and Dithiothreitol (50 mM final concentration) is added and following incubation at 37°C for 30 min the mixture is resolved by SDS-PAGE. The gel is stained with Coomassie blue and the polypeptide bands of interacting proteins are excised, subjected to in-gel trypsin hydrolysis and the sample prepared for MALDI-TOF mass spectrometric analysis for identification of the protein. This procedure has been used for analysis of proteins, interacting with Hsp80 of *Neurospora*, in extracts of germinating conidia and mycelial homogenates.

5.4   Role of cross-linking enzymes in meat systems

5.4.1   Meat protein modification

In addition to meat tenderisation by proteolytic enzymes, meat proteins are modified by creating new covalent bonds in the system rather than by degrading them. Hitherto TG has been the main enzyme studied and applied in meat protein modification. Although the most obvious source of TG for the meat industry would be blood, microbial enzymes have superseded the mammalian ones due to their Ca^{2+}-independence, feasible pH and temperature profiles and commercial availability for industrial use. TG was examined in covalent cross-link formation of myofibrillar proteins already more than four decades ago when Derrick and Laki used the enzyme to label actin and tropomyosin with 14C-putrescine. TGs of different origins that have been studied ever since in myofibrillar protein modification have differed in their cross-linking specificity. Ca^{2+}-dependent mammalian TGs, i.e. bovine plasma TG, human placental factor XIII or pig plasma TG, are capable of modifying myosin and actin, whereas microbial enzymes originating from *Streptomyces* have shown limited activity on actin. Although actin has not exhibited major changes in the presence of microbial TG, an intramolecular cross-link catalyzed by this enzyme has been identified in rabbit actin.

It has been suggested that TGs from different origins recognize different peptide sequences in the same proteins, thus leading to differences in cross-linking efficiency. TG-catalyzed modification of myofibrillar proteins are shown in Table 15.3. In addition to myofibrillar proteins, TG is capable of cross-linking collagen, the main protein component of the connective tissue.

Although no detailed reports on how TG acts with intramuscular collagen or about the subsequent consequences on meat quality have been published, it is known that TG cross-links at least porcine skin type I collagen, resulting in greater resistance to proteolysis, increased tensile strength and shift of thermal denaturation towards higher temperatures compared with the native collagen. Although meat protein cross-linking has hitherto been carried out almost solely by TG, mushroom tyrosinase is also known to be capable of cross-linking collagen and increasing the firmness of gels of unheated pork. In addition, laccase has been claimed to cross-link Bovine serum albumin (BSA) and gelatin and to improve the textural properties of pork sausage and kamaboko gels.

Table 15.3 TG-catalyzed modification of myofibrillar proteins.

Myofibrillar protein substrate	Origin of the protein	Origin of TG	Modification
Myosin	Beef	Bovine plasma Human placental factor XIIIa	Polymerization
Actomyosin	Turkey	Guinea pig liver	Polymerization
	Beef	Guinea pig liver	
	Tilapia	Tilapia (*Oreochromis niloticus*) muscle	
Actin andmyosin	Rabbit	Human plasma factor XIII	Coupling of actin to myosin
Actin	Rabbit	*Streptomyces mobaraense*	Intramolecular isopeptide bond
	Beef	Human placental factor XIIIa	Partial polymerization of actin
Myofibrillar proteins	Mackerel	*Streptomyces ladakanum*	Polymerization of MHC
	Beef heart	Microbial	Polymerization of myosin, possible intramolecular bond in MHC
	Chicken breast		Intramolecular cross-links inMHC and actin in salt-free conditions. Suggested proteolytic fragmentation
	Spent hen breast	Pig plasma	Polymerization of MHC and actin
	Beef	*Streptomyces mobaraense*	Polymerization of myosin

5.4.2   Effects on meat protein gels

By adding a cross-linking enzyme to a meat protein system the gel formation, texture and water-holding properties of a gel can be modified. Due to the efficient cross-linking of myosin TG is generally capable of positively affecting gelation and the texture of meat protein gels. As well as by enzymatic protein modification, gelation is affected by the protein type and quantity, protein solubility, pH, temperature and the amount and types of salts in the system.

Gelation of solubilized myofibrillar proteins is a functional property that is necessary for restructuring of fresh meat, but particularly in processing meat products such as sausages or hams.

TG has been used in meat systems to improve the binding of fresh meat pieces together, low-temperature gelation of fish, heat-induced gelation of meat and fish and to strengthen meat gels. Ramirez-Suarez and others reported about 2–3-fold increase in storage modulus (G') in the presence of TG for heated gels containing 3% or 5% of beef heart protein with poor binding and water binding characteristics.

Improvement of G' was dependent on TG but independent on the TG-pretreatment time (0.5–15 h) and temperature (5 and 15°C) within the given range of these variables. TG-aided myosin cross-linking was observed in these pretreatment conditions whereas actin did not show major modifications. The result indicated that myosin polymerization occurring during the TG-pretreatment was sufficient to cause clear improvement in heat-induced gelation. Smith and others obtained 3.9-fold improvement in gel strength when fish surimi was set at 25°C for 80 min in the presence of TG and then heated to 90°C. Lantto and others recently reported a 2-fold increase in firmness of unheated pork gels by TG and 52% increase by mushroom tyrosinase. The enzymatic treatment was carried out at 4°C for 22 h. In this study gel formation during heating was improved by TG but not by tyrosinase.

Enzyme-catalyzed formation of additional covalent bonds in structural proteins leads, by definition, to firmer gel structures as has been clearly shown in many studies and is often desired in various meat product applications. However, excessive increase in gel firmness due to additional transverse bonds in the protein network may cause constraint in protein mobility and flexibility of myofibrils, leading to undesired decrease in water holding.

Decreased water holding is mostly an unwanted phenomenon and has been observed to take place particularly in heated comminuted meat systems when salt concentrations are low.

TG has not been shown unequivocally to be an adequate salt replacer, but in high enough salt levels by ensuring sufficient protein solubility, it is capable of improving water-holding properties.

When salt levels are low (<2% NaCl), introduction of strong bonds in proteins by TG has not generally been found to improve water-holding of heated meat or fish systems, although textural properties have been improved. The reason is most probably the increased protein–protein and reduced protein–water interaction due to TG catalyzed covalent bonds and subsequent limited protein solubility and swelling ability resulting in reduced WHC. Smith and others postulated that the combined effect of TG and low amount of salt (1.5%) in a comminuted and heated meat system leads to increased cooking loss because of the reduced WHC of solubilized meat proteins cross-linked with TG. This conclusion is very much in line with what has been known for years about water-holding in unheated and heated meat systems. Extra cross-links between myofibrillar proteins may prohibit the swelling of the myofibrils, which is a requirement for good water-holding.

However, contrasting results indicating improved water-holding caused by TG in low salt concentrations have also been reported at least by Tseng and others, Pietrasik and Li-Chan were able to increase cooking yield as a function of increasing TG dosage in chicken meatballs containing 1% salt, 0.2% tripolyphosphate and 25% pork fat. The high fat amount, phosphate and no added water might have contributed to the result. Pietrasik and Li-Chan studied the effects of TG and various salt levels (0.2%) on pork batter gel properties. The authors found that low salt levels, as expected, caused decreased gel firmness and cooking yield, but TG was able to increase both properties in the low-salt batters but not to the same level as with the normal salt amount (2%). Trespalacios and Pla reported significant reduction of the expressible moisture of low-salt chicken batters treated with TG. However, in their study ground chicken meat was mixed with fresh egg yolk and dehydrated egg white, both at 10% of the meat mass. Although the reported effects of egg proteins on hydration properties of meat gels are said to be somewhat contradictory, they may have played a role in water-holding in the studied chicken meat batters.

High quality but moderate prices of meat products demanded by consumers have been the driving forces to develop methods to restructure low-value cuts and muscles of poorer quality to improve their market value by making them palatable steaks resembling intact muscle and to maximize the efficiency of carcass utilization. Traditionally salt and phosphates in conjunction with heat treatment have been used to bind meat pieces together. Unheated comminuted products are usually frozen to enhance meat binding. Nowadays when consumers demand fresh, unfrozen meat as well as lower salt contents, technologies have been developed to eliminate the need for freezing and to enable the use of less salt. One of these technologies is TG-aided restructuring, which has been adopted in commercial scale production already some time ago and is still the main TG application in the meat sector. TG has been found to improve the strength of restructured meat protein gels with or without low levels of added salt and phosphates. Kuraishi and others reported that restructured meat products that are traditionally prepared using salt and phosphates to promote extraction of proteins can be prepared without added salt using TG, and that binding strength can be boosted with caseinate added to the system as an extender. Caseinate is an excellent substrate for microbial TG. It polymerizes during the enzymatic reaction, it becomes viscous and thus acts like a glue binding the meat pieces together. The TG/caseinate technology has also been tested in fish products. Textural parameters of fish gels, i.e. firmness and springiness were best when TG and caseinate were present. In addition, all the measured textural parameters increased along with the increasing salt concentration (0%, 1% and 2%). The amount of expressible water was drastically increased (indicating reduced WHC) by TG in all tested salt concentrations, particularly in non- and low-salt gels. Adding caseinate into the system reduced expressible water to some extent but not to the level of the normal salt control.

As already stated, WHC is a feature that is affected by many technological factors, but also by many intrinsic features of meat proteins themselves. Due to the complexicity of WHC *par se* and the inconsistent findings about the effects of TG in the development of WHC, no generalizations concerning the role of TG in the water-holding of meat systems can be made. Despite this, the positive effect of TG on the textural properties is beyond dispute.

5.4.3   Future applications of protein cross-linking

Although protein cross-linking is often considered to be detrimental to the quality of food, it is increasingly clear that it can also be used as a tool to improve food properties. The more we understand of the chemistry and biochemistry that take place during processing, the better placed we are to exploit it–minimizing deleterious reactions and maximizing beneficial ones.

Food chemists are faced with the task of understanding the vast array of reactions that occur during the preparation of food. Food often has complex structures and textures and through careful manipulation of specific processes that occur during food preparation, these properties can be enhanced to generate a highly marketable product. The formation of a structural network within food is critical for properties such as food texture. From a biopolymer point of view, this cross-linking process has been successfully applied in the formation of protein films, ultimately for use as packaging. Enhancing textural properties, emulsifying and foaming properties, by protein cross-linking has been a subject of interest of many working in this area, as exemplified by the number of studies detailed above. Cross-linking using chemical reagents remains challenging, in terms of both controlling the chemistry and gaining consumer acceptance. However, as the extensive recent use of transglutaminase dramatically illustrates, protein cross-linking using enzymes has huge potential for the improvement of traditional products and the creation of new ones. Transglutaminase itself will no doubt find yet more application as its precise mode of action becomes better understood, especially if variants of the enzyme with a broader substrate specificity are found. Whether other enzymes, which cross-link by different mechanisms, can find equal applicability remains open to debate. The thiol exchange enzymes, such as protein disulphide isomerase, may offer advantages to food processors if their mechanisms can be unravelled and then controlled within a foodstuff. Other enzymes, such as lysyl oxidase, have not yet been used, but have potential to improve foods, especially in the light of recent work characterizing this class of enzymes, which may allow their currently unpredictable effects to be better understood.

6

Enzymes as biocatalysts

6.1 Introduction

Enzymes are very efficient catalysts for biochemical reactions. They speed up reactions by providing an alternative reaction pathway of lower activation energy. Like all catalysts, enzymes take part in the reaction that is how they provide an alternative reaction pathway. But they do not undergo permanent changes and so remain unchanged at the end of the reaction. They can only alter the rate of reaction, not the position of the equilibrium. Most chemical catalysts catalyze a wide range of reactions. They are not usually very selective. In contrast enzymes are usually highly selective, catalyzing specific reactions only. This specificity is due to the shapes of the enzyme molecules. Many enzymes consist of a protein and a non-protein (called the cofactor). The proteins in enzymes are usually globular. The intra- and intermolecular bonds that hold proteins in their secondary and tertiary structures are disrupted by changes in temperature and pH. This affects shapes and so the catalytic activity of an enzyme is pH and temperature sensitive.

Enzymes break down our food. Enzymes are the workhorses of the body. When we eat, enzymes break down the food into tiny particles which can be converted into energy in the body. The process starts in the mouth, where an enzyme called amylase attacks all incoming food particles. Like a well-drilled team of engineers, different enzymes continue to break down the food all the way to the stomach and intestines.

The breakdown of food is an essential part of the conversion of food into energy. Undigested food is unable to pass on the energy stored within it. The enzymes involved in the digestion process carry out the final cutting of the food particles so that they can be easily converted into the essential energy needed by all parts of our body. Without enzymes we would die from starvation. Enzymes are catalysts. This means that they make biochemical reactions happen faster than they would otherwise.

6.2 Characteristics of enzymes

1. Enzymes catalyze the conversion of one or more compounds called substrates into one or more compounds called products.

2. Enzymes accelerate (speed up) the rate of reaction by a factor of at least 106.
3. Like all catalysts, enzymes are neither, consumed or altered after catalysis.
4. Enzymes are highly selective catalysts; they are specific for the type of reaction catalyzed and for a single substrate or a set of closely related substrates.

6.3   Enzyme catalysis

The enzyme (E) has a reactive site (called active centre) which binds the reactant [or substrate (S)] by non-covalent interactions. The reaction starts by substrate binding to the enzyme to form ES complex, then reaction proceeds with the formation of the product (P):

$$E + S \rightleftharpoons ES \rightleftharpoons EP \rightleftharpoons E + P$$

6.3.1   Factors that affect the rate of enzyme-catalyzed reactions

1. pH: A change in pH changes the rate of enzyme-catalyzed reaction. A bell shaped curve is obtained when the rate of the reaction is plotted against pH with an optimum pH at which the rate is optimum. Changes in pH (Fig. 6.1) can change the ionization of the substrate or the catalytic site of the enzyme.

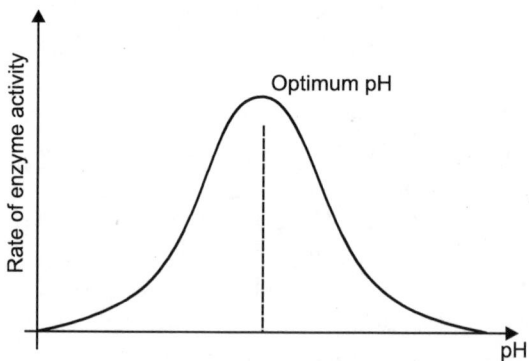

Figure 6.1 pH and enzyme activity.

2. Temperature: The rate of an enzyme-catalyzed reaction increases with increasing temperature (Fig. 6.2) up to an optimum point (called optimum temperature), then it decreases because of denaturation of the enzyme protein.

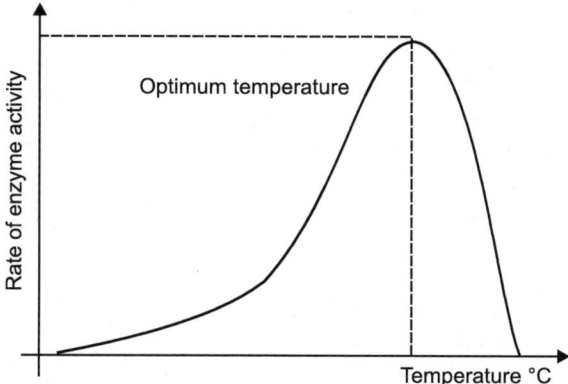

Figure 6.2 Temperature and enzyme activity.

3. Substrate concentration: The rate of enzyme-catalyzed reaction is increased with the increase of substrate concentration [S] till a point is reached (point of saturation), when there is no further increase in the reaction rate with the increase in substrate concentration. This observation can be explained as follows:

   (a) At low substrate concentration [S]: The active sites (centers) of the enzyme are still not saturated with substrate molecules and addition of substrate increases [ES] complex gradually.

   (b) At the point of saturation with the substrate: As the substrate concentration increases, the sites become saturated and as substrate is added no further increase in the [ES] with no further increase in the rate of reaction and the result is shown in Fig. 6.3.

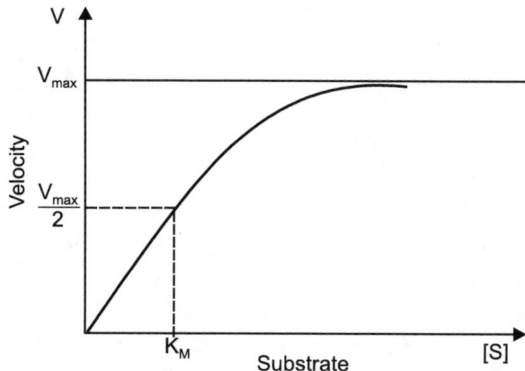

Figure 6.3 Rate of enzyme-catalyzed reaction and substrate concentration.

V_{max} = Maximal velocity or Maximal rate of the reaction.

K_M = The substrate concentration that gives half maximal velocity (or $1/2\ V_{max}$). K_M is a good measure of enzyme affinity towards its substrate. This means that, the smaller the K_M the higher the affinity and vice versa.

4. Concentration of the product (P) of the reaction: Most enzyme-catalyzed reactions are reversible; therefore, if the products are not properly removed from the cell, they will inhibit the reaction rate.
5. Physical agents: Enzyme activity is seriously affected by some physical agents that cause denaturation of proteins, e.g. ultrasonic vibrations, X-ray, ultraviolet light (UV), repetitive freezing and thawing, etc.
6. Inhibitors: Many compounds can combine with enzymes although they are not substrates, therefore, they may block the catalytic function of the enzyme. These compounds are called inhibitors. There are 2 main types of inhibitors: (i) competitive inhibitors and (ii) non-competitive inhibitors.

6.3.2   Classification of enzymes

Enzymes are divided into six (6) major classes:
1. *Oxidoreductases:* Involved in oxidation and reduction reactions, e.g. oxidases, dehydrogenases, oxygenases, peroxidases.
2. *Transferases:* Transfer functional groups, e.g. amino or phosphate groups, e.g. aminotransferases.
3. *Hydrolases:* Catalyse hydrolysis of the substrate, e.g. lipase, maltase, protease.
4. *Lyases:* Add or remove elements of water, ammonia, or CO_2 to form double bonds, e.g. decarboxylases.
5. *Isomerases:* Catalyze the rearrangements of atoms within a molecule to give its isomer, e.g. glucose to fructose.
6. *Ligases:* Joins two molecules, e.g. carboxylases and synthetases.

6.3.3   Enzyme cofactors

The majority of enzymes require the presence of some components to help them catalyze the reaction; these components are called cofactors.

Cofactors can be classified into 3 groups as follows:
1. *Coenzymes:* These are organic compounds with low molecular weight, heat stable, loosely attached to the enzyme molecule, therefore can be separated easily by simple dialysis. Examples: NAD^+, $NADP^+$, B_6P.

2. *Prosthetic groups:* These are also low molecular weight organic compounds, which are firmly attached to the enzyme protein; therefore they are not separated by dialysis. Examples: FAD, heme.
3. *Metal activators:* These are inorganic monovalent and divalent cations such as K^+, Mn^{2+}, Ca^{2+}, Zn^{2+}, and Mg^{2+}; these cations may be either loosely or firmly attached to the enzyme protein.

6.3.4   Holoenzyme, apoenzyme and coenzyme

Holoenzyme is the term used to describe the whole enzyme molecule which may be composed of an enzyme protein and a coenzyme or a prosthetic group. In this case, the enzyme protein is called apoenzyme. Therefore:

Holoenzyme = Apoenzyme + Coenzyme (or Prosthetic group)

Remember that: Neither apoenzyme nor coenzyme alone is catalytically active. Only when they are both combined together they become catalytically active.

6.3.5   Proenzymes (Zymogens)

Proenzymes (or zymogens) are inactive form of some enzymes produced by some cells, to become active in a second site like intestine and stomach.

Example: Pepsinogen (inactive form) is produced by gastric mucosal cells, and in the stomach, it is activated to pepsin by gastric acidity by active pepsin (Auto-activation):

$$\text{Pepsinogen} \xrightarrow{\text{HCl}} \text{Pepsin}$$
$$\text{(Inactive)} \qquad\qquad \text{(Active)}$$

6.3.6   Isoenzymes

Isoenzymes mean a group of enzymes which catalyze one and the same reaction but they differ in physical, chemical, immunological and electrophoretic properties.

Examples:
1. The enzyme lactate dehydrogenase (LDH) occurs in different forms, which can be separated from each other by electrophoresis into 5 types: LDH_1, LDH_2, LDH_3, LDH_4 and LDH_5 which come from different organs; LDH_1 is found in the heart and LDH_5 occurs in the liver. Determination of the level of either of these isoenzymes in blood is of value in the diagnosis of liver and heart diseases.
2. Creatine phosphokinases (CK) also have different isoenzyme forms. These isoenzymes are of value in the diagnosis of muscle, heart and brain diseases.

6.4   Enzyme inhibitors

An enzyme inhibitor is a molecule that binds to an enzyme and decreases its activity. Since blocking an enzyme's activity can kill a pathogen or correct a metabolic imbalance, many drugs are enzyme inhibitors. They are also used in pesticides. Not all molecules that bind to enzymes are inhibitors; enzyme activators bind to enzymes and increase their enzymatic activity, while enzyme substrates bind and are converted to products in the normal catalytic cycle of the enzyme.

6.4.1   Competitive inhibitors

These are compounds which compete with the natural substrate for the biding with the active site of the enzyme, therefore decreasing the catalytic activity of the enzyme, e.g. the enzyme succinate dehydrogenase (SDH) oxidizes succinic acid to fumaric acid. Addition of malonic acid, which has a chemical structure similar to succinic acid, will cause inhibition of this enzyme. Such inhibition can be reversed by increasing the concentration of the substrate succinic acid. This shows that there is competition between the inhibitor (i.e. malonic acid) and the substrate (i.e. succinic acid) to bind the active site of the enzyme (Fig. 6.4). This competition is due to similarity in chemical structure between succinic acid and malonic cid.

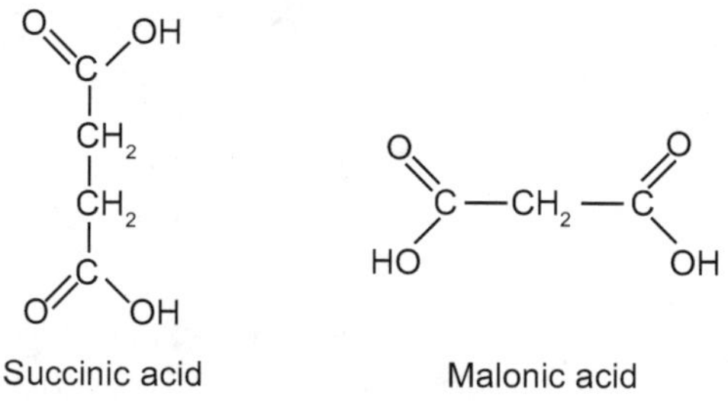

Succinic acid   Malonic acid

Figure 6.4 Inhibitor (i.e. malonic acid) and the substrate (i.e. succinic acid).

6.4.2   Non-competitive inhibitors

These are compounds which combine irreversibly with the active center of the enzyme and are not displaced with increasing substrate concentration.

Examples: Iodoacetate react with SH – groups essential for the catalytic activity of some enzymes:

Enzyme – SH + ICH$_2$–COOH → Enzymes–S–CH$_2$–COOH + HI
Active enzyme   Iodoacetic   Inactive enzyme

pH levels

An enzyme loses its ability to function when it loses its shape. Because ionic and hydrogen bonds are important factors in enzymes retaining their specific shapes, pH levels that interrupt this bonding, negatively impact enzyme activity. Extreme pH levels may also impact the substrate or its bonds, hindering the ability of the substrate to react with the enzyme, causing a further decrease in activity.

Temperature levels

Temperature levels also play an important role in how effectively enzymes function. Very high temperatures, however, run the risk of denaturing enzymes by destroying their bonds and their shapes. Low temperatures, on the other hand, can cause enzymes to slow down and decrease their rate of interaction with substrates.

Substrate concentration

Substrates, as the 'key' that unlocks enzyme activity, have a significant effect on how efficiently enzymes function. Higher substrate concentrations will increase an enzymes activity, simply because more substrate present means more substrate available for binding. At low substrate concentrations, increases in substrate concentration result in a steep increase in enzyme activity. At very high substrate concentrations, however, the increase in enzyme activity flattens out, and, at very high concentrations, no increase in enzyme activity will be realized with an increase in substrate concentration.

6.5   Biocatalysis

Biocatalysis may be broadly defined as the use of enzymes or whole cells as biocatalysts for industrial synthetic chemistry. They have been used for hundreds of years in the production of alcohol via fermentation, and cheese via enzymatic breakdown of milk proteins. Over the past few decades, major advances in our understanding of the protein structure–function relationship have increased the range of available biocatalytic applications. In particular, new developments in protein design tools such as rational design and directed evolution have enabled scientists to rapidly tailor the properties of biocatalysts for particular chemical processes. Biocatalysis has become an important tool for industrial chemical synthesis and processes and is on the verge of significant growth. Presently, approximately 100 different biocatalytic processes are implemented in pharmaceutical, chemical, agricultural, and food industries. Similar to other catalysts, biocatalysts increase the speed in which a reaction takes place but do not affect the thermodynamics of the reaction. However, they offer some unique characteristics over conventional catalysts (Table 6.1). The most important advantage of a biocatalyst is its high selectivity. Biocatalytic processes are similar to conventional chemical processes in many ways. However, when considering a biocatalytic process one must account for enzyme reaction kinetics and enzyme stability for single-step reactions, or metabolic pathways for multiple-step reactions. Figure 6.4 shows the key steps in the development of a biocatalytic process. It usually starts with the identification of a target reaction, followed by biocatalyst discovery, characterization, engineering, and process modelling. In many cases, biocatalyst engineering is the most time consuming step, often involving two major approaches: rational design and directed evolution. In addition to biocatalyst development, product isolation is an important step. The overall process economics depends on all these factors, which needs to be demonstrated in a pilot-scale plant before scale-up. Biocatalysts can constitute a significant portion of the operating budget; however, their cost can be reduced by reusing them when immobilized.

Table 6.1 Advantages and disadvantages of biocatalysis in comparison with chemical catalysis.

Advantages	Disadvantages
Generally more efficient (lower concentration of enzyme needed)	Susceptible to substrate or product inhibition
Can be modified to increase selectivity, stability, and activity	Solvent usually water (high boiling point and heat of vaporization)
More selective (types of selectivity: chemo-selectivity, regio-selectivity, diastereo-selectivity, and enantio-selectivity)	Enzymes found in nature in only one enantiomeric form
Milder reaction conditions (typically in a pH range of 5–8 and temperature range of 20–40°C)	Limiting operating region (enzymes typically denatured at high temperature and pH)
Environment friendly (completely degraded in the environment)	Enzymes can cause allergic reactions

6.5.1   Enzyme-based biocatalysis vs. whole-cell biocatalysis

Both isolated enzymes and whole cells can be used as biocatalysts. Compared to whole cells, isolated enzymes offer several benefits, including simpler reaction apparatus, higher productivity owing to higher catalyst concentration, and simpler product purification. Until recently, only enzymes that were abundantly produced by cells could be used in industrial applications. Now it is possible to produce large amounts of an enzyme through the use of recombinant DNA technology. In brief, the DNA sequence encoding a given enzyme is cloned into an expression vector and transferred into a production host such as *Escherichia coli* or *Saccharomyces cerevisiae* for gene expression. The over expressed enzymes are purified from the cell extracts based on their chemical and physical properties.

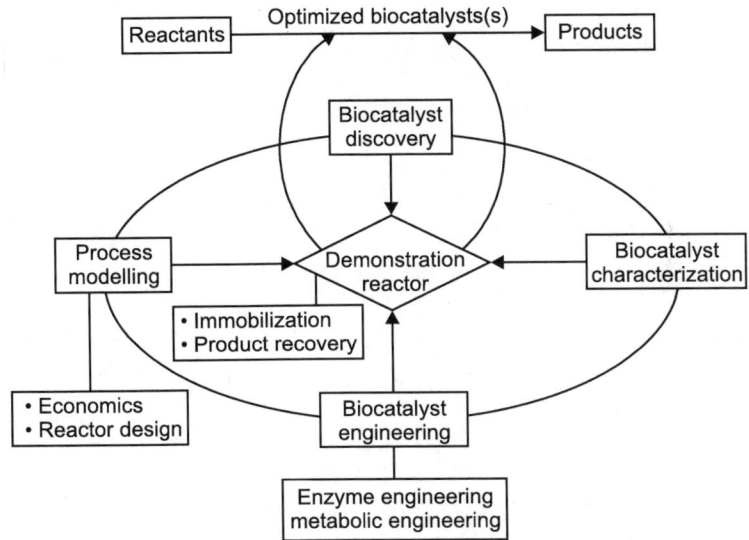

Figure 6.5 Flowchart of the development of a biocatalytic process.

The most commonly used enzyme purification techniques include electrophoresis, centrifugation, and chromatography. Centrifugation separates enzymes based on their differences in mass or shape, whereas electrophoresis separates enzymes based on their differences in charge.

Liquid chromatography separates enzymes based on their differences in charge (ion-exchange chromatography), in mass (gel filtration chromatography), or in ligand-binding property (affinity chromatography).

The whole-cell biocatalysis approach is typically used when a specific biotransformation requires multiple enzymes or when it is difficult to isolate the enzyme.

6.5.2   Nonaqueous biocatalysis

Enzymes are well suited to their natural aqueous environment; however, biotransformations in industrial synthesis often involve organic molecules insoluble in water. More importantly, because of its high boiling point and high heat of vaporization, water is usually the least desired solvent of choice for most organic reactions. Thus, shifting enzymatic reactions from an aqueous to an organic medium is highly desired. Recent advances in protein engineering and directed evolution have aided in the development of enzymes that show improved activity in organic solvents.

6.5.3   Biocatalyst immobilization

Immobilization is the process of adhering biocatalysts (isolated enzymes or whole cells) to a solid support. Immobilized biocatalysts offer several potential advantages over soluble biocatalysts, such as easier separation of the biocatalysts from the products, higher stability of the biocatalyst, and more flexible reactor configurations. More than one hundred techniques for immobilizing enzymes have been developed which can be divided into five major groups summarized in Table 6.2. Recently the emerging field of nanotube biotechnology has created another possible means of immobilizing biocatalysts.

6.5.4   Biocatalyst engineering

Nature has supplied us with a vast array of biocatalysts capable of catalyzing numerous biological reactions. Unfortunately, naturally occurring biocatalysts are often not optimal for many specific industrial applications, such as low stability and activity. Moreover, naturally occurring biocatalysts may not catalyze the reaction with the desired non-natural substrates or produce the desired products. To address these limitations, molecular techniques have been developed to create improved or novel biocatalysts with altered industrial operating parameters. It should be noted that, for enzyme-based biocatalysts, many molecular techniques have been developed for engineering enzymes with novel or improved characteristics.

Metabolic engineering is a rapidly growing area with great potential to impact biocatalysis. It has been broadly defined as 'the directed improvement of product formation or cellular properties through modifications of specific biochemical reaction(s) or the introduction of new one(s) with the use of recombinant DNA technology.'

In the past two decades, metabolic engineering has been successfully used to engineer microorganisms to produce a wide variety of products, including polymers, aromatics, carbohydrates, organic solvents, proteins, antibiotics, amino acids, and organic acids. According to the approach taken or the aim, these applications can be classified into seven groups: (i) expression of heterologous genes for protein production, (ii) extension of the range of substrate for cell growth and product formation, (iii) design and construction of pathways for the production of new products, (iv) design and construction of pathways for degradation of xenobiotics, (v) engineering of cellular physiology for process improvement, (vi) elimination or reduction of by-product formation and (vii) improvement of yield or productivity.

Table 6.2 Methods of enzyme immobilization.

Covalent attachment

Isolated enzymes usually attached through amino or carboxyl groups to a solid support; variety of supports such as porous glass, cellulose, ceramics, metallic oxides.

Adsorption

Ion-exchangers frequently used in industry because of simplicity.

Industrial applications include anion-exchangers diethylaminoethyl cellulose (DEAE-cellulose) and the cation-exchanger carboxymethyl cellulose (CM-cellulose).

Entrapment in polymeric gels

Enzyme becomes trapped in gel volume by changing temperature or adding gel-inducing chemical.

Enzymes may be covalently bound to gel (for instance, polyacrylamide cross-linked with N,N'-methylenebisacrylamide) or noncovalently linked (calcium alginate).

Intermolecular cross-linking

Enzyme cross-linked with bifunctional reagents.

Popular cross-linkers are glutaraldehyde, dimethyl adipimidate, dimethyl suberimidate, and aliphatic diamines.

Encapsulation

Enzymes enveloped in semipermeable membrane, which allows low molecular weight substrates and products to pass through the membrane.

Enclosed in a variety of devices: hollow fibers, cloth fibers, microcapsules, film.

In particular, with the recent advances in genomics, proteomics, and bioinformatics, many new genes and pathways will be discovered and the regulation of metabolic network will also be better understood, all of which will accelerate the development of more commercially viable bioprocesses through metabolic engineering.

Enzymes in food industry

7.1   Introduction

Enzymes are produced by all living cells as catalysts for specific chemical reactions. In food industry, enzyme has been used to produce and to increase the quality and the diversity of food. Some examples of products that use enzyme are cheese, yoghurt, bread syrup, etc. Ancient traditional arts such as brewing, cheese making, meat tenderization with papaya leaves and condiment preparation (e.g. soya sauce and fish sauce) rely on proteolysis, albeit the methods were developed prior to our knowledge of enzymes. Early food processes involving proteolysis were normally the inadvertent consequence of endogenous or microbial enzyme activity in the foodstuff.

7.2   Production of food enzymes from microorganisms

In the twentieth century, enzymes began to be isolated from living cells, which led to their large-scale commercial production and wider application in the food industry. Today, microorganisms are the most important source of commercial enzymes. Although microorganisms do not contain the same enzymes as plants or animals, a micro-organism can usually be found that produces a related enzyme that will catalyze the desired reaction. Enzyme manufacturers have optimized microorganisms for the production of enzymes through natural selection and classical breeding techniques.

Direct genetic modification (biotechnology) encompasses the most precise methods for optimizing microorganisms for the production of enzymes. These methods are used to obtain high-yielding production organisms. Biotechnology also provides the tools to have a genetic sequence from a plant, animal, or a micro-organism, from which commercial scale enzyme production is not adequate, to be transferred to a micro-organism that has a safe history of enzyme production for food use.

Although the production organism is genetically modified, the enzyme it produces is not. Enzymes produced through biotechnology are identical to those found in nature. Additionally, enzymes produced by microorganisms are extracted and purified before they are used in food manufacturing. Genetically modified microorganisms are useful from a commercial standpoint but would not survive in nature.

Some major applications of various types of enzymes are discussed below.

7.3   Rennet

The use of rennet in cheese manufacture was among the earliest applications of exogenous enzymes in food processing, dating back to approximately 6000 BC. The use of rennet, as an exogenous enzyme, in cheese manufacture is perhaps the largest single application of enzymes in food processing. In recent years, proteinases have found additional applications in dairy technology, for example in acceleration of cheese ripening, modification of functional properties and preparation of dietic products.

Rennet is the stomach extract that contains the enzyme chymosin in a stabilized form that is usable for cheese making. While the amount of chymosin that is required for cheese making is very small, this enzyme industry has undergone an interesting transformation over time.

For the manufacture of traditional rennet, calves, lambs or kids that are no more than about 2 weeks old and fed only milk are used. As calves become older and begin eating other feeds, the proportion of bovine pepsin in relation to chymosin increases. Extracts from milk-fed calves that are 3 weeks old contain over 90 per cent chymosin and the balance is pepsin. As the calves age and are fed other feeds such as concentrates, the ratio of chymosin to pepsin drops to 30:70 by 6 months of age. In a full-grown cow there are only traces of chymosin.

Powdered rennet is manufactured by saturating a rennet suspension with sodium chloride or acidifying it with a food-grade acid. Chymosin precipitates and secondary enzymes such as pepsin remain in the original suspension. The chymosin-containing precipitate is dried to rennet powder.

A method has been developed for manufacturing rennet without slaughtering calves. A hole (fistula) is surgically bored in the side of live calves and at milk feeding time, excreted rennet is removed. After the calf matures, the hole is plugged and the animal is returned to the herd. This method has not been commercialised but may be of value where religious practices do not allow calf slaughter.

Rennet paste, a rich source of lipolytic enzyme, has been a major factor in the flavour development of ripened Italian cheeses such as Provolone and Romano. Years ago, Romans applied rennet paste to cheese milks to develop a typical 'picante' flavour in Romano and related cheeses, for even then calf rennet was used to set their cheese milks. In more modern times many countries, including the United States, applied calf rennet paste to induce flavour, largely for Italian ripened cheeses.

Animal rennet (bovine chymosin) is conventionally used as a milk-clotting agent in dairy industry for the manufacture of quality cheeses with good flavour and texture. At present, microbial rennet is used for one-third of all the cheese produced worldwide. Rennin acts on the milk protein in two stages, by enzymatic and by nonenzymatic action, resulting in coagulation of milk. In the enzymatic phase, the resultant milk becomes a gel due to the influence of calcium ions and the temperature used in the process. Many microorganisms are known to produce rennet-like proteinases which can substitute the calf rennet. Micro-organisms like *Rhizomucor pusillus, R. miehei, Endothia parasitica, Aspergillus oryzae* and *Irpex lactis* are used extensively for rennet production in cheese manufacture.

Different strains of species of Mucor are often used for the production of microbial rennets. Whereas best yields of the milk-clotting protease from *Rhizomucor pusillus* are obtained from semisolid cultures containing 50% wheat bran, *R. miehei* and *Endothia parasitica* are well suited for submerged cultivation. Using the former, good yields of milk-clotting protease may be obtained in a medium containing 4% potato starch, 3% soyabean meal, and 10% barley. During growth, lipase is secreted together with the protease. Therefore, the lipase activity has to be destroyed by reducing the pH, before the preparation can be used as cheese rennet.

Types of rennets are shown in Table 7.1.

Table 7.1 Types of rennets.

Type	Source	Active enzymes	Examples
Animal	Calf	Chymosin (88–94%) Bovine pepsin (6–12%)	Naturen
Animal	• Ovine • Porcine • (Chicken)	Chymosin Secondary proteases	–
Plant	Dried flowers *Cynara cardunculus*	Plant proteases	–
Microbial	Fungal species e.g. *Rhizomucor miehei*	Microbial proteases	Hannilase Fromase Kesozym
Fermentation Produced Chymosin	Fungus (Bovine chymosin gene cloned into fungus)	Chymosin (100%)	Chy-max Maxiren

Advantages and disadvantages of different types of rennets are shown in Table 7.2.

Table 7.2 Advantages and disadvantages of rennets.

Rennet type	Advantages	Disadvantages
Animal (Calf)	• Traditional use • Historical acceptance • Flavour can vary slightly (Farm house)	• Can have slight variations in manufacture • Animal origin • Not suitable for vegetarians
Fermentation-produced Chymosin	• Very consistent in manufacture • Suitable for vegetarians	• Not GMO but GE implications
Microbial	• Non-animal, non GE • Suitable for vegetarians • More Thermoliable–Mozzarella	• Perception may cause bitterness • Curd sets slightly softer

7.3.1   Rennet substitutes

Various proteolytic enzymes from plant, microbial and animal sources were identified and developed for commercial applications.

These enzymes should possess certain key characteristics to be successful rennet substitutes: (i) the clotting-to-proteolytic ratio should be similar to that of chymosin (i.e. the enzyme should have the capacity to clot milk without being excessively proteolytic), (ii) the proteolytic specificity for β-casein should be low because otherwise bitterness will occur in cheese, (iii) the substitute product should be free of contaminating enzymes such as lipases and (iv) cost should be comparable to or lower than that of traditional rennet.

Animal: Pepsin derived from swine shows proteolytic activity between pH 2 and 6.5, but by itself it has difficulty in satisfactorily coagulating milk at pH 6.6. For this reason, it is used in cheese making as a 1:1 blend with rennet. Pepsin, used alone as a milk coagulator, shows high sensitivity to heat and is inclined to create bitter cheese if the concentrations added are not calculated and measured exactly.

Plant: Rennet substitutes from plant sources are the least widely used because of their tendency to be exclusively proteolytic and to cause formation of bitter flavours. Most such enzymes are also heat stable and require higher setting temperatures in milk. Plant sources include ficin from the fig tree, papain from the papaya tree and bromelin from pineapple. A notable exception to these problems is the proteolytic enzymes from the flower of thistle *(Cynara cardunaculus)* and Cynara humilis. These enzymes have been used successfully for many years in Portugal to make native ripened Serra cheese with excellent flavour and without bitterness.

7.4   Lactases

Lactose can be obtained from various sources like plants, animal organs, bacteria, yeasts (intracellular enzyme), or molds. Some of these sources are used for commercial enzyme preparations. Lactase preparations from *A. niger, A. oryzae,* and *Kluyveromyces lactis* are considered safe because these sources already have a history of safe use and have been subjected to numerous safety tests. The most investigated *E. coli* lactase is not used in food processing because of its cost and toxicity problems.

Lactose, the sugar found in milk and whey, and its corresponding hydrolase, lactase or β-galactosidase, have been extensively researched during the past decade. This is because of the enzyme immobilization technique which has given new and interesting possibilities for the utilization of this sugar. Because of intestinal enzyme insufficiency, some individuals, and even a population, show lactose intolerance and difficulty in consuming milk and dairy products.

Hence, low-lactose or lactose-free food aid programme is essential for lactose-intolerant people to prevent severe tissue dehydration, diarrhoea, and, at times, even death. Another advantage of lactase-treated milk is the increased sweetness of the resultant milk, thereby avoiding the requirement for addition of sugars in the manufacture of flavoured milk drinks.

Manufacturers of ice cream, yoghurt and frozen desserts use lactase to improve scoop and creaminess, sweetness, and digestibility, and to reduce sandiness due to crystallization of lactose in concentrated preparations. Cheese manufactured from hydrolyzed milk ripens more quickly than the cheese manufactured from normal milk.

Technologically, lactose crystallizes easily which sets limits to certain processes in the dairy industry, and the use of lactase to overcome this problem has not reached its fullest potential because of the associated high costs. Moreover, the main problem associated with discharging large quantities of cheese whey is that it pollutes the environment. But, the discharged whey could be exploited as an alternate cheap source of lactose for the production of lactic acid by fermentation. The whey permeate, which is a byproduct in the manufacture of whey protein concentrates, by ultra filtration could be fermented efficiently by *Lactobacillus bulgaricus.*

The properties of the enzyme depend on its source. Temperature and pH optima differ from source to source and with the type of particular commercial preparation. Immobilization of the enzymes, method of immobilization, and type of carrier can also influence these optima values. In general, fungal lactase have pH optima in the acidic range 2.5–4.5, and yeast and bacterial lactases in the neutral region 6–7 and 6.5–7.5, respectively. The variation in pH optima of lactases makes them suitable for specific applications, for example fungal lactases are used for acid whey hydrolysis, while yeast and bacterial lactases are suitable for milk (pH 6.6) and sweet whey (pH 6.1) hydrolysis. Product inhibition, e.g. inhibition by galactose, is another property which also depends on the source of lactase. The enzyme from *A. niger* is more strongly inhibited by galactose than that from *A. oryzae*. This inhibition can be overcome by hydrolyzing lactose at low concentrations by using immobilized enzyme systems or by recovering the enzyme using ultra filtration after batch hydrolysis. Lactases from *Bacillus* species are superior with respect to thermostability, pH operation range, product inhibition, and sensitivity against high-substrate concentration. Thermostable enzymes, able to retain their activity at 60°C or above for prolonged periods, have two distinct advantages viz. they give higher conversion rate or shorter residence time for a given conversion rate, and the process is less prone to microbial contamination due to higher operating temperature. *Bacillus* species have a pH optima of 6.8 and temperature optima of 65°C. Its high activity for skim milk and less inhibition by galactose has made it suitable for use as a production organism for lactase.

The enzymatic hydrolysis of lactose can be achieved either by free enzymes, usually in batch fermentation process, or by immobilized enzymes or even by immobilized whole cells producing intracellular enzyme. Although numerous hydrolysis systems have been investigated, only few of them have been scaled up with success and even fewer have been applied at an industrial or semi-industrial level. Several acid hydrolysis systems have been developed to industry-scale level. Large-scale systems which use free enzyme process have been developed for processing of UHT-milk and processing of whey, using *K. lactis* lactase.

7.5   Catalases

Catalases (EC 1.11.1.6) are some of the most efficient enzymes found in cells. The enzyme is composed of four identical subunits, each with its own active site buried deep inside.

Catalase is a form of protein. In fact, all enzymes are protein. Proteins are large globular molecules made of amino acid subunits. These amino acids are arranged in a chain to form the protein molecule.

The catalase found in animals is slightly different than the catalase found in plants. But the general composition remains the same.

, Catalases are produced by aerobic organisms ranging from bacteria to man. Catalases are haem-containing proteins that catalyze the conversion of hydrogen peroxide (H_2O_2) to water and molecular oxygen, thereby protecting cells from the toxic effects of hydrogen peroxide:

$$2H_2O_2 \rightarrow 2H_2O + O_2$$

Some haem-containing catalases are bifunctional, acting as a catalase and a peroxidase (EC 1.11.1.7). In these bifunctional catalase-peroxidases, a variety of organic substances can be used as a hydrogen donor, for example alcohol, which can be oxidised in the liver. These bifunctional catalases are closely related to plant peroxidases. There are also non-haem manganese-containing catalases, which occur in bacteria.

The catalase enzyme is so critical to our health that it is found in nearly every living organism on the planet that is exposed to oxygen. This antioxidant enzyme can catalyze the conversion of hydrogen peroxide into water and oxygen. Hydrogen peroxide is a byproduct of cell metabolism, which serves some useful functions including healthy immune response.

Catalase has one of the highest rates of turnover when compared to all other enzymes. In other words, one catalase enzyme can change 40 million molecules of hydrogen peroxide into water and oxygen in just one second. In fact, catalase enzymes act to protect our cells, counteracting and balancing the continual production of hydrogen peroxide. Because of its undeniable, scientifically proven powerful antioxidant properties, catalase is very beneficial to the organs and body processes. In addition to acting as a super antioxidant, catalase also has the ability to use hydrogen peroxide to oxidize toxins including methanol, ethanol, formic acid, formaldehyde, and nitrite. This type of dual activity makes it a crucial cellular enzyme.

The best supplemental source of catalase is the vegetarian form derived from the fermentation process of the fungus *Aspergillus niger*.

7.5.1   Health benefits of catalase

1. *Powerful antioxidant support:* Catalases are perhaps the single most efficient enzymes found in the cells of the human body. Catalase has been shown to create a speedy reaction against hydrogen peroxide free radicals, turning them into water and oxygen.

2. *Possible anti-ageing and anti-degenerative effects:* Catalase is currently being studied for its applications on extending life span and vitality. Research scientists from the University of Washington in

Seattle conducted a lab study on rats and the augmentation of natural catalase in their bodies. By supplementing with increased catalase, the life span of these laboratory rats increased by almost 20%. This is the equivalent of nearly 25 human years.

3. *Catalase may increase lifespan:* There is a direct link between the catalase enzyme, free radical damage and extending our life span. This also suggests that the catalase enzyme may help ward off degenerative diseases. Spanish scientists found that very high servings of apple polyphenols boosted the gene expression of natural catalase in the body.

4. *Fat reduction:* Very exciting research from Japan shows a link between catalase and lowered amounts of organ fat in lab rats. This study also showed a link between the enzyme and an increase in muscle strength.

5. *Helps prevent DNA damage:* A 2012 study from the Institute of Cytology and Genetics found that oxidative stress, accumulation of protein and DNA damage could be reduced in the presence of antioxidant enzymes, including catalase and glutathione in the cytosol and mitochondrial extracts from liver cells of rats. This study also found that dietary supplements for increasing the activity of catalase in the liver mitochondria in rats led to reduced mitochondrial dysfunction and slowed the process of ageing in these animals.

7.6   Lipases

Lipase is an enzyme the body uses to break down fats in food so they can be absorbed in the intestines. Lipase is produced in the pancreas, mouth, and stomach. Most people produce enough pancreatic lipase, but people with cystic fibrosis, Crohn disease, and celiac disease may not have enough lipase to get the nutrition they need from food. Along with lipase, the pancreas secretes insulin and glucagon, two hormones the body needs to break down sugar in the bloodstream. Other pancreatic enzymes include amylase, which breaks down a certain starch into its sugar building blocks, and protease, which breaks down protein into single amino acids.

A lipase is a water–soluble enzyme that catalyzes the hydrolysis of ester bonds in water-insoluble, lipid substrates. Lipases (triacylglycerol acylhydrolases) are produced by micro-organism in individual or together with esterase. Micro-organisms that produce lipases are *Pseudomonas aeruginosa, Serratia marcescens, Staphylococcus aureus* dan *Bacillus subtilis*. Lipase is used as biocatalyst to produce free fatty acid, glycerol and various esters, part of glycerides and fat that is modified or esterified from cheap substrate i.e. palm oil. These products are extensively used in pharmacy, chemical and food industry.

Various animal or microbial lipases gave pronounced cheese flavour, low bitterness and strong rancidity, while lipases in combination with proteinases and/or peptidases give good cheese flavour with low levels of bitterness. In a more balanced approach to the acceleration of cheese ripening using mixtures of proteinases and peptidases, attenuated starter cells or cell-free extracts (CFE) are being favoured.

How lipase deficiency can affect our health:

- Since lipase digests fat and fat-soluble vitamins, lipase deficient people may have a tendency towards high cholesterol, high triglycerides, difficulty losing weight and diabetes or a tendency towards glucosuria (sugar in the urine without symptoms of diabetes).

- Because lipase requires the coenzyme, chloride, lipase deficient people have a tendency towards hyphochlorhydria (low chlorides in our electrolyte balance). This can be easily remedied with lipase, but often nutritionists recommend using betaine HCl, which may place an acidic stress on the blood, leading to an inability to provide the alkalinity required to activate the body's pancreatic enzymes. Lipase requires a high pH for its activation among food enzymes. That is why fats are the most difficult of all foods to digest.

- Fat intolerant people can be helped by taking a lipase supplement, but the fat intolerance problem still exists. The lipase will help prevent an aggravated condition only if the fat intolerant person minimizes fat consumption.

- Lipase deficient people have decreased cell permeability, meaning nutrients cannot get in and the waste cannot get out. Lipase modulates cell permeability so that nutrients can enter and wastes exit. Waste-eating enzymes (such as protease) may also be taken to help cleanse the blood of the unwanted debris.

- A common symptom of lipase deficiency is muscle spasms. This is not the 'muscle cramp' (tetany) resulting from low ionized blood calcium. It commonly occurs as trigger point pain in the muscles across the upper shoulders, but it can occur in other muscles, such as those in the neck or anywhere in the small or large intestines including the muscles of the rectal tissues.

- People with 'spastic colon' may be lipase deficient. They are often given toxic muscle relaxant drugs to control the symptoms, but a simple food enzyme called lipase may provide relief.

7.7   Proteases

Protein is one of the three major food groups needed for proper nutrition. Protease is the digestive enzyme needed to digest protein.

Every animal, including humans, must have an adequate source of protein in order to grow or maintain itself. Proteins, which yield amino acids, are the fundamental structural element of every cell in the body. Specific proteins are now recognized as the functional elements in certain specialized cells, glandular secretions, enzymes and hormones.

Proteins, which are among the most complex organic compounds found in nature, are made up of nitrogen containing compounds known as amino acids. During the digestion of proteins, hydrochloric acid and proteolytic enzymes break down the intact protein molecule into amino acids which are then absorbed through the intestinal wall. If proteins are not properly broken down before they are absorbed, various health consequences may occur. Assuring optimal protein digestion and proper blood flow is necessary for effective nutrient delivery, a healthy immune response, and detoxification. Protease enzymes are therefore a must.

Proteases refer to a group of enzymes whose catalytic function is to hydrolyze (breakdown) proteins. They are also called proteolytic enzymes or systemic enzymes.

Proteolytic enzymes are very important in digestion as they breakdown the peptide bonds in the protein foods to liberate the amino acids needed by the body. Additionally, proteolytic enzymes have been used for a long time in various forms of therapy. Their use in medicine is notable based on several clinical studies indicating their benefits in oncology, inflammatory conditions, blood rheology control, and immune regulation.

Protease is able to hydrolyze almost all proteins as long as they are not components of living cells. Normal living cells are protected against lysis by the inhibitor mechanism.

Parasites, fungal forms, and bacteria are protein. Viruses are cell parasites consisting of nucleic acids covered by a protein film. Enzymes can break down undigested protein, cellular debris, and toxins in the blood, sparing the immune system this task. The immune system can then concentrate its full action on the bacterial or parasitic invasion.

The proteolytic system of lactic acid bacteria is essential for their growth in milk, and contributes significantly to flavour development in fermented milk products. The proteolytic system is composed of proteinases which initially cleaves the milk protein to peptides; peptidases which cleave the peptides to small peptides and amino acids; and transport system responsible for cellular uptake of small peptides and amino acids. Lactic acid bacteria have a complex proteolytic system capable of converting milk casein to the free amino acids and peptides necessary for their growth.

These proteinases include extracellular proteinases, endopeptidases, aminopeptidases, tripeptidases, and proline-specific peptidases, which are all serine proteases. Apart from lactic *streptococcal proteinases*, several other proteinases from nonlactostreptococcal origin have been reported. There are also serine type of proteinases, e.g. proteinases from *Lactobacillus acidophillus, L. plantarum, L. delbrueckii* sp. *bulgaricus, L. lactis,* and *L. helveticus*. Aminopeptidases are important for the development of flavour in fermented milk products, since they are capable of releasing single amino acid residues from oligopeptides formed by extracellular proteinase activity.

How protease deficiency can affect our health:

- Acidity is created through the digestion of protein. Therefore a protease deficiency results in an alkaline excess in the blood. This alkaline environment can cause anxiety and insomnia.

- In addition, since protein is required to carry protein-bound calcium in the blood, a protease deficiency lays the foundation for arthritis, osteoporosis and other calcium-deficient diseases.

- Because protein is converted to glucose upon demand, inadequate protein digestion leads to hypoglycemia, resulting in moodiness, mood swings and irritability.

- Protease also has an ability to digest unwanted debris in the blood including certain bacteria and viruses. Therefore, protease deficient people are immune compromised, making them susceptible to bacterial, viral and yeast infections and a general decrease in immunity.

7.8   Amylases

Amylase is a digestive enzyme that aids in the breaks down of carbohydrates by breaking the bonds between sugar molecules in polysaccharides through a hydrolysis reaction. It can be found in animals, plants and bacteria. Amalase can be classified into three types: alpha-amylase, beta-amylase and gamma-amylase.

The three types differ in how they hydrolyze the polysaccharide bonds. Neither beta nor gamma-amylase is found in animal tissue.

Amylase is critical in the digestion of starch into sugars to make them available energy sources for the body. Amylase is found in two primary places within the human body, and the two types are classified according to where they are found. Salivary Amylase is a component of saliva, and breaks starch into glucose and dextrin. It hydrolyzes the bonds between long-chain polysaccharides found in food, breaking compounds such as glycogen and starch into their useful monomers, glucose and maltose. Pancreatic Amylase is added to the small intestine to further digest starches; amylase is denatured in the acidic stomach. Amylase is also present in blood where it digests dead white blood cells.

7.8.1   Applications of amylases

Liquefaction

Liquefaction is a process of dispersion of insoluble starch granules in aqueous solution followed by partial hydrolysis using thermostable amylases. In industrial processes, the starch suspension for liquefaction is generally in excess of 35% (w/v). Therefore the viscosity is extremely high following gelatinization. Thermostable α-amylase is used as a thinning agent, which brings about reduction in viscosity and partial hydrolysis of starch. Retrogradation of starch is thus avoided during subsequent cooling.

Liquefaction is the first and most important step in starch processing. The purpose is to provide a partially hydrolyzed starch suspension of relatively low viscosity which is free from by products, stable to retro gradation and suitable for further processing, i.e. saccharification. If the liquefaction process does not go well, problems like poor filtration and turbidity of the processed solution occurs. The most important factor for ideal liquefaction of starch is that the starch slurry which contains suitable amount of α-amylase is treated at 105 to 107°C as quickly and uniformly as possible.

Thermostable amylases are not sufficiently heat stable to be used during liquefaction process, but they can be used as saccharifying enzymes. The most widely used enzymes in this group are the maltogenic enzymes.

7.8.2   Manufacturing of maltose

Maltose is a naturally occurring disaccharide. Maltose is widely used as sweetener and also as intravenous sugar supplement. It is used in food industries because of low tendency to be crystallized and is relatively nonhygroscopic. Corn, potato, sweet potato and cassava starches are used for maltose manufacture. The concentration of starch slurry is adjusted to be 10–20% for production of medical grade maltose and 20–40% for food grade. Thermostable α-amylase from *B. licheniformis* and *B. amyloliquefaciens* are used. Maltose is widely used as sweetener and also as intravenous sugar supplement. It is used in food industries because of low tendency to be crystallized and is relatively nonhygroscopic. Corn, potato, sweet potato, and cassava starches are used for maltose manufacture. The concentration of starch slurry is adjusted to be 10–20% for production of medical grade maltose and 20–40% for food grade. Thermostable α-amylase from *B. licheniformis* and *B. amyloliquefaciens* are used.

Manufacture of high fructose containing syrups

High fructose containing syrups (HFCS) 42°F (Fructose content = 42%) is prepared by enzymic isomerization of glucose with glucose isomerase. The starch is first converted to glucose by enzymic liquefaction and saccharification.

Manufacture of oligosaccharides mixture

Oligosaccharides mixture (maltooligomer mix) is obtained by digestion of corn starch with α-amylase, β-amylase and pullulanase. Maltooligomer mix is a new commercial product. Its composition is usually as follows: glucose, 2.2%, maltose, 37.5%, maltotriose, 46.4% and maltotetraose and larger maltooligo-saccharides, 14%.

Maltooligomer mix powder obtained by spray drying is highly hygroscopic. Therefore it serves as a moisture regulator of the food with which it is mixed. Maltooligomer mix tastes less sweet than sucrose.

Manufacture of maltotetraose syrup

Maltotetraose syrup (G4 syrup) is produced by subjecting starch to the action of maltotetraose forming amylase (G4 amylase). The sweetness of the syrup is as low as 20% of sucrose.

Production of anomalously linked oligosaccharides mixture (Alo mixture) 'Alo mixture' is a mixture of isomaltose, panose, isomaltotriose and branched oligosaccharide composed of 4 and 5 glucose residues. The 'Alo mixture' has properties that are favourable to food industry. It is mildly sweet, has low viscosity, high moisture retaining capacity and low water activity convenient in controlling microbial contamination. For the manufacturing of 'Alo mixture', starch is dextrinized using thermostable bacterial α-amylase. The degree of hydrolysis (DE) of starch is kept between 6 and 10. The simultaneous reaction of sacharification and transglucosidation of dextrin is done by using soyabean β-amylase and *Aspergillus niger* transglucosidase. The reaction mixture is finally purified and concentrated to 25% moisture.

Manufacturing of high molecular weight branched dextrins

Branched dextrins of high molecular weight are prepared by hydrolysis of corn starch with α-amylase. The extent of starch degradation to be carried out depends on the type of starch and the physical properties desired. The branched dextrins are obtained as powder after chromatography and spray drying. These are used as extender and a glozing agent for production of powdery foods and rice cakes, respectively.

Direct fermentation of starch to ethanol

The amylolytic activity rate and amount of starch utilization and ethanol yields increase in several fold in cocultures. Molds amylases are used in alcohol production and brewing industries. The advantages of such system are uniform enzyme action in mashes, increase rate of saccharification, alcohol yield and yeast growth.

Starch is also present in waste produced from food processing plants. Starch waste causes pollution problems. Biotechnological treatment of food processing wastewater can produce valuable products such as microbial biomass protein and also purifies the effluent.

Other applications

Amylases, especially alkaline amylases are used in detergents. To some extent amylases are also used as digestive aids to supplement the diastatic activity of flour and to improve digestibility of some of the animal feed ingredients.

Section III
Milk, dairy and bakery products

Dairy products

8.1 Introduction

Dairy products include milk and any of the foods made from milk, including butter, cheese, ice cream, yogurt, and condensed and dried milk. Milk has been used by humans since the beginning of recorded time to provide both fresh and storable nutritious foods. In some countries almost half the milk produced is consumed as fresh pasteurized whole, low fat, or skim milk.

Cow milk (bovine species) is by far the principal type used throughout the world. Other animals utilized for their milk production include buffalo (in India, China, Egypt, and the Philippines), goats (in the Mediterranean countries), reindeer (in northern Europe), and sheep (in southern Europe). This section focuses on the processing of cow milk and milk products unless otherwise noted. In general, the processing technology described for cow milk can be successfully applied to milk obtained from other species.

8.2 Properties of milk

8.2.1 Nutrient composition

Many factors influence the composition of milk, including breed, genetic constitution of the individual cow, age of the cow, stage of lactation, interval between milkings, and certain disease conditions. Since the last milk drawn at each milking is richest in fat, the completeness of milking also influences a sample. In general, the type of feed only slightly affects the composition of milk, but feed of poor quality or insufficient quantity causes both a low yield and a low percentage of total solids. Current feeding programs utilize computer technology to achieve the greatest efficiency from each animal.

The composition of milk varies among mammals, primarily to meet growth rates of the individual species. The proteins contained within the mother's milk are the major components contributing to the growth rate of the young animals. Human milk is relatively low in both proteins and minerals compared with that of cows and goats.

Goat milk has about the same nutrient composition as cow milk, but it differs in several characteristics. Goat milk is completely white in colour because all the beta-carotene (ingested from feed) is converted to vitamin A. The fat globules are smaller and therefore remain suspended, so the cream does not rise and mechanical homogenization is unnecessary. Goat milk curd forms into small, light flakes and is more easily digested, much like the curd formed from human milk. It is often prescribed for persons who are allergic to the proteins in cow milk and for some patients afflicted with stomach ulcers. Sheep milk is rich in nutrients, having 18 per cent total solids (5.8 per cent protein and 6.5 per cent fat). Reindeer milk has the highest level of nutrients, with 36.7 per cent total solids (10.3 per cent protein and 22 per cent fat). These high-fat, high-protein milks are excellent ingredients for cheese and other manufactured dairy products. The major components of milk are water, fat, protein, carbohydrate (lactose), and minerals (ash).

However, there are numerous other highly important micronutrients such as vitamins, essential amino acids, and trace minerals. Indeed, more than 250 chemical compounds have been identified in milk. Table 8.1 shows the nutrient composition of dairy products (per 100 g).

Table 8.1 Nutrient composition of dairy products (per 100 g).

Dairy product	Energy (kcal)	Water (g)	Protein (g)	Fat (g)	Carbohyd-rate (g)	Choleste-rol (mg)	Vitamin A (IU)	Ribofla-vin (mg)	Calcium (mg)
Fresh milk									
Whole	61	88	3.29	3.34	4.66	14	126	0.162	119
Low-fat*	50	89	3.33	1.92	4.80	8	205	0.165	122
Skim*	35	91	3.41	0.18	4.85	2	204	0.140	123
Evaporated milk	134	74	6.81	7.56	10.04	29	243	0.316	261
Evaporated skim milk*	78	79	7.55	0.20	11.35	4	392	0.309	290
Sweetened condensed milk	321	27	7.91	8.70	54.40	34	328	0.416	284
Nonfat dry milk*	358	4	35.10	0.72	52.19	18	2,370	1.744	1,231
Butter	717	16	0.85	81.11	0.06	219	3,058	0.034	24
Ice cream (vanilla)	201	61	3.50	11.00	23.60	44	409	0.240	128
Ice milk (vanilla)	139	68	3.80	4.30	22.70	14	165	0.265	139
Sherbet (orange)	138	66	1.10	2.00	30.40	5	76	0.068	54
Frozen yogurt, nonfat	128	69	3.94	0.18	28.16	2	7	0.265	134
Buttermilk	40	90	3.31	0.88	4.79	4	33	0.154	116
Sour cream	214	71	3.16	20.96	4.27	44	790	0.149	116
Yogurt, plain, low-fat	63	85	5.25	1.55	7.04	6	66	0.214	183
Yogurt, fruit, low-fat	102	74	4.37	1.08	19.05	4	46	0.178	152
Cheese									
Blue	353	42	21.40	28.74	2.34	75	721	0.382	528
Brie	334	48	20.75	27.68	0.45	100	667	0.520	184
Cheddar	403	37	24.90	33.14	1.28	105	1,059	0.375	721
Cottage	103	79	12.49	4.51	2.68	15	163	0.163	60
Cream	349	54	7.55	34.87	2.66	110	1,427	0.197	80
Mozzarella**	280	49	27.47	17.12	3.14	54	628	0.343	731
Parmesan, grated	456	18	41.56	30.02	3.74	79	701	0.386	1,376
Emmentaler (Swiss)	376	37	28.43	27.54	3.38	92	845	0.365	961

*Fortified with vitamin A

**Low moisture, part skim

Fat

The fat in milk is secreted by specialized cells in the mammary glands of mammals. It is released as tiny fat globules or droplets, which are stabilized by a phospholipid and protein coat derived from the plasma membrane of the secreting cell. Milk fat is composed mainly of triglycerides—three fatty acid chains attached to a single molecule of glycerol. It contains 65 per cent saturated, 32 per cent monounsaturated, and 3 per cent polyunsaturated fatty acids. The fat droplets carry most of the cholesterol and vitamin A. Therefore, skim milk, which has more than 99.5 per cent of the milk fat removed, is significantly lower in cholesterol than whole milk (2 milligrams per 100 grams of milk, compared with 14 milligrams for whole milk) and must be fortified with vitamin A.

Protein

Milk contains a number of different types of proteins, depending on what is required for sustaining the young of the particular species. These proteins increase the nutritional value of milk and other dairy products and provide certain characteristics utilized for many of the processing methods. A major milk protein is casein, which actually exists as a multi subunit protein complex dispersed throughout the fluid phase of milk. Under certain conditions the casein complexes are disrupted, causing curdling of the milk. Curdling results in the separation of milk proteins into two distinct phases, a solid phase (the curds) and a liquid phase (the whey).

Lactose

Lactose is the principal carbohydrate found in milk. It is a disaccharide composed of one molecule each of the monosaccharides (simple sugars) glucose and galactose. Lactose is an important food source for several types of fermenting bacteria. The bacteria convert the lactose into lactic acid, and this process is the basis for several types of dairy products.

In the diet lactose is broken down into its component glucose and galactose subunits by the enzyme lactase. The glucose and galactose can then be absorbed from the digestive tract for use by the body. Individuals deficient in lactase cannot metabolize lactose, a condition called lactose intolerance. The unmetabolized lactose cannot be absorbed from the digestive tract and therefore builds up, leading to intestinal distress.

Vitamins and minerals

Milk is a good source of many vitamins. However, its vitamin C (ascorbic acid) content is easily destroyed by heating during pasteurization. Vitamin D is formed naturally in milk fat by ultraviolet irradiation but not in sufficient quantities to meet human nutritional needs. Beverage milk is commonly fortified with the fat-soluble vitamins A and D. In the United States the fortification of skim milk and low-fat milk with vitamin A (in water-soluble emulsified preparations) is required by law.

Milk also provides many of the B vitamins. It is an excellent source of riboflavin (B_2) and provides lesser amounts of thiamine (B_1) and niacin. Other B vitamins found in trace amounts are pantothenic acid, folic acid, biotin, pyridoxine (B_6), and vitamin B_{12}.

Milk is also rich in minerals and is an excellent source of calcium and phosphorus. It also contains trace amounts of potassium, chloride, sodium, magnesium, sulphur, copper, iodine, and iron. A lack of adequate iron is said to keep milk from being a complete food.

8.2.2   Physical and biochemical properties

Milk contains many natural enzymes, and other enzymes are produced in milk as a result of bacterial growth. Enzymes are biological catalysts capable of producing chemical changes in organic substances. Enzyme action in milk systems is extremely important for its effect on the flavour and body of different milk products. Lipases (fat-splitting enzymes), oxidases, proteases (protein-splitting enzymes), and amylases (starch-splitting enzymes) are among the more important enzymes that occur naturally in milk. These classes of enzymes are also produced in milk by microbiological action. In addition, the proteolytic enzyme (i.e. protease) rennin, produced in calves' stomachs to coagulate milk protein and aid in nutrient absorption, is used to coagulate milk for manufacturing cheese.

The coagulation of milk is an irreversible change of its protein from a soluble or dispersed state to an agglomerated or precipitated condition. Its appearance may be associated with spoilage, but coagulation is a necessary step in many processing procedures. Milk may be coagulated by rennin or other enzymes, usually in conjunction with heat. Left unrefrigerated, milk may naturally sour or coagulate by the action of lactic acid, which is produced by lactose-fermenting bacteria. This principle is utilized in the manufacture of cottage cheese. When milk is pasteurized and continuously refrigerated for two or three weeks, it may eventually coagulate or spoil owing to the action of psychrophilic or proteolytic organisms that are normally present or result from post pasteurization contamination. Milk fat is present in milk as an emulsion in a water phase. Finely dispersed fat globules in this emulsion are stabilized by a milk protein membrane, which permits the fat to clump and rise. The rising action is called creaming and is expected in all unhomogenized milk.

Homogenization is practiced in many dairy processes in order to improve the physical properties of products.

Milk and other dairy products are very susceptible to developing off-flavours. Some flavours, given such names as 'feed,' 'barny,' or 'unclean,' are absorbed from the food ingested by the cow and from the odours in its surroundings. Others develop through microbial action due to growth of bacteria in large numbers. Chemical changes can also take place through enzyme action, contact with metals (such as copper), or exposure to sunlight or strong fluorescent light.

8.3   Fresh fluid milk

Fresh fluid milk requires the highest quality raw milk and is generally designated as Grade 'A.' This grade requires a higher level of sanitation and inspection on the farm than is necessary for 'manufacturing grade' milk.

8.3.1   Quality concerns

Raw milk is a potentially dangerous food that must be processed and protected to assure its safety for humans. While most bovine diseases, such as brucellosis and tuberculosis, have been eliminated, many potential human pathogens inhabit the dairy farm environment. Therefore, it is essential that all milk be either pasteurized or (in the case of cheese) held for at least 60 days if made from raw milk. While milk from healthy cows is often totally bacteria-free, that condition quickly changes when milk is exposed to the farm environment.

Milk received at the processing plant is tested before being unloaded from either farm-based tank trucks or over-the-road tankers. The milk is checked for odour, appearance, proper temperature, acidity, bacteria, and the presence of drug residues. These tests take no longer than 10–15 minutes. If the tank load passes these tests, the milk is pumped into the plant's refrigerated storage tanks. The milk is then stored for the shortest possible time. The process steps are shown in Fig. 8.1.

8.3.2   Processing

Pasteurization

Pasteurization is most important in all dairy processing. It is the biological safeguard which ensures that all potential pathogens are destroyed. Extensive studies have determined that heating milk to 63°C (145°F) for 30 minutes or 72°C (161°F) for 15 seconds kills the most resistant harmful bacteria. In actual practice these temperatures and times are exceeded, thereby not only ensuring safety but also extending shelf-life.

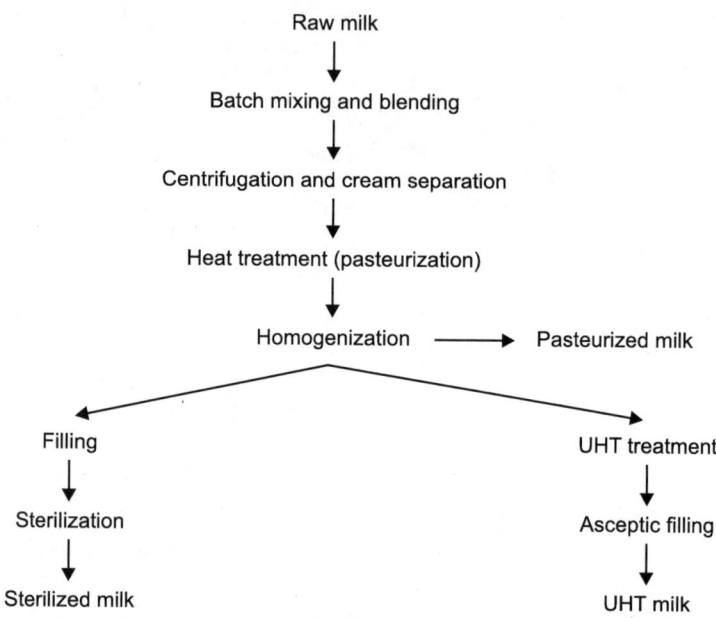

Figure 8.1 Milk-processing operations

Most milk today is pasteurized by the continuous high-temperature short-time (HTST) method (72°C or 161°F for 15 seconds or above). The HTST method is conducted in a series of stainless steel plates and tubes, with the hot pasteurized milk on one side of the plate being cooled by the incoming raw milk on the other side. This 'regeneration' can be more than 90 per cent efficient and greatly reduces the cost of heating and cooling. There are many fail-safe controls on an approved pasteurizer system to ensure that all milk is completely heated for the full time and temperature requirement. If the monitoring instruments detect that something is wrong, an automatic flow diversion valve will prevent the milk from moving on to the next processing stage. Higher temperatures and sometimes longer holding times are required for the pasteurization of milk or cream with a high fat or sugar content.

Pasteurized milk is not sterile and is expected to contain small numbers of harmless bacteria. Therefore, the milk must be immediately cooled to below 4.4°C (40°F) and protected from any outside contamination. The shelf-life for high-quality pasteurized milk is about 14 days when properly refrigerated.

Extended shelf life can be achieved through ultra pasteurization. In this case, milk is heated to 138°C (280°F) for two seconds and aseptically placed in sterile conventional milk containers. Ultra pasteurized milk and cream must be refrigerated and will last at least 45 days. This process does minimal damage to the flavour and extends the shelf life of slow-selling products such as cream, eggnog, and lactose-reduced milks.

Ultrahigh-temperature (UHT) pasteurization is the same heating process as ultra pasteurization (138°C or 280°F for two seconds), but the milk then goes into a more substantial container—either a sterile five-layer laminated 'box' or a metal can. This milk can be stored without refrigeration and has a shelf life of six months to a year. Products handled in this manner do not taste as fresh, but they are useful as an emergency supply or when refrigeration is not available.

Separation

Most modern plants use a separator to control the fat content of various products. A separator is a high-speed centrifuge that acts on the principle that cream or butterfat is lighter than other components in milk. (The specific gravity of skim milk is 1.0358, specific gravity of heavy cream 1.0083.) The heart of the separator is an airtight bowl with funnel like stainless steel disks. The bowl is spun at a high speed (about 6,000 revolutions per minute), producing centrifugal forces of 4,000 to 5,000 times the force of gravity. Centrifugation causes the skim, which is denser than cream, to collect at the outer wall of the bowl. The lighter part (cream) is forced to the centre and piped off for appropriate use.

An additional benefit of the separator is that it also acts as a clarifier. Particles even heavier than the skim, such as sediment, somatic cells, and some bacteria, are thrown to the outside and collected in pockets on the side of the separator. This material, known as 'separator sludge,' is discharged periodically and sometimes automatically when buildup is sensed.

Most separators are controlled by computers and can produce milk of almost any fat content. Current standards generally set whole milk at 3.25 per cent fat, low fat at 1–2 per cent, and skim at less than 0.5 per cent. (Most skim milk is actually less than 0.01 per cent fat.)

Homogenization

Milk is homogenized to prevent fat globules from floating to the top and forming a cream layer or cream plug. Homogenizers are simply heavy-duty, high-pressure pumps equipped with a special valve at the discharge end. They are designed to break up fat globules from their normal size of up to

18 micrometers to less than 2 micrometers in diameter (a micrometer is one-millionth of a meter). Hot milk (with the fat in liquid state) is pumped through the valve under high pressure, resulting in a uniform and stable distribution of fat throughout the milk.

Two-stage homogenization is sometimes practiced, during which the milk is forced through a second homogenizer valve or a breaker ring. The purpose is to break up fat clusters or clumps and thus produce a more uniform product with a slightly reduced viscosity.

Homogenization is considered successful when there is no visible separation of cream and the fat content in the top 100 milliliters of milk in a one-liter bottle does not differ by more than 10 per cent from the bottom portion after standing 48 hours.

In addition to avoiding a cream layer, other benefits of homogenized milk include a whiter appearance, richer flavour, more uniform viscosity, better 'whitening' in coffee, and softer curd tension (making the milk more digestible for humans). Homogenization is also essential for providing improved body and texture in ice cream, as well as numerous other products such as half-and-half, cream cheese, and evaporated milk.

8.3.3   Speciality milks

Many speciality milks are now available (even in remote areas) as a result of the 45-day refrigerated shelf life of ultra pasteurized milk. One of the most useful products, lactose-reduced milk, is available in both nonfat and low-fat composition as well as in many flavoured versions. The lactose (milk sugar) is reduced by 70–100 per cent, making it possible for lactose-intolerant individuals to enjoy the benefits of milk in their diets. Lactose reduction is accomplished by subjecting the appropriate milk to the action of the enzyme lactase in a refrigerated tank for approximately 24 hours. The enzyme breaks down the lactose to more readily digestible glucose and galactose. The reaction is halted when the lactose is consumed or when the milk is heat-treated. The resulting beverage is sweeter than regular milk but acceptable for most uses.

Other speciality milks include calcium-fortified, special and seasonal flavours (e.g. eggnog), and high-volume flavoured milk shakes.

8.4   Condensed and dried milk

8.4.1   Condensed and evaporated milk

Whole, low-fat, and skim milks, as well as whey and other dairy liquids, can be efficiently concentrated by the removal of water, using heat under vacuum. Since reducing atmospheric pressure lowers the temperature at which liquids boil, the water in milk is evaporated without imparting a cooked flavour. Water can also be removed by ultra filtration and reverse osmosis, but this membrane technology is more expensive. Usually about 60 per cent of the water is removed, which reduces storage space and shipping costs. Whole milk, when concentrated, usually contains 7.5 per cent milk fat and 25.5 per cent total milk solids. Skim milk can be condensed to approximately 20–40 per cent solids, depending on the buyer's needs.

Condensed milk is often sold in refrigerated tank-truck loads to manufacturers of candy, bakery goods, ice cream, cheese, and other foods. When preserved by heat in individual cans, it is usually called 'evaporated milk.' In this process the concentrated milk is homogenized, fortified with vitamin D (A and D in evaporated skim milk), and sealed in a can sized for the consumer. A stabilizer, such as disodium phosphate or carrageenan, is also added to keep the product from separating during processing and storage. The sealed can is then sterilized at 118°C (244°F) for 15 minutes, cooled, and labelled. Evaporated milk keeps indefinitely, although staling and browning may occur after a year.

Sweetened condensed milk: Sweetened condensed milk is also made by partially removing the water (as in evaporated milk) and adding sugar. The final product contains about 8.5 per cent milk fat and at least 28 per cent total milk solids. Sugar is added in sufficient amount to prevent bacterial action and subsequent spoilage. Usually, at least 60 per cent sugar in the water phase is required to provide sufficient osmotic pressure for prevention of bacterial growth. Because sweetened condensed milk (or skim milk) is preserved by sugar, the milk merely needs to be pasteurized before being placed in a sanitary container (usually a metal can).

8.4.2   Dry milk products

Milk and by-products of milk production are often dried to reduce weight, to aid in shipping, to extend shelf-life, and to provide a more useful form as an ingredient for other foods. In addition to skim and whole milk, a variety of useful dairy products are dried, including buttermilk, malted milk, instant breakfast, sweet cream, sour cream, butter powder, ice cream mix, cheese whey, coffee creamer, dehydrated cheese products, lactose, and caseinates. Many drying plants are built in conjunction with a butter-churning plant. These plants utilize the skim milk generated from the separated cream and the buttermilk produced from churning the butter. Most products are dried to less than 4 per cent moisture to prevent bacterial growth and spoilage. However, products containing fat lose their freshness rather quickly owing to the oxidation of fatty acids, leading to rancidity.

Two types of dryers are used in the production of dried milk products—drum dryers and spray dryers. Each dryer has certain advantages.

Drum dryers

The simplest and least expensive is the drum, or roller, dryer. It consists of two large steel cylinders that turn toward each other and are heated from the inside by steam. The concentrated product is applied to the hot drum in a thin sheet that dries during less than one revolution and is scraped from the drum by a steel blade. The flake like powder dissolves poorly in water but is often preferred in certain bakery products. Drum dryers are also used to manufacture animal feed where texture, flavour, and solubility are not a major consideration.

Spray dryers

Spray dryers are more commonly used since they do less heat damage and produce more soluble products. Concentrated liquid dairy product is sprayed in a finely atomized form into a stream of hot air. The air may be heated by steam-heated 'radiators' or directly by sulphur-free natural gas. The drying chamber may be rectangular (the size of a living room), conical, or silo-shaped (up to five stories high). The powder passes from the drying chamber through a series of cyclone collectors and is usually placed in plastic-lined, heavy-duty paper bags.

Spray-dried milk is also difficult to reconstitute or mix with water. Therefore, a process called agglomeration was developed to 'instantize' the powder, or make it more soluble. This process involves rewetting the fine, spray-dried powder with water to approximately 8–15 per cent moisture and following up with a second drying cycle. The powder is now granular and dissolves very well in water. Virtually all retail packages of nonfat dry milk powder are instantized in this manner.

8.5   Butter

8.5.1   Composition

Butter is one of the most highly concentrated forms of fluid milk. Twenty liters of whole milk are needed to produce one kilogram of butter. This process leaves approximately 18 liters of skim milk and buttermilk, which at one time were disposed of as animal feed or waste. Today the skim portion has greatly increased in value and is fully utilized in other products.

Commercial butter is 80–82 per cent milk fat, 16–17 per cent water, and 1–2 per cent milk solids other than fat (sometimes referred to as curd). It may contain salt, added directly to the butter in concentrations of 1–2 per cent. Unsalted butter is often referred to as 'sweet' butter. This should not be confused with 'sweet cream' butter, which may or may not be salted. Reduced-fat, or 'light,' butter usually contains about 40 per cent milk fat.

8.5.2   Production

Butter is produced when the cream emulsion in unhomogenized milk is destabilized by agitation, or churning. Breaking the emulsion produces butterfat granules the size of rice grains. The granules mat together and separate from the water phase or serum, which is known as buttermilk. (This milky liquid is drained away and is concentrated or dried, later to become an ingredient in ice cream, candy, or other foods.) The butterfat is then washed with clean water and 'worked' (kneaded) until more buttermilk separates and is removed. Ultimately, only about 16 per cent of the water and milk solids present in the original milk remains trapped in the butter.

The churning process can take 40–60 minutes to complete in a traditional churn, but butter is more commonly made by high-speed continuous 'churns' in factories. Although the basic principle is the same, in the continuous churn cream is pumped into a cylinder and mixed by high-speed blades, forming butter granules in seconds. The butter granules are forced through perforated plates while the buttermilk is drained from the system. A salt solution may be added if salted butter is desired. The butter is then worked in a twin screw extruder and emerges ready to be packaged.

Quality concerns

The quality of butter is based on its body, texture, flavour, and appearance. In the United States, the Department of Agriculture (USDA) assigns quality grades to butter based on its score on a standard quality point scale. Grade AA is the highest possible grade; Grade AA butter must achieve a numerical score of 93 out of 100 points based on its aroma, flavour, and texture. Salt (if present) must be completely dissolved and thoroughly distributed. Grade A butter is almost as good, with a score of 92 out of 100 points. Grade B butter is based on a score of 90 points, and it usually is used only for cooking or manufacturing. The flavour of Grade B is not as fresh and sweet, and its body may be crumbly, watery, or sticky.

Additions and treatment

The addition of salt to butter contributes to its flavour and also acts as a preservative. Added in concentrations of approximately 2 per cent, all the salt goes into solution in the water phase. Since the water content of butter is less than 16 per cent of the total volume, each water droplet can contain more than 10 per cent salt. Such a concentration in the water phase limits bacterial growth overall, since the fat portion of butter is generally safe from microbial degradation. Butter may contain added colouring. Butter from cows that are eating dry, stored feed during the winter may not contain enough beta-carotene for proper colouring, as it does when cows are pasture-fed. In such cases small amounts of a yellow vegetable colouring from the seed of the annatto tree may be added to enhance the colour.

Because butter is so firm when first removed from the refrigerator, it is sometimes whipped to improve spreadability. Generally, volume is increased by 50 per cent by whipping in air—or, better still, nitrogen or an inert gas in order to prevent oxidation of the fat. Whipped butter, both salted and sweet, is sold in small plastic-coated tubs.

8.6   Ice cream manufacture

The essential ingredients in ice cream are milk, cream, sugar, flavouring, and stabilizer. Cheaper ingredients such as dry whey, corn syrup, and artificial flavourings may be substituted to create a lower-cost product.

The first step in ice cream making is formulating a suitable mix. The mix is composed of a combination of dairy ingredients, such as fresh milk and cream, frozen cream, condensed or dried skim, buttermilk, dairy whey, or whey protein concentrate. Sugars may include sucrose, corn syrup, honey, and other syrups. Stabilizers and emulsifiers are added in small amounts to help prevent formation of ice crystals, particularly during temperature fluctuations in storage.

The ice cream mix is pasteurized at no less than 79°C (175°F) for 25 seconds. The heated mix is typically homogenized in order to assure a smoother body and texture. Homogenizing also prevents churning (i.e. separating out of fat granules) of the mix in the freezer and increases the viscosity. (Since smaller fat globules have more surface area, the associated milk protein can hydrate more water and produce a more viscous fluid.)

After homogenization, the hot mix is quickly cooled to 4.4°C (40°F). The mix must age at this temperature for at least four hours to allow the fat to solidify and fat globules to clump. This ageing process results in quicker freezing and a smoother product.

The next step, freezing the mix, is accomplished by one of two methods: continuous freezing, which uses a steady flow of mix, or batch freezing, which makes a single quantity at a time. For both methods, the objective is to freeze the product partially and, at the same time, incorporate air. The freezing process is carried out in a cylindrical barrel that is cooled by a refrigerant. The barrel is equipped with stainless steel blades, called dasher blades, which scrape the frozen mixture from the sides of the freezing cylinder and incorporate or whip air into the product. The amount of air incorporated during freezing is controlled by a pump or the dasher speed. Depending on individual conditions, freezing can be instantaneous in the continuous freezer or require approximately 10 minutes in the batch freezer.

Semi-frozen ice cream leaves the freezer at a temperature between –9°C and –5°C (16°F and 23°F). It is placed in a suitable container and conveyed to a blast freezer, where temperatures are in the range of –29°C to –34°C (–20°F to –30°F). While in this room, the ice cream continues to freeze without agitation. Rapid freezing at this stage prevents the formation of large ice crystals and favours a smooth body and texture. The length of time in the hardening room depends on the size of the package, but usually 6–12 hours are required to bring the internal ice cream temperature to –18°C (0°F) or below. In larger manufacturing plants, final freezing takes place in a hardening tunnel, where packages are automatically conveyed on a continuous belt to the final storage area.

Much of the appeal of ice cream comes from the variety of standard and holiday flavours available throughout the year. Most ice cream manufacturers make a standard mix consisting of milk, cream, sugars, and stabilizers, to whi ch flavours may be added just prior to freezing. High-volume flavours such as vanilla, chocolate, and strawberry may be blended in their own large batch tanks. For flavours with large particles, such as fruit, nuts, cookies, or candy parts, a 'feeder' on the outlet of the freezer is used to inject the material. Ingredients such as fruits and nuts are carefully selected and specially treated to avoid introducing unwanted bacteria into the already pasteurized mix.

Ice cream and other frozen desserts require no preservatives and have long shelf lives if they are kept below –23°C (–10°F) and are protected from temperature fluctuations. Airtight packaging materials have made it possible to consider frozen storage of six months or longer without loss of flavour or body and texture. When ice cream is finally dipped, composition and overrun will determine ideal scooping temperature. This can vary from –16°C to –9°C (3°F to 15°F), with lower temperatures resulting in less dipping loss.

Ice cream can also be freeze-dried by the removal of 99 per cent of the water. Freeze-drying eliminates the need for refrigeration and provides a high-energy food for hikers and campers and a 'filling' centre for candy and other confections.

8.7   Cultured dairy foods

With the development of microbiological and nutritional sciences in the late 19th century came the technology necessary to produce cultured dairy products on an industrial or commercial basis. Fermented milks had been made since early times, when warm raw milk from cows, sheep, goats, camels, or horses was naturally preserved by common strains of *Streptococcus* and *Lactobacillus* bacteria. (The 'cultures' were obtained by including a small portion from the previous batch.) These harmless lactic acid producers were effective in suppressing spoilage and pathogenic organisms, making it possible to preserve fresh milk for several days or weeks without refrigeration. Cultured products eventually became ethnic favourites and were introduced around the world as people migrated.

Central to the production of cultured milk is the initial fermentation process, which involves the partial conversion of lactose (milk sugar) to lactic acid. Lactose conversion is accomplished by lactic-acid–producing *Streptococcus* and *Lactobacillus* bacteria. At temperatures of approximately 32°C (90°F), these bacteria reproduce very rapidly, perhaps doubling their population every 20 minutes. Many minute by-products that result from their metabolic processes assist in further ripening and flavouring of the cultured product. Subsequent or secondary fermentations can result in the production of other compounds, such as diacetyl (a flavour compound found in buttermilk) and alcohol (from yeasts in kefir), as well as butyric acid (which causes bitter or rancid flavours).

Cultured buttermilk, sour cream, and yogurt are among the most common fermented dairy products in the Western world. Other, lesser-known products include kefir, koumiss, acidophilus milk, and new yogurts containing *Bifidobacteria*. Cultured dairy foods provide numerous potential health benefits to the human diet. These foods are excellent sources of calcium and protein. In addition, they may help to establish and maintain beneficial intestinal bacterial flora and reduce lactose intolerance.

8.7.1   Buttermilk

Because of its name, most people assume buttermilk is high in fat. Actually, the name refers to the fact that buttermilk was once the watery end-product of butter making. Modern buttermilk is made from low-fat or skim milk and has less than 2 per cent fat and sometimes none. Its correct name in many jurisdictions is 'cultured low-fat milk' or 'cultured nonfat milk.'

The starting ingredient for buttermilk is skim or low-fat milk. The milk is pasteurized at 82°C to 88°C (180°F to 190°F) for 30 minutes or at 90°C (195°F) for two to three minutes. This heating process is done to destroy all naturally occurring bacteria and to denature the protein in order to minimize wheying off (separation of liquid from solids). The milk is then cooled to 22°C (72°F), and starter cultures of desirable bacteria, such as *Streptococcus lactis, S. cremoris, Leuconostoc citrovorum,* and *L. dextranicum,* are added to develop buttermilk's acidity and unique flavour. These organisms may be used singly or in combination to obtain the desired flavour. The ripening process takes about 12–14 hours (overnight). At the correct stage of acid and flavour, the product is gently stirred to break the curd, and it is cooled to 7.2°C (45°F) in order to halt fermentation. It is then packaged and refrigerated.

8.7.2   Sour cream

Sour cream is made according to the same temperature and culture methods as used for buttermilk. The main difference is the starting material—sour cream starts with light 18 per cent cream.

8.7.3   Yogurt

Yogurt is made in a similar fashion to buttermilk and sour cream, but it requires different bacteria and temperatures. Whole, low-fat, or skim milk is fortified with nonfat dry milk or fresh condensed skim milk, in order to raise the total solids to 14–16 per cent. The mixture is heat-treated as for buttermilk and then cooled to 45.6°C to 46.7°C (114°F to 116°F). At this point a culture of equal parts *Lactobacillus bulgaricus* and *Streptococcus thermophilus* is added to the warm milk, followed by one of two processing methods. For set, or sundae-style, yogurt (fruit on the bottom), the cultured mixture is poured into cups containing the fruit, held in a warm room until the milk coagulates (usually about four hours), and then moved to a refrigerated room. For blended (Swiss- or French-style) yogurt, the milk is allowed to incubate in large heated tanks. After coagulation occurs, the mixture is cooled, fruit or other flavours are added, and the product is placed in containers and immediately made ready for sale.

Many yogurt manufacturers have added *Lactobacillus acidophilus* to their bacterial cultures. *L. acidophilus* has possible health benefits in easing yeast infections and restoring normal bacterial balance to the intestinal tract of humans after antibiotic treatment.

8.7.4   Cheese

Cheese is the most complex of the dairy products, involving chemical, biochemical and microbiological processes. The steps in all cheese making include milk acidification, milk coagulation, whey removal, packaging and storage. Most cheese making also includes heating the cheese curd and salting the curd. Even slight changes in these processes can lead to significant differences in the final cheese. Control of these has been crucial in the transformation of cheese making from an art or craft to a skilled large-scale technological operation that is constantly undergoing minor shifts to accommodate the changing raw material (milk) and the increasing range of custom-made cheese types and styles. Flow diagram for the production of hard and semi-hard cheeses is shown in Fig. 8.2.

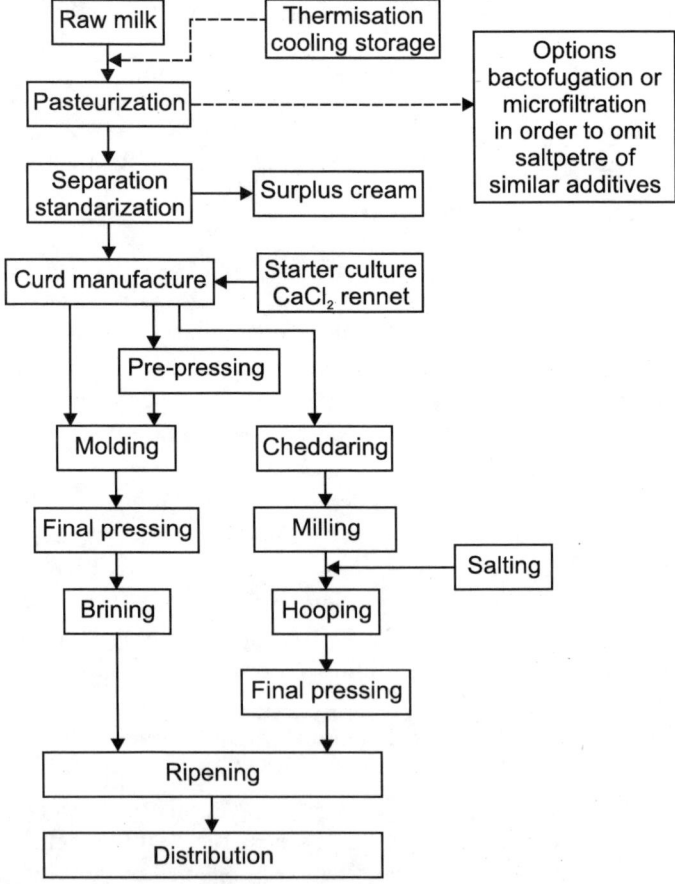

Figure 8.2 Flow diagram for the production of hard and semi-hard cheeses

There are more than 400 varieties of cheese produced throughout the world, created by differences in milk source (geographic district or mammalian species), fermentation and ripening conditions as well as pressing, size and shape. Most of the cheese types that are produced today originated many centuries ago within smaller communities and are thus named, for example, Camembert and Brie from France, Gouda and Edam from the Netherlands, Cheddar and Cheshire from England, Emmentaler and GruyÉre from Switzerland, Parmesan and Gorgonzola from Italy, and Colby from the USA; others are named for some aspect of their manufacture, e.g. Feta from Greece, processed cheese (best known as the cheese slices that go on hamburgers) from the USA, and Mozzarella from Italy; other names are more generic, e.g. cottage cheese from the USA.

8.7.5   Fundamentals of cheese making

The cheese-making process consists of removing a major part of the water contained in fresh fluid milk while retaining most of the solids. Since storage life increases as water content decreases, cheese making can also be considered a form of food preservation through the process of milk fermentation.

The fermentation of milk into finished cheese requires several essential steps: preparing and inoculating the milk with lactic-acid-producing bacteria, curdling the milk, cutting the curd, shrinking the curd (by cooking), draining or dipping the whey, salting, pressing, and ripening. These steps begin with four basic ingredients: milk, microorganisms, rennet, and salt.

Inoculation and curdling

Milk for cheese making must be of the highest quality. Because the natural microflora present in milk frequently include undesirable types called psychrophiles, good farm sanitation and pasteurization or partial heat treatment are important to the cheese-making process. In addition, the milk must be free of substances that may inhibit the growth of acid-forming bacteria (e.g. antibiotics and sanitizing agents). Milk is often pasteurized to destroy pathogenic microorganisms and to eliminate spoilage and defects induced by bacteria. However, since pasteurization destroys the natural enzymes found in milk, cheese produced from pasteurized milk ripens less rapidly and less extensively than most cheese made from raw or lightly heat-treated milk.

During pasteurization, the milk may be passed through a standardizing separator to adjust the fat-to-protein ratio of the milk. In some cases the cheese yield is improved by concentrating protein in a process known as ultra filtration. The milk is then inoculated with fermenting microorganisms and rennet, which promote curdling.

The fermenting microorganisms carry out the anaerobic conversion of lactose to lactic acid. The type of organisms used depends on the variety of cheese and on the production process. Rennet is an enzymatic preparation that is usually obtained from the fourth stomach of calves. It contains a number of proteolytic (protein-degrading) enzymes, including rennin and pepsin. Some cheeses, such as cottage cheese and cream cheese, are produced by acid coagulation alone. In the presence of lactic acid, rennet, or both, the milk protein casein clumps together and precipitates out of solution; this is the process known as curdling, or coagulation. Coagulated casein assumes a solid or gel like structure (the curd), which traps most of the fat, bacteria, calcium, phosphate, and other particulates. The remaining liquid (the whey) contains water, proteins resistant to acidic and enzymatic denaturation (e.g. antibodies), carbohydrates (lactose), and minerals.

Lactic acid produced by the starter culture organisms has several functions. It promotes curd formation by rennet (the activity of rennet requires an acidic pH), causes the curd to shrink, enhances whey drainage (syneresis), and helps prevent the growth of undesirable microorganisms during cheese making and ripening. In addition, acid affects the elasticity of the finished curd and promotes fusion of the curd into a solid mass. Enzymes released by the bacterial cells also influence flavour development during ripening. Salt is usually added to the curd. In addition to enhancing flavour, it helps to withdraw the whey from the curd and inhibits the growth of undesirable microorganisms.

Cutting and shrinking

After the curd is formed, it is cut with fine wire 'knives' into small cubes approximately one centimeter (one-half inch) square. The curd is then gently heated, causing it to shrink. The degree of shrinkage determines the moisture content and the final consistency of the cheese. Whey is removed by draining or dipping. The whey may be further processed to make whey cheeses (e.g. ricotta) or beverages, or it may be dried in order to preserve it as a food ingredient.

Ripening

Most cheese is ripened for varying amounts of time in order to bring about the chemical changes necessary for transforming fresh curd into a distinctive aged cheese. These changes are catalyzed by enzymes from three main sources: rennet or other enzyme preparations of animal or vegetable origin added during coagulation, microorganisms that grow within the cheese or on its surface, and the cheese milk itself. The ripening time may be as short as one month.

The ripening of cheese is influenced by the interaction of bacteria, enzymes, and physical conditions in the curing room. The speed of the reactions is determined by temperature and humidity conditions in the room as well as by the moisture content of the cheese. In most cheeses lactose continues to be fermented to lactic acid and lactates, or it is hydrolyzed to form other sugars. As a result, aged cheeses such as Emmentaler and cheddar have no residual lactose.

In a similar manner, proteins and lipids (fats) are broken down during ripening. The degree of protein decomposition, or proteolysis, affects both the flavour and the consistency of the final cheese. It is especially apparent in Limburger and some blue-mold ripened cheeses. Surface-mold ripened cheeses, such as Brie, rely on enzymes produced by the white *Penicillium camemberti* mold to break down proteins from the outside. When lipids are broken down, the process is called lipolysis.

The eyes, or holes, typical of Swiss-type cheeses such as Emmentaler and Gruyère come from a secondary fermentation that takes place when, after two weeks, the cheeses are moved from refrigerated curing to a warmer room, where temperatures are in the range of 20°C to 24°C (68°F to 75°F). At this stage, residual lactates provide a suitable medium for propionic acid bacteria *(Propionibacterium shermanii)* to grow and generate carbon dioxide gas. Eye formation takes three to six weeks. Warm-room curing is stopped when the wheels develop a rounded surface and the echo of holes can be heard when the cheese is thumped. The cheese is then moved back to a cold room, where it is aged at about 7°C (45°F) for 4–12 months in order to develop its typical sweet, nutty flavour.

The unique ripening of blue-veined cheeses comes from the mold spores *Penicillium roqueforti* or *P. glaucum,* which are added to the milk or to the curds before pressing and are activated by air. Air is introduced by 'needling' the cheese with a device that punches about 50 small holes into the top. These air passages allow mold spores to grow vegetative cells and spread their greenish blue mycelia, or threadlike structures, through the cheese. *Penicillium* molds are also rich in proteolytic and lipolytic enzymes, so that during ripening a variety of trace compounds also are produced, such as free amines, amino acids, carbonyls, and fatty acids—all of which ultimately affect the flavour and texture of the cheese.

Surface-ripened cheeses such as Gruyère, brick, Port Salut, and Limburger derive their flavour from both internal ripening and the surface environment. For instance, the high-moisture wiping of the surface of Gruyère gives that cheese a fuller flavour than its Emmentaler counterpart. Specific organisms, such as *Brevibacterium* linens, in Limburger cheese result in a reddish brown surface growth and the breakdown of protein to amino nitrogen. The resulting odour is offensive to some, but the flavour and texture of the cheese are pleasing to many.

Not all cheeses are ripened. Cottage, cream, ricotta, and most mozzarella cheeses are ready for sale as soon as they are made. All these cheeses have sweet, delicate flavours and often are combined with other foods.

8.7.6   Varieties of cheese

As a result of the many combinations of milks, cultures, enzymes, molds, and technical processes, literally hundreds of varieties of cheese are made throughout the world. The different types of cheese can be classified in many ways; the most effective is probably according to hardness or ripening method. In recent years different types of cheese have been combined in order to increase variety and consumer interest. For example, soft and mildly flavoured Brie is combined with a more pungent semisoft cheese such as blue or Gorgonzola.

The resulting 'Blue-Brie' has a bloomy white edible rind, while its interior is marbled with blue *Penicillium roqueforti* mold. The cheese is marketed under various names such as Bavarian Blue, Cambazola, Lymeswold and Saga Blue.

Another combination cheese is Norwegian Jarlsberg. This cheese results from a marriage of the cultures and manufacturing procedures for Dutch Gouda and Swiss Emmentaler. Varieties of cheese, classified by hardness and ripening method are shown in Table 8.2.

Table 8.2 Varieties of cheese, classified by hardness and ripening method.

	Ripening method	Cheese variety
Very hard	Bacteria/enzymes	Asiago, Parmesan, Romano, Sapsago, Sonoma Dry Jack
Hard	Bacteria/enzymes	Cantal, cheddar, Colby
	Eye producing	Emmentaler (Swiss), Gruyère,
	Bacteria/enzymes	Fontina, Jarlsberg
Semihard/ semisoft	Bacteria/enzymes	Brick, Edam, Gouda, Monterey Jack mozzarella, Munster, Provolone
	Bacteria/enzymes and surface microorganisms	Bel Paese, brick, Limburger, Port Salut, Trappist
	Bacteria/enzymes and blue mold	Blue, Gorgonzola, Roquefort, Stilton
Soft	Bacteria/enzymes and surface microorganisms	Brie, Camembert, Neufchâtel (France), Pont l'Évêque
	Unripened	Baker's, cottage, cream, feta, Neufchâtel (United States), pot

8.7.7   Pasteurized process cheese

Some natural cheese is made into process cheese, a product in which complete ripening is halted by heat. The resulting product has an indefinite shelf life. Most process cheese is used in food service outlets and other applications where convenient, uniform melting is required.

Pasteurized process cheese is made by grinding and mixing natural cheese with other ingredients, such as water, emulsifying agents, colouring, fruits, vegetables, or meat. The mixture is then heated to temperatures of 74°C (165°F) and stirred into a homogeneous, plastic mass. Process cheese foods, spreads, and products differ from process cheese in that they may contain other ingredients, such as nonfat dry milk, cheese whey, and whey protein concentrates, as well as additional amounts of water.

American cheddar is processed most frequently. However, other cheeses such as washed-curd, Colby, Swiss, Gruyère, and Limburger are similarly processed. In a slight variation, cold pack or club cheese is made by grinding and mixing together one or more varieties of cheese without heat. This cheese food may contain added flavours or ingredients.

9

Designer milk

9.1   Introduction

Nutritional and genetic interventions to alter the milk composition for specific health and/or processing opportunities are gaining importance in dairy biotechnology. Altered fatty acid and amino acid profiles, more protein, less lactose and absence of β-lactoglobulin are some challenges of 'designing' milk for human health benefits. Alteration of primary structure of casein and lipid profile, increased protein recovery, milk containing nutraceuticals and replacement for infant formula are some of the processing advantages envisaged. Final acceptability of the newly designed products will depend on animal welfare, safety and enhanced health properties of the products and increased *profitability vis-à-vis conventional practices.*

Advances in biotechnology and genetic engineering have hinted at possibilities that were hitherto not fathomed in the field of dairying. It is now firmly established that a new generation of value-added products can be harvested from milk and milk products.

Milk composition can be altered by nutritional management or through the exploitation of naturally occurring genetic variation among cattle. By a thorough understanding of the biochemistry, genetic traits and changes in the cow's diet that affect milk synthesis and composition, ways and means to manipulate milk composition to suit specific needs can be found. By combining the two approaches of nutritional and genetic interventions, researchers are now hoping to develop 'designer milk' tailored to consumer preferences or rich in specific milk components that have implications in health as well as processing. This chapter discusses the potential that exists in altering the milk composition or 'designing' milk by nutritional and genetic approaches so as to achieve specific health and/or processing opportunities.

However, various ethical, legal and social aspects of biotechnological research need to be addressed through public education to bring about greater understanding of the issues involved and to find classical solutions for the existing issues. Until recently, emphasis had been on breeding bigger animals that produced more milk. Now we are turning our attention to adding more value to the milk we produce and studying its health implications. An interest in understanding how a cow makes milk, what controls the casein's synthesis, what controls the synthesis of fat in mammary glands, how the biochemical pathways work has led to find ways to produce milk with lower fat and greater amounts of produce milk with lower fat and greater amounts of protein or different types of proteins. By combining the two approaches—the genetic work and the diet studies researchers are hoping to develop 'designer milk' tailored to consumer preferences or rich in specific milk components that have health implications. There has been a decline in livestock, product consumption viz. egg and egg product-50 per cent, liquid milk-20 per cent, butter-70 per cent and meat-11 per cent over the last two decades. As consumer preference for animal products is likely to continue, it would be important to modify animal products in such a way that dietary risks are minimized while there would be maximum benefits in respect of processing of products. Rapid development of genetic technology has placed the dairy processors open to improvement by modern biotechnology, while novel horizons beckon in nutrition, food technology and pharmacology.

9.2   Modification of milk composition by genetic technology

Milk composition can be dramatically altered using gene transfer. Classical studies have revealed association between major lacto-protein variants and milk production traits. Introduction of DNA technology in the dairy science field has enabled to identify new genetic polymorphism and revealed molecular background of lacto-protein gene expression. Subsequently, several emerging transgenic technologies have focused on the mammary gland.

A transgenic animal has been defined as an animal that is altered by the introduction of recombinant DNA through human intervention. The process involves that a DNA construct is designed and built to express the desired protein in animal, the construct then introduced into a single cell embryo to allow incorporation of the transgene into animal genome. There are several methods available for this purpose, including retroviral transmission, stem cell transfection, and microinjection into the pronucleus or cytoplasm.

9.3   Designer milk and nutritional significance

9.3.1   Change in milk fat

Composition of milk could be altered by feeding lactating dairy cows a control diet to obtain a naturally modified component. This method is eyed towards altering milk fat. Fatty acids of less than 12 carbon atoms are neutral or actually may decrease cholesterol. Stearic acid (C18:0) acts similarly to oleic acid (*cis*-C18:1) to decrease cholesterol. Only three saturated fatty acids (lauric, C12:0; myristic, C14:0; and palmitic, C16:0) now are considered to be hypercholesterolemic.

These fatty acids constitute about 44 per cent of total milk fatty acids. According to a group of nutritionists from industry and academic, the 'ideal' milk fat for human health would contain <10 per cent polyunsaturated fatty acids, <8 per cent saturated fatty acids, and >82 per cent monosaturated fatty acids. Although it is not likely that the 'ideal' milk fat composition could be achieved, manipulation of composition of milk fat is possible through feeding practices for dairy cows.

Studies have demonstrated significant and prompt changes in the iodine number of milk fat when the diet fat was changed from a high to a low degree of unsaturation. The modest increase in unsaturation was probably due to an increase in the oleic acid (C18:1) with little increase in linoleic acid (C18:2). The rumen microflora is apparently able to hydrogenating only one double bond. A 'Designer Cow' which can produce semi-skimmed (half-fat milk) milk and butter that spreads straight from the refrigerator (soft-spreading butter) has been bred by Britain's Agricultural Development and Advisory Services (ADAS) in collaboration with Samsbury's, the retail grocer chain. Cows fed with fish oil, fishmeal or plankton can produce milk rich in $\bar{\omega}-3$ fatty acids. Restricted quantities of fish products are added to the cow's diet of grass or silage to produce $\bar{\omega}-3$ fatty acids enriched milk. Similarly for half-fat milk, the supplement is dehusked oats and for a spreadable butter is rapeseed oil.

In the United States, researches have a way to boost the whole milk natural cancer fighting ability. Dietary fats such as corn oil are fed to cows in protected form. The cows produce milk with substantially increased levels of conjugated linoleic acid (CLA).

A number of laboratory studies have shown that CLA suppresses carcinogens, inhibits proliferation of leukemia and colon, prostate, ovarian, and breast cancers. CLA inhibits cancer growth even in extremely low dietary concentration of 0.5 per cent and is among the mostly potent of all naturally occurring anticarcinogens.

9.3.2   Changes in milk protein

One of the major products of the mammary glands is protein, it is thus legitimate to consider the production of additional proteins for which there is a demand and that introduce desirable compositional changes in the milk, several potential changes are listed in Table 9.1.

Table 9.1 Potential changes in milk through genetic engineering.

Potential modification	Change in milk
Increase casein content	Increased protein, better manufacturing properties particularly for cheese making
Engineered Casein	Better manufacturing properties
Remove a β-lactoglobulin	Better manufacturing properties, decreased milk allergies
Remove fat	Easier to produce low fat milk products, decrease the butter surplus. Increase solids content, remove lactase lactose
Produce b-galactosidase,Lactose, produce antibodies of pathogens	Safer food, mastitis prevention

9.3.3   Human-like milk

A human-like milk enriched with human lysozyme and lactoferrin might enhance defense against gastrointestinal infections and promote iron transport in the digestive tract. Given the much higher level of lactoferrin in the human milk (10 times more than in cow's milk); it may be desirable to supplement the level in baby formula to bring it up to the human milk level. A construct comprising human lysozyme cDNA driven by the bovine alpha S_1-casein promoter yielded a high level of lysozyme in mice. Obtention of a transgenic bull carrying a human lactoferrin construct was also reported. These studies drive towards humanization of non-human mammals' milk.

9.4   Changes in the individual casein content of milk

9.4.1   Increasing the α-S_1-or α-S_2-casein Content

A higher content of α S_2-casein, which contains 10–13 phosphoserine and two cysteine residues in the bovine species, would obviously enhance their nutritional value of casein, which is deficient in sulphur-containing amino acids, and would presumably increase micelle stability. The bovine αS_1-casein gene and hybrid genes driven by the α S_1-casein promoter have been well expressed in mice but attempts to express the bovine α S_2-casein gene have been unsuccessful.

9.4.2   Producing milk devoid of β-lactoglobulin

Cow's milk is an allergic trigger in a significant fraction of infants and β-lactoglobulin, which is not found in human milk, is believed to be one of the culprits. Elimination of this protein from cow's milk is unlikely to have any detrimental effects, on either to cow or human formula, and might actually overcome many of the major allergy problems associated with cow's milk. Presently, a reasonable genetic approach is the systematic search for individuals, which might carry a null β-lactoglobulin allele, which proved successful for the α-and β-casein loci in the goat species. The use of antisense or ribozyme anti-β-lactoglobulin mRNA is appealing and is very potent to prevent synthesis of 3 g of β-lactoglobulin in bovine species. The methodology of choice is obviously the knockout of β-lactoglobulin gene, which relies on the availability of the ES cells in dairy species.

9.4.3   Changes in milk sugar

Reduction of lactose in milk

Lactose is the major sugar present in milk and is an important osmotic regulator of milk secretion. It is synthesized in the mammary gland by the lactose synthetase complex composed of uridine diphosphate-galactosyl transferase and of the mammary specific α-lactalbumin. The adult population suffers from intestinal disorders after milk ingestion, as a consequence of lactose maldigestion that results from the normal drop of the intestinal lactose-hydrolyzing enzyme. Lactase-phlorizin associated with intestinal pathologies and with rare cases of intolerance can also be limited primarily by dietary changes either by avoidance of dairy products or through the use of low lactose (post harvest) or lactase replacement products. Each of these management strategies requires dietary supplementation and they vary in their efficiencies.

9.5   Designer milk and technological aspects

9.5.1   Modification of protein

Increasing the κ-casein content

In the mammary gland, calcium-induced precipitation of αS_1-, αS_2-, and β-casein is prevented by their association with κ-casein as sub micelles stably bound together through calcium bridges. Higher κ-casein content was found to be associated with smaller micelle size. Furthermore, heat stability of milk was markedly improved by addition of κ-casein. Because of the key roles of κ-casein, the relevant gene is one of the most obvious candidates for trangenesis. Transgenic mice carrying a caprine β-casein: β-casein fusion gene produced milk containing up to 3 mg caprine κ-casein per ml. This hybrid gene is thus quite suitable for microinjection into eggs from farm animals.

Modification of β-casein in milk

An increase in (β-casein reduces RCT and increases syneresis rate, and also increases curd firmness by 50 per cent. One modification that has been made to β-casein is the deletion of plasmin cleavage site causing prevention of bitter flavour in cheese due to plasmin cleavage. The second alternation is removal of cleavage site for chymosin. The last modification made is the addition of glycosylation sites to the molecule increasing its hydrophilicity (the above modification done to mice milk). Also the native caprine (β-casein gene successfully tested in mice, which produced up to 25 mg exogenous (β-casein per ml of milk, is being used as such for generating transgenic goats producing β-casein. β-lactoglobulin causes aggregation and gelation of milk at higher temperature, and also is potent allergen. This gives the reason to remove this gene by ES cells.

9.5.2   Modification of fat

The most significant changes in milk fat quality relate to rheological and organoleptic properties, which influence numerous aspects of character and quality of manufactured dairy products. The type of fatty acids present in milk fat can influence the flavour and physical properties of dairy products. Sensory evaluation indicated that butter produced from cows fed high oleic sunflower seeds and regular sunflower seeds were equal or superior in flavours to the control butter.

The high oleic sunflower seed and regular sunflower seed treatment butter were softer, more unsaturated and exhibited acceptable flavour, manufacturing, and storage characteristics. Sensory evaluation of Cheddar cheese indicated that extruded soyabean and sunflower diets yielded a product of quality similar to that of control diet.

Cheese made from milk obtained with extruded soyabean and sunflower diets contained higher concentrations of unsaturated fatty acids while maintaining acceptable flavour, manufacturing, and storage characteristics. It is anticipated that the total unsaturated fatty acids content could be increased beyond the amounts achieved to date when feeding unsaturated fat sources and without adversely affecting flavour or product processing properties.

Milk containing naturally modified fat, obtained either by feeding extruded soybean or sunflower seeds produced blue and Cheddar cheeses enhanced the safety against *L. monocytogenes* and *S.typhimurium* due to accumulation of fatty acids, namely C_{12}, C_{14}, $C_{18:1}$ and $C_{18:2}$.

9.5.3   Other modifications

Increasing total solids and reducing lactose and water content of milk

Mice that produce milk with 33 per cent more total solids and 17 per cent less lactose than normal control mice have been generated by transgene. Mouse milk normally has approximately 30 per cent total solids, while these transgenic mice produce milk that is 40–50 per cent total solids when milk reaches 45–50 per cent total solids it lacks fluidity and becomes difficult to remove from mammary gland. The increase in the total solids is associated with a decrease in total milk volume and a decrease in amount of water in the milk. The activity of lactose synthetase enzyme within the mammary gland decreased by 35 per cent in the transgenic mice. These data indicate that due to a decrease in lactose synthetase activity, less lactose is being produced; less water is being transferred into milk causing a reduction in milk volume. So it appears that the same amount of total milk fat and protein are being produced in a lesser total milk volume.

Using the technology described above and increasing our understanding of these mechanisms may lead to cattle with similar characteristics to the mice that have been produced. Having milk from a dairy cow that contain 6.5 per cent protein, 7 per cent fat, 2.5 per cent lactose and 50 per cent less water would have a number of economic and processing benefits. The main direct economic benefit would be the 50 per cent reduction in the cost of shipping milk. In addition, since the cow would be producing one-half her normal volume of milk there would be less stress on the cow and on her udder. On the manufacturing side, after the removal of fat from this type of milk, a skim milk having twice protein content and have half the lactose content of normal milk could be produced. This type of milk would also make it easier to produce low lactose or lactose free 'hard dairy' products. The concentrated milk should lead to better product yields form the same amount of initial input. The lowering of milk volume and lactose content will reduce the total whey output produced during processing. The reduction of stress on the mammary gland of the cow and the more viscous milk may also decrease the susceptibility to obtaining mastitis infection. Organisms that cause mastitis use lactose as their energy source and since lactose would be reduced in the system there would be a decrease in the available food source for these bacteria.

9.5.4   Designer milk with higher levels of β-casein and κ-casein-A

To enhance milk composition and milk processing efficiency by increasing the casein concentration in milk, researchers have introduced additional copies of the genes encoding bovine β- and κ-casein (*CSN2* and *CSN3*, respectively) into female bovine fibroblasts. Nuclear transfer with four independent donor cell lines resulted in the production of 11 transgenic calves. The analysis of hormonally induced milk showed substantial expression and secretion of the transgene-derived caseins into milk.

Nine cows, representing two high-expressing lines, produced milk with an 8–20 per cent increase in β-casein, a twofold increase in κ-casein levels, and a markedly altered κ-casein to total casein ratio. These results show that it is feasible to substantially alter a major component of milk in high producing dairy cows by a transgenic approach and thus to improve the functional properties of dairy milk.

Substantial gains

The researchers, led by Goetz Laible, New Zealand biotech company AgResearch, engineered cells in the laboratory to overproduce casein proteins. The cells were then fused with cow eggs. The resulting embryos were transferred into recipient cows, and 11 transgenic calves were born. Nine were found to produce the enhanced milk. One protein, called κ-casein, increases heat stability in the cheese-making process. The other, β-casein, improves the process. The other, β-casein, improves the process by reducing the clotting time of the rennet, which curdles the milk. It also increases the expulsion of whey; the watery part of milk, which remains after the cheese, has formed. The cows are now producing milks with 8–20 per cent more β-casein, and double the normal amount of κ-casein. Reporting their findings, the scientists said that controlling levels of the two proteins could offer big savings for cheese manufacturers. Cows genetically modified to produce high-protein milk for the cheese industry have been created in New Zealand. It is the first time that cow's milk has been engineered to improve its quality, rather than to contain profitable pharmaceuticals. The cows possess additional copies of genes for two proteins: β-casein and κ-casein. As a result, their milk contains between up to 20 per cent more β-casein and twice the amount of κ-casein as milk from ordinary cows. This should allow cheese-makers to produce more cheese from the same volume of milk. The manufacturing process should also be quicker, due to the faster clotting times associated with the higher protein levels.

Designer milk to cut cheese costs

Protein-rich designer milk from modified cows could speed dairy processing. Protein-rich milk from cloned, genetically modified cows could cut cheese-making costs. Dairy manufacturers would need less milk to make cheddar firm and ice cream creamy. Two years old and living in New Zealand, the clones produce about 13 per cent more milk protein than normal cows. They carry extra copies of the genes for two types of the protein casein, key for cheese and yoghurt manufacture. They allow milk to have high protein content, but to remain watery, says study leader Gotz Laible of New Zealand biotech company AgResearch. His team must now find out whether the increase improves milk's calcium content or its ability to coagulate before they seek approval to sell the clones to dairy farmers. Most scientists believe that milk from cloned cows is no different than normal milk. But they are less certain about the safety of milk from genetically modified cows.

Genetically modified cheese

10.1 Introduction

Cheese is a food derived from milk that is produced in a wide range of flavors, textures, and forms by coagulation of the milk protein casein. It comprises proteins and fat from milk, usually the milk of cows, buffalo, goats, or sheep. During production, the milk is usually acidified, and adding the enzyme rennet causes coagulation. The solids are separated and pressed into final form. Some cheeses have molds on the rind or throughout. Most cheeses melt at cooking temperature. Considerable developments have been made to shorten the long ripening period, increase the cheese yield, reduce the bitterness in cheese, produce desired flavour level within short maturation period, etc. The unprecedented accomplishments of biotechnology and genetic engineering since the past decade have added a new margin in designing transgenic cheese.

The cheese technology had been modified from time to time in order to surmount the processing challenges and to gift the mankind with novel type of cheese with improved physico-chemical and functional properties. With the triumphal achievements in genetic engineering in the field of food, agriculture and health sector, recent developments in cheese technology have emerged with a new margin as the unprecedented consequence of genetic games alike the conventional and gradual amelioration in cheese making technique as usually expected with the passing generation by the virtue of evolution in order to satisfy the more and more requirements. The controversy concerning the biosafety of genetically modified (GM) food, development of genetically modified cheese (GME) as the outcome of good cheese-biotech sorority seems to curtail or minimize a number of processing challenges for cheese manufacture.

10.2 Genetic modification

Genes are organized into chromosomes which are found in all living cells. They are a coded form of instructions to make proteins. Most of the proteins manufactured by living cells are enzymes. Enzymes regulate the functional activities. Genetic information mainly is stored in a coded form in deoxyribonucleic acid (DNA) molecules present in genes. So, DNA is a kind of molecular blue print where it stores all the information needed to make a new cell. Therefore, the phrase, 'genetically modified' is commonly used to describe the application of recombinant deoxyribonucleic acid (rDNA) technology for the genetic alteration of microorganisms, plants and animals. This advanced molecular technology allows for effective and efficient transfer or alter genetic material from one organ to another. An identified single gene responsible for a particular trait can be inserted/transferred among all living organisms. The technology is now well developed and can be applied to any living organism.

10.3   Genetically modified cheeses

Cheese obtained or manufactured by the application of genetic engineering by adopting the following three approaches can be called as genetically modified cheese. These three approaches may be adopted individually and or in combination to obtain cheese with approved vegetarian sentiment, improved functional attributes, increased nutritional/therapeutic value, accelerated ripening activities and enhanced sensory quality. These approaches are:

- Modification in milk composition.
- Addition of recombinant coagulating enzymes.
- Application of modified starter culture.

Among the three approaches, considerable amount of work have been carried out on application of genetically modified micro-organism to obtain improved starter culture and recombinant coagulating enzymes for the cheese production. Work on alteration of milk composition by Genetic manipulation in animals results in the improvement of milk production system which can even produce dairy products with functional characteristics. In this context, main emphasis for genetically modified cheese is now on the processing aspects, i.e. on starter culture and coagulating enzymes.

10.4   Modification in milk composition

It is established that genetic variation in animal causes the different composition in milk. Transgenic cows with altered genetic make up to produce milk with 2 per cent fat with a greater proportion of unsaturated fatty acids in milk fat, higher levels of milk protein, β- and κ-casein and reduced lactose content in milk have been developed. Such milk is suited ideally for people suffering from lactose intolerance and also to produce improved varieties of specialty cheeses with effective cost price efficiency.

Genetically modified bovine somatotrophin (BST) also play a role in the regulation of milk yield, growth rate and protein to fat ratio of milk which results in milk composition alteration.

10.4.1   Recombinant coagulating enzymes

One of the success stories for the application of genetic engineering is the manufacture of recombinant chymosin and its use as milk coagulating enzyme for commercial cheese production. Concern over the supply of chymosin from traditional source (suckling calves) has led to efforts over the past three decades to develop a recombinant source. Cheese industry has been the major beneficiary of this technological work. The gene coding for the chymosin enzyme has been cloned in the bacteria *Escherichia coli*, the yeast *Kluyveromyces lactis* and the mold *Aspergilus niger*. The enzymatic properties of the recombinant enzymes are indistinguishable from those of calf chymosin. The cheese making properties of recombinant chymosin produces very satisfactory results and its use in commercial plant have been approved by many countries.

All the three chymosin available are identical to calf chymsoin, considered as vegetarian source and accepted by religion group. More knowledge on genetic engineering combined with better understanding of protein structure might one day give us chymosin with higher activities, lower cost and with flavour enhancing properties of cheese during ripening.

10.4.2   Modified starter culture

The application of gene cloning technology to lactic acid bacteria is the potential process in the generation of enhanced starter cultures for the manufacture of cheese and yoghurt. These starter cultures are mainly made of species of *Lactococcus, Lactobacillus* and *Streptococcus*.

Modifications of these microorganisms were achieved mainly on three directions for cheese making process. These are:
- Development of phage resistant cheese culture.
- Organisms with probiotic activity for cheeses.
- Acceleration of cheese ripening.

Phage resistant cheese culture: The major cause of slow acid production in cheese plants today is bacteriophage (phage). This can significantly upset manufacturing schedules and, in extreme cases, result in complete failure of acid production or 'dead vats'. Phages are viruses that can multiply only within a bacterial cell. They have a head, which contain the DNA and a tail, which is compared of protein. Morphologically, there are three types of phage for *latococci* as mentioned below:

- Small isometric headed (spherical headed) phage most common.
- Prolate—headed (oblong—headed) phage.
- Large isometric—headed phage.

Phage multiplication occurs in one or two ways, called the lytic and lysogenic cycle. Multiplication of phage is very fast in their hosts. Microbiologists have enhanced cheese culture performance by genetically engineering bacteria increasing the viability of the culture during cheese making. The new strain of bacteria resist phase contamination and are suitable for prolonged use in milk fermentation. Several phase resistance mechanism, including inhibition of phage adsorption, restriction—modification mechanism and abortive infection mechanism are found in LAB. All of these are commonly encoded on plasmids.

A phage resistant starter culture of cheddar cheese, *L. lactis* DPC 5000 have been developed which was shown to embody three effective phage resistant mechanisms. Cheddar cheese manufactured with DPC 5000 compared favourably in term of composition with cheeses manufactured using commercial starter. Two phage-resistant thermophilic starter strains DPC 1842 and DPC 5099 have been developed by Cork and others. Ireland performed well in commercial plants for Mozzarella cheese preparation. The new cultures provide more predictable performance and reduce the chance of vats failure.

10.4.3   Cheese with increased yield

Scientists in New Zealand have created the world's first cow clones that produce special milk that can increase the speed and ease of cheese making. The researchers in Hamilton say their herd of nine transgenic cows makes highly elevated levels of milk proteins, i.e. casein-with improved processing properties and heat stability. Cows have previously been engineered to produce proteins for medical purposes, but this is the first time the milk itself has been genetically enhanced. The scientists hope the breakthrough will transform the cheese industry, and if widened, the techniques could also be used to 'tailor' milk for human consumption. But opponents of GM cheese continue to doubt whether such products will be safe.

Probiotic organisms for cheese: Ripened variety of cheeses may offer certain advantages as a carrier of probiotic microorganisms with specified health benefits. Growing public awareness of diet-related health benefits has fuelled the demand for probiotic foods. Dairy foods including fermented milks and in particular yoghurt are the best accepted food, which can carry probiotic cultures. Among the most important criteria when considering cheddar cheese as a probiotic food is that the micro-organism be able to survive for relatively long ripening time of at least 6 months and/or that they grow in the cheese over this period although the pH is higher in comparison to traditional probiotic dairy foods. The potential health promoting effects achieved by the consumption of dairy foods containing probiotic organisms, such as *Lactobacillus* and *Bifidobacterium* spp., have resulted in intensive research efforts in recent years. Even though small number of probiotic cheese is currently in the market worldwide, a few reports are available concerning cheese as a carrier of probiotic organisms. European manufactures have introduced different varieties of cheese with added *Lactobacillus paracasei* NFBE 338. Dinakar and Mistry incorporated *Bifidobacterium bifidum* into cheddar cheeses as starter adjunct. This strain survived well and retained a viability of 2×10^7 cfu/gm of cheese even after 6 months of ripening, without adversely affecting cheese flavour, texture or appearance. This suggested that cheddar cheese can provide suitable environment of probiotic organisms.

Bifidobacteria were used in combination with *L. acidophilus* strain Ki as starter in Gouda cheese manufacture. There was a significant effect on flavour development in the cheese after 9 weeks of ripening. Cheddar cheese was manufactured with either *Lactobacillus salivarius* NFBC 310, NFBC 321 and NFBC 348 or *L. paracasei* NFBC 338 or NFBC 364, isolated from small intestine, as the dairy starter adjunct. Using randomly amplified polymorphic DNA method, it was found that both *L. Paracasel* strains grew and sustained high viability in cheese during ripening, while each of the *L. salivarius* species declined over the ripening period. These data demonstrate that Cheddar Cheese can be an effective vehicle for delivery of some probiotic organisms to the consumer.

Therefore, though the reports on probiotic cheeses are few but there is promising scope to exert probiotic effect on the cheese by enhancing the existing characteristic of probiotic organisms with genetic engineering. They may be called as second-generation probiotics. These might include the incorporation of attachment molecules to facilitate colonization, increased production of antimicrobial compounds directed against common food pathogen, cholesterol metabolism, enhanced immune stimulation and viability which contribute beneficial effect on human health. Attempts are also being made to identify new strains of probiotic culture by identifying the genes involved in such attributes and exploring the advanced molecular technique such as PCR, RFLP, RAPD and DNA finger printing, etc. All these modifications in dairy starters can be brought also with the help of genetic engineering with the sole objective of improving their functionality so that cheese industry could benefit from value addition in cheeses/dairy foods through the intervention of these genetically modified organisms. Another desirable property might be the generation of probiotic genetically modified organism to digest especially prebiotic carbohydrates. The combinations of probiotic and a prebiotic are known a synbiotic and the prebiotic is thought to enhance the survival of the probiotic. Several strains of *Lactobacillus* and *Bifidobacterium* sp. have been engineered to metabolize unusual carbohydrate, but the potential to form enhanced synbiotic has not yet been evaluated.

Acceleration of cheese ripening: The objective of acceleration of cheese ripening is to accelerate the proteolytic process and related events that occur in naturally ripened cheese as closely as possible. The methods used to accelerate ripening fall into six categories:

- Elevated ripening temperature
- Exogenous enzymes
- Chemically or physically modified bacterial cells
- Genetically modified starters
- Adjunct culture
- Cheese slurries and enzyme-modified cheeses

In this context genetically modified starter cultures is of special interest.

Genetically modified starters: Genetically modified starter cultures with enhanced complements of proteinase and/or peptidase, which could be released early and evenly distributed in the curd would be an ideal method of accelerating cheese ripening. Modified/genetically tailored microorganisms, genetically engineered proteolytic and lipolytic enzymes are now being used for enhancing flavour production in cheese. Enzyme addition is now one of the few preferred methods of accelerated ripening of cheese. Enzyme may be immobilized or encapsulated for long-term action on the production for quick action and homogenous distribution in the product.

Cheese manufactured with an amino peptidase N-negative clone strain of *Lactococcus* produced bitter off flavour. The possible role of amino peptidase as a debittering agent confirmed by Prost and Chamba after making the Emmetal cheese with *Lb. helveticus* strain L_1 (high amino peptidase activity), L_2 or strain L_3 (clones selected for lack of amino peptidase activity). They also explained the bitterness in ripened cheese made with *Lb. delbrueckii subsp. lactis* to be due to their very low amino peptidase activity.

The gene for the neutral proteinase (neutrase) of *B. subtilis* has been cloned in *Lc. lactis* UC317. Cheddar cheese manufactured with this engineered culture as the sole starter showed very extensive proteolysis, and the texture became very soft within 2 weeks at 8°C. Cheddar cheese made with *Lc. lactis* subsp. *cremoris* Sk 11 (cloned with proteinase) revealed that starter proteinases are required for the accumulation of small peptides and free amino acids in cheddar cheese. The strain in which the proteinase remained attached to the cell wall appeared to contribute more to proteolysis than the strain that secreted the enzyme. Cheeses made with proteinase positive starter produce more pronounced flavour than those with proteinase negative strain during ripening. The inactivation of genes in a metabolic pathway can be used to alter end products that accumulate from a given pathway. Application of regulated promoters (in genetic engineering) is the controlled expression of lytic genes resulting in autolysis of the starter culture. This would result in rapid release of enzymes, (i.e. peptidase) into the cheese matrix and potentially accelerate cheese flavour development.

Therefore, through genetic engineering, specific genes can be implanted/removed to increase or decrease the activity of specific property of the existing strain of LAB. It can be stated from the utilization of genetically modified starter as: (i) cloning of exogenous proteinases in starter cells leads to enhanced proteolysis, (ii) debittering action of amino peptidase is now well recognized and (iii) starter peptidases and proteinases produce small peptides and amino acids in cheese but may not have a direct impact on flavour.

Bakery and cereal products

11.1   Introduction

The bakery industry comprises mainly of bread, biscuits, cakes and pastries manufacturing units. The contributing factors for the popularity of bakery products are urbanization resulting in increased demand for ready to eat convenient product, availability at reasonable cost, greater nutritional quality, availability of varieties with different textural and taste profiles and better taste. The bakery products have become popular among all cross section of populations irrespective of age group, and economic conditions.

11.2   Bread formulation

Many different types of bread formulations have been developed so far. These formulations are developed in different regions based on the traditional food habits of the people.

The main bread types can be classified as under:

1. *Pan bread:* This type of bread is popular in the economically developed countries including USA, Canada, UK and European nations.
2. *Hearth bread or sour bread:* This category of bread is produced with or without lactic acid fermentation. Hearth breads are baked in an open hearth. These breads are becoming popular in France.
3. *Flat bread or roti/chappati:* This category of bread is popular in Asian countries. The product is unfermented and flat. This is baked on a flat hot pan.
4. *Rolls and other small fermented breads:* These products generally have higher levels of sugar and fat in the formulation and thus typically have sweeter taste and softer bite characteristics.

The basic recipes for bread making include wheat flour, yeast, salt and water. If any one of these basic ingredients is missing, the acceptable product cannot be prepared. Other ingredients are known as optional, for example, fat, sugar, milk and milk product, malt and malt product, oxidants (such as ascorbic acid and potassium bromate), surfactants and anti-microbial agents. Each of these ingredients has specific role to play in bread making. The wheat flour is the main ingredient in bread production. It is primarily responsible for bread structure and bite characteristics. Water transforms flour into a viscoelastic dough that retains gas produced during fermentation and water also provides medium of all chemical reaction to occur. Yeast ferments sugar and produces carbon dioxide gas and ethanol. Thus it gives us porous and leavened bread. Sugar is the source of fermentable carbohydrate for yeast and it also provides sweet taste. Salt enhances flavour of all other ingredients and adds taste to the bread. It also strengthens the gluten network in the dough. Fat makes the bread texture softer and improves its freshness and shelf-life. The oxidizing agents such a ascorbic acid, potassium bromate, potassium iodate and azodicarbonamide are used at parts per million levels to enhance dough strength loaf volume and softness. Surfactants are used as antistaling agents. Calcium propionate is used as prevent mold growth.

11.3   Bread-making procedure

The following steps are generally considered essential for the production of good quality bread.

11.3.1   Sieving

The flour is generally sieved before using in bread primarily for following reasons:

1. To aerate the flour
2. To remove coarse particles and other impurities
3. To make flour more homogeneous

11.3.2   Weighing

The next step is weighing of different ingredients as per formulation. Minor ingredients have to be weighed more precisely. Salt, sugar, oxidizing agents and yeast are added in solution form. Yeast is added as a suspension, which is mixed well each time before dispensing. Sequence of addition of ingredients also affects the dough characteristics. Generally shortening and salt are added after the clean up stage.

11.3.3   Mixing

Mixing of flour and ingredients involves, i.e. hydration and blending, dough development and dough breakdown. The process of mixing begins with hydration of the formula ingredients. The mixing, whilst the flour is hydrating, brings about development of the gluten network in dough, which is evidenced as an ascending part of the mixing curve. The dough system subsequently becomes more coherent, losing its wet and lumpy appearance, and it achieves a point of maximum consistency or minimum mobility. This is the point to which dough should be mixed for producing bread of superior loaf quality. At this stage the dough is converted into a viscoelastic mass from thick and viscous slurry. At this stage the gluten forms a continuous film or sheet suitable for processing into bakery applications. If mixing is continued beyond this point, mechanical degradation of the dough occurs resulting in the breaking down of the dough network. Eventually the dough becomes wet, sticky and extremely extensible, and is capable of being drawn out into long strands. This is generally referred to as the dough being 'broken down'. Such dough will create problem in dough handling and frequent break down in the plants and ultimately results into processing losses. Various stages of mixing are explained below:

1. *Initial hydration stage:* At this stage ingredients are blended and homogenized. The dough begins to wet and sticky.

2. *Pick up stage:* At this stage the hydration of ingredients is advanced and they are aggregated into wet mass. The wet mass is uneven and wet. The gluten begins to develop in the dough system.

3. *Clean up stage:* Further mixing develops the gluten network in the dough. Dough becomes extensible and elastic. Dough forms a cohesive mass and ceases to stick to mixing blades and walls of mixer.

4. *Development stage:* The dough becomes more viscoelastic in nature. It gives silky and shine character.

5. *Optimum stage:* This is the optimum mixing stage. Dough at this stage is elastic, silky and smooth. Forms thin membrane of uniform thickness when stretched without breaking. It is the right stage to process dough for bread making.

6. *Break down stage:* Beyond optimum stage the dough becomes increasingly soft, smooth and highly extensive. Dough also becomes sticky and demonstrates poor machinability.

Mixing time

The mixing time varies with the type of flour, type of mixer, speed of mixing presence of salt or shortening, additive, particle size as well as damaged starch content of flour. A flour of good bread quality flour should have medium to long dough development and mixing time. If the mixing time is very short, the flour can easily be over mixed, and if the mixing time is very long it might never reach its optimum. On the other hand, biscuit quality flour will develop in to dough rapidly. Flour with weak gluten develops more quickly, whereas strong and 'extra strong' flours need a longer time to mix to peak dough resistance, suggesting that the mixing requirement of a flour is related to its gluten protein composition. It has been observed that gliadin and its subgroups decreases the mixing time of flour. The mixing time is related to the size of gliadin proteins.

11.3.4   Fermentation

Optimally mixed dough is subjected to fermentation for a suitable length of time to obtain light aerated porous structure of fermented product. Fermentation is achieved by yeast *(Saccaromyces cerevisiae)*. The yeast in dough breaks down the sugars to carbon dioxide and ethanol. The gas produced during fermentation leavens the dough into foam. The foam structure of dough is discrete and has stability during fermentation. When fermented dough is baked, the foam structure gets converted into sponge structure that is responsible for aerated structure of breadcrumb. The conditions under which fermentation occurs affect the rate of carbon dioxide production and flavour development in the dough. The temperature and relative humidity conditions are particularly important for yeast activity and gas production. In the temperature range of 20–40°C, the yeast fermentation rate is doubled for each 10°C rise in temperature. Above 40°C yeast cells are started to get killed. The yeast performs well at 30–35°C and relative humidity of 85% and above. The optimum pH range for yeast is 4–6. Below pH 4 the yeast activity begins to diminish and it is inactivated below pH 3. Osmotic pressure also affects the activity of yeast.

11.3.5   Knock back

Punching of dough in between the fermentation periods increases gas retaining capacity of the dough. The knock back has the objectives of equalizing dough temperature throughout the mass, reducing the effect of excessive accumulation of carbon dioxide within the dough mass and introduces atmospheric oxygen for the stimulation of yeast activity. The knock back also aids in the mechanical development of gluten by the stretching and folding action. Usually knock back is done when 2/3 of the normal fermentation time is over.

11.3.6   Dough make-up

The function of dough make-up is to transform the fermented bulk dough into properly sealed and molded dough piece, when baked after proofing it yields the desired finished product.

Dough make-up includes: (a) scaling and (b) rounding, intermediary proof and molding.

Scaling or dividing

The dough is divided into individual pieces of predetermined uniform weight and size. The weight of the dough to be taken depends on the final weight of the bread required. Generally, 12% extra dough weight is taken to compensate for the loss. Dividing should be done within the shortest time in order to ensure the uniform weight. If there is a delay in dividing, corrective steps should be taken either by degassing the dough or increasing the size of the dough. The degassers are essentially dough pumps which feed the dough into the hopper and in the process remove most of the gas. The advantages of using degassers are: (i) more uniform scaling, (ii) uniform pan flows and (iii) uniform grain and texture of bread.

Rounding

When the dough piece leaves the divider, it is irregular in shape with sticky cut surfaces from which the gas can readily diffuse. The function of the rounder is to impart a new continuous surface skin that will retain the gas as well as reduce the stickiness thereby increasing its handling. Rounder are of two types, i.e. umbrella and bowl type.

Intermediate proofing

When the dough piece leaves the rounder, it is rather well degassed as a result of the mechanical it received in that machine and in the divider. The dough lacks extensibility and tears easily. It is rubbery and will not mold easily. To restore more flexible, pliable structure which will respond well to the manipulation of molder, it is necessary to let the dough piece rest while fermentation proceeds.

Intermediate proofer contains a number of trays that are chain driven. The dough piece is deposited in the tray with completed number of laps at predetermined rate. Average time at this stage ranges from 5 min to 20 min.

Molding

The molder receives pieces of dough from the inter-mediate proofer and shapes them into cylinders ready to be placed in the pans. Molding involves three separate steps: (i) sheeting, (ii) curling and (iii) scaling.

Sheeter degasses the dough and sheeted dough can be easily manipulated in the later stages of molding. Sheeting is accomplished by passing the dough through 2 or 3 sets of closely spaced rolls that progressively flatten and degas the dough. The first pair of rolls is spaced about 0.25″ apart where the degassing takes place. The successive two rollers are spaced 0.125″ and 0.06″ apart for optimum grain and texture development in the finished products. The sheeted dough piece next enters the curling section. A belt conveyor under a flexible woven mash chain that rolls into a cylindrical form carries the sheeted dough. The rolling operation should produce a relatively tight curl that will avoid air entrapment. The curled dough piece finally passes under a pressure board to eliminate any gas pockets within and to seal the same.

Panning

The molded dough pieces are immediately placed in the baking pans. Panning should be carried out so that the seam of the dough is placed on the bottom of the pan. This will prevent subsequent opening of the seam during proofing and baking. Optimum pan temperature is 90°F.

11.3.7   Proving or proofing process

Proving or proofing refers to the dough resting period during fermentation after molding has been accomplished and molded dough pieces are placed in bread pans or tins. During this resting period the fermentation of dough continues. The dough finally proofed or fermented in baking pan for desired dough height. It is generally carried out at 30–35°C and at 85% relative humidity. Proofing takes about 55–65 minutes. During proofing the dough increases remarkably in volume. The dough expands by a factor of 3–4 during proofing. During proofing care has to be taken that the skin of dough remains wet and flexible so that it does not tear as it expands. A high humid condition is also required to minimize weight loss during proving.

Temperature, humidity and time influence proofing depends on the variety of factors such as flour strength, dough formulation with respect to oxidants, dough conditioners, type of shortening, degree of fermentation and type of product desired. During proofing lower humidity gives rise to dry crust in the dough. Excessive humidity leads to condensation of moisture. Dough is generally proved to a constant time or constant height.

11.3.8   Baking process

After proofing the dough is subjected to heat in a baking oven. Baking temperature generally varies depending up on oven and product type but it is generally kept in the range of 220–250°C. During baking the temperature of dough centre reaches to about 95°C in order to ensure that the product structure is fully set. When the dough is placed in the oven, heat is transferred through dough by several mechanisms such as convection, radiation, conduction, and condensation of steam and evaporation of water. Heat transfer inside dough is said to occur through the mechanism of heat conduction and evaporation/ condensation. The baking time of bread may range from 25 to 30 minutes depending up on size of bread loaf. After baking, bread is cooled prior to packaging to facilitate slicing and to prevent condensation of moisture in the wrapper. Desirable temperature of bread during slicing is 95–105°F.

11.4   Functions of mixing

There is some difference between mixing and kneading of flour to dough. Mixing refers to homogenization of formula ingredients, whereas kneading is the development of the dough or gluten network by mechanical means.

Mixing of flour and other formula ingredients is carried out to serve following functions:

- To disperse the formula ingredients uniformly.
- To blend and hydrate the dough ingredients.
- To develop the gluten structure or network in the dough in order to enable the dough to retain gas without rupture.
- To aerate the dough to serve two purposes: (i) to provide gas nuclei for carbon dioxide produced during fermentation and (ii) to provide atmospheric oxygen for oxidation of dough and yeast activity.

11.5   Types of mixers

Mixers used to develop dough vary widely in size and intensity of mixing action. Many mixing machines are available those still work similar to hand mixing. A series of other mixers have also been developed, in which a very high speed mixing is practiced and dough is mechanically developed within few minutes.

The mixers commonly used for mixing of wheat dough are classified as under:

Low speed mixer: This type of mixer takes longer time to develop the dough for bread making.

Spiral mixers: It is commonly used mixer in the industry. The machine is fitted with a spiral-shaped mixing hook that rotates on a vertical axis. This mixer can operate on slow and fast speed. The slow speed mode is used for weaker dough and high speed mode is used for mixing of strong flour.

High speed and twin-spiral mixers: This category of mixer imparts high level of mechanical energy to the dough in a short period of time. It can mix a dough within 5 minutes.

11.6   Functions of molding and dividing

Important functions of dividing and molding are:
- To divide the bulk dough in desired shape and size.
- To divide the dough into individual pieces of uniform weight.
- To improve grain and texture of bread.
- To improve appearance and acceptability of bread.

11.7   Functions of proving

Final proof has the following functions:
- To relax the dough from the stress received during previous operations.
- To facilitate production of gas in order to give desired volume to the dough.
- To mellow gluten to extensible character for oven rise.

11.8   Changes during mixing, fermentation and baking

11.8.1   Changes during mixing

1. Formation of three-dimensional net work of protein: Mixing of bread formula ingredients brings about physico-chemical changes in the dough and its components. The dough constituents particularly gluten proteins interact to form a three dimensional structure to the dough. As a result of this change, the dough becomes extensible and elastic that is responsible for gas retention and bread quality.

2. Sulphydryl-disulphide interchange reactions: Mixing brings about sulphydryl (SH) and disulphide (SS) interchange, which matures and develops the dough for further processing into bread.

3. Oxygen incorporation into the dough, which helps in oxidation as well as the formation of nuclei for the formation of gas cells.

4. Starch, lipid and protein complex formation, which is responsible for gas retention during baking.

5. Mixing also increases the temperature of dough: The temperature of the dough has great influence on the overall quality of bread. The temperature of the dough out of the mixer should be within the range of 25–28°C. The dough temperature depends on the temperature of different ingredients, friction factor of the machine, heat of hydration of ingredients and their specific heat. The temperature of the mixed dough can be controlled by altering the temperature of water used for preparing the dough.

11.8.2   Changes during fermentation

Dough after optimum mixing is subjected to fermentation at around 30°C and 85% relative humidity for a suitable length of time. Fermentation is essential for obtaining light aerated loaf of bread. During fermentation several desired physico-chemical changes occur in the dough system, which is explained below.

Physical changes

- Increase in volume due to production of CO_2
- Increase in temperature
- Increase in the number of yeast cells
- Loss of moisture
- Changes in the consistency of dough. The dough becomes soft, elastic as well as extensible

Chemical changes

- Reduction in pH: The pH of dough is reduced from 5.5 to 4.7 due to formation of acids like acetic acid yeast activity.
- Formation of maltose sugar by diastatic enzymes acting on starch.
- Development of dough due to the cessation of S-S bonds and formation of new SH and SS groups, which improve the gas retention property of dough.
- Conversion of starch into simple sugar by diastatic enzymes which is then converted into CO_2 and alcohol by the following group of enzymes present by yeast:

$$Starch \xrightarrow{\text{Diastase}} Maltose + Dextrin$$

$$Maltose \xrightarrow{\text{Maltase}} Dextrin$$

$$Dextrose \xrightarrow{\text{Zymase}} Carbon\ dioxide + Alcohol$$

11.8.3   Changes during baking

As the dough enters into baking oven, it undergoes several physical and biochemical changes. These changes are described below:

Physical

1. *Oven-spring:* The dough expands rapidly in first few minutes in the oven. This sudden rise is called oven-spring. Several factors are held responsible for oven-spring. The gases heat and increase in volume, water, carbon dioxide and ethanol evaporates. All this causes increase in internal pressure of dough and the dough rise rapidly in the initial stage of baking. Yeast activity decreases as the dough warms and the yeast is inactivated at 55°C.
2. *Crust formation:* The dough that is exposed to oven temperature develops skin and forms a crust as moisture evaporates from surface of dough evaporates very rapidly. The crust provides the strength of the loaf.

Chemical

1. *Yeast activity:* Activity of yeast depends on the temperature. Yeast activity increases very rapidly initially as the dough is placed in the oven but it is inactivated at 55°C.
2. *Starch gelatinisation:* Starch begins to gelatinise at about 60°C. The dough contains limited water to gelatinise the starch completely. This limited gelatinisation of dough helps in gas retention and setting of bread texture.
3. *Gluten coagulation:* Starch gelatinisation is associated with absorption of water while gluten denaturalisation is associated with the removal of water. Gelatinisation sets in when the temperature is around 74°C, and continues till the end of baking. In this process, gluten matrix surrounding the individual cells is transformed into a semi-rigid film structure. Thus, a major change that takes place during the oven process is the redistribution of water from the gluten phase to the starch phase.
4. *Enzyme activity:* The action of amylase on starch increases with temperature approximately doubling for every 10°C rise. At the same time, heat inactivation of the enzymes also commences. β-amylase denature at lower temperature (57–71°C) as compared α-amylase which denatures at temperatures ranging from 65°C to 95°C. Insufficient amylase activity can restrict loaf volume because the starch becomes rigid soon, whereas excess amylase activity may cause collapse of loaf.

5. *Browning reaction:* The browning reaction starts at around 160°C. It is the result of heating reducing sugars with proteins or other nitrogen containing substance to form coloured compounds, known as melanoidins. This reaction also imparts colour and flavour to the bread.

11.9   Major bread-making processes

The processes used for commercial production of bread differ principally in achieving dough development. These may be classified into three broad processing groups although there are numerous variations and also elements of overlap between each of the individual groups.

11.9.1   Long fermentation processes

Straight dough bulk fermentation process and Sponge and dough process are example, which falls under this group. In these processes resting periods (floor time) for the dough in bulk after mixing and before dividing are longer. In the case of straight dough method all the ingredients are mixed in one step, whereas in sponge and dough process, a part of the dough formulation receives a prolonged fermentation period before being added back to the remainder of the ingredients for further mixing to form the final dough.

11.9.2   Rapid processes

In these methods a very short or no period of bulk fermentation is given to the dough after mixing and before dividing.

11.9.3   Mechanical dough development process

Here a primary function of mixing is to impart significant quantities of energy to facilitate dough development, and the dough moves without delay from mixer to divider. The dough is developed by high level of energy imparted at the stage of mixing.

11.9.4   Straight dough bulk fermentation process

The dough is fermented in bulk. This is the most traditional and most 'natural' of the bread making process. Essential features of bulk fermentation processes are summarized as follows:

1. Mixing of all the ingredients to form homogeneous dough.
2. Fermentation of the dough so formed in bulk for a prescribed time (floor time), depending on flour quality, yeast level, dough temperature and the bread variety being produced.
3. Dough formation for bulk fermentation is usually achieved by low speed mixers or may be carried out by hand.

In general, the stronger flour will require longer fermentation to achieve optimum dough development. So higher protein flours require longer bulk fermentation times than lower protein flours.

The supplementation of flours with dried, vital wheat gluten to increase the protein content of weaker base flours is a common practice in many parts of the world.

Gluten supplementation is less successful in bread-making processes where mixer has lower speed as seen in the case of long processes. Flours used for bulk fermentation processes should be low in cereal α-amylase because of the potential softening effects on the dough handling character with extended bulk resting time.

During bulk fermentation the dough develops by enzymatic action. Since enzymatic actions are time and temperature dependent, therefore, adjustment of added water levels will have to be made to compensate for these changes. An outline of a straight dough method is shown in Fig. 11.1.

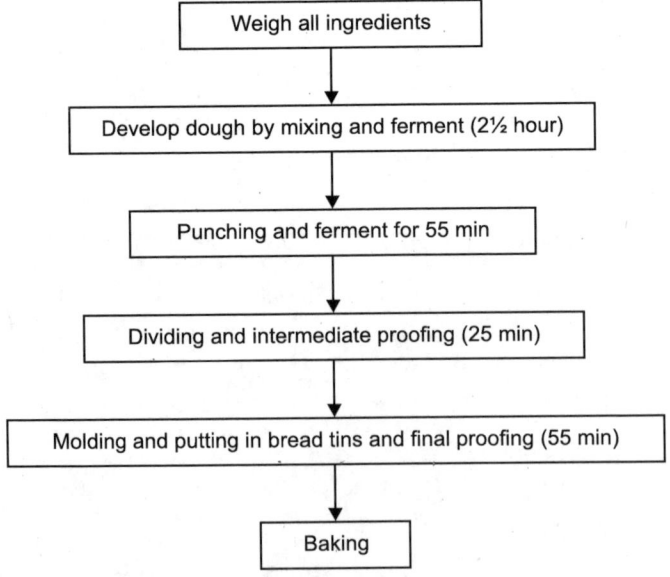

Figure 11.1 A straight dough bulk fermentation bread process.

11.9.5   Sponge and dough bulk fermentation process

The key features of sponge and dough processes are:

1. In this process a part flour (generally two-third), part of water and yeast are mixed just to form loose batter or dough (sponge).
2. Sponge is allowed to ferment for up to 5 hours.
3. Mixing of the sponge with the remainder of the ingredients to develop the dough optimally.
4. Immediate processing of the developed dough with a short period of bulk fermentation period.

An outline of a sponge and dough method is shown in Fig. 11.2.

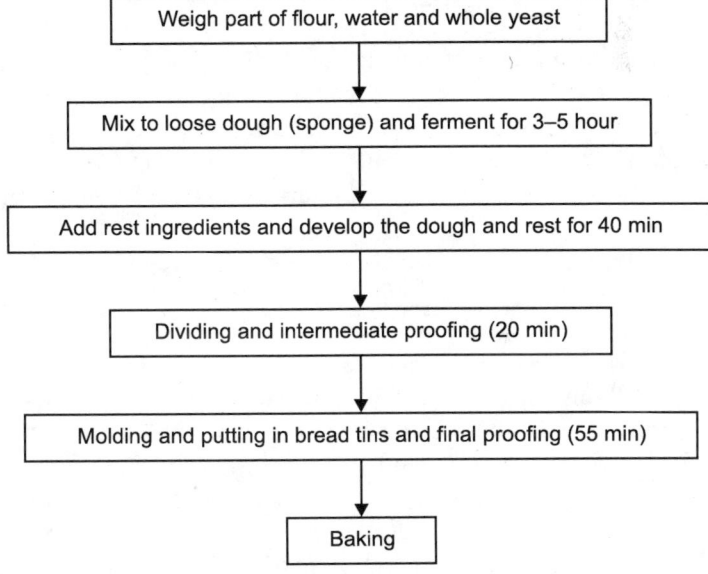

Figure 11.2 Sponge and dough bread process

Role of the sponge

The primary role of the sponge is to modify the flavour and to contribute to the development of the final dough through the modification of its rheological properties. The flavour is developed by yeast in the sponge. The sponge fermentation conditions should be closely controlled and care should be taken to avoid a build-up of unwanted flavours by thorough cleaning of storage containers after use.

During the sponge fermentation period, the pH decreases with increasing fermentation time. The rheological character of the gluten formed during initial sponge mixing will change as fermentation progresses, with the sponge becoming very soft and losing much of its elastic character. As standing time increases, the condition of the sponge increasingly resembles over fermented dough. The low pH of the sponge and its unique rheological character are carried through to the dough where they have the effect of producing a softer and more extensible gluten network after the second mixing. In many cases the addition of the sponge changes the rheological character of the final dough enough to render further bulk resting time unnecessary, so that dividing and molding can proceed without further delay. The sponge and dough process produces soft bread with uniform crumb grain structure. The sponge and dough process has tolerance to time and other conditions.

11.9.6   Rapid processing

Under this category, the processes are evolved based on different combinations of active ingredients and processing methods. All processes covered under this group include improvers to assist in dough development and the reduction fermentation period. The prominent examples include activated dough development, no-time doughs with spiral mixers and the Dutch green dough process.

11.10   Packaging of bread

A wide variety of bakery products are available including bread, biscuits, cookies, cakes, buns, rusk, etc. Bakery industry is one of the largest foods processing industries in India and is rapidly growing because of increased demand for baked goods as a result of industrialization and urbanization. Packaging of baked goods constitutes 10–30% of the entire cost of the pack.

The main functions of a package are: (i) to contain the product, (ii) to protect the product and (iii) to help in selling of the product. To perform these functions satisfactorily, the packaging materials should have:

1. Necessary strength properties to withstand conditions of processing, storage and transport.
2. Should protect from environmental factors such as humidity, oxygen, light and heat.
3. Should have desired machinability, heat sealability and printability
4. Should preferably be economic and easily available.

Baked goods are very susceptible to physico-chemical changes under adverse climatic conditions e.g. desiccation in bread, loss of crispness and rancidity in biscuits and microbiological spoilage. So a package designed for these products must take these factors also into consideration.

11.11   Variety of bread products

This section explains of the formulations and manufacturing process of various types of breads. It also discusses the methods of assessment of bread quality.

11.11.1   Multigrain bread

Multigrain breads are made with wheat flour and cereal grains as well as oil seeds. Many types of multigrain breads have been made to sustain consumer interest in number of developed countries. The grains and vegetables that have been used include corn, flax, millet, triticale, buckwheat, barley, oats, alfalfa, soya, potato, rye, rice and sauerkraut.

11.11.2   High fiber bread

In view of the importance of dietary fiber in reducing diet related diseases, increased emphasis is being laid on high fiber products. The sources of fibers include wheat bran, corn bran, rice bran, rye bran, barley bran, triticale bran and oat bran. However, the beneficial effect of bran has been found only in coarse brans. Among the above sources of bran, barley and oat brans have been proved to reduce the level cholesterol in the human body.

11.11.3   Cracked wheat bread

Adding about 15% cracked wheat in the formulation makes this bread. Even clear flours could be used. It is better to soak cracked wheat in water for 2–3 hours.

11.11.4   Sour dough bread

Sour dough breads and rolls have become popular in recent years particularly in European countries. These breads resemble white pan or hearth breads in formulation and overall quality except that these are acidic in taste. Mostly, starter is used in the initial stage, which is taken from ripe sour. It can also be produced by spontaneous fermentation process by bacteria present in flour and air. Generally, rye bread is prepared by sourdough process.

11.11.5   Milk bread

Milk bread should contain at least 6.0% milk solids. Milk solids could be added as skimmed or whole milk solids, or milk, which is sterilized or condensed milk.

11.11.6   Composite flour bread

Considerable proportion of bread produced in Germany and other European countries includes mixture of wheat and rye. The concept of composite flour was developed: (i) to utilize indigenously available non-wheat cereals, tubers, millets, etc. (ii) to improve the nutritional quality by using protein rich raw materials, (iii) to extend the wheat availability and (iv) to produce products with different flavour/taste profiles. Refined flours from maize, jowar, barley, tapioca, potato or any other cereal or tuber flours could be used in bread. These flours do not contain gluten and hence their incorporation into wheat flour naturally bring down the gluten content and hence the quality of bread.

11.11.7   High protein bread

Wheat flour has a good carrying capacity and hence it could be fortified with respect to calories, protein, salt, carbohydrates, vitamins and minerals required for special target groups. There is a greater scope to produce high protein breads using protein rich oilseed meals. Normal white pan bread contains only 6–10% proteins. Any of the defatted oilseed meals contain protein ranging from 40% to 60%.

11.11.8   Wheat germ bread

Wheat germ bread must contain at least 10% of germ. Wheat germ, which is a by-product of flour milling, is nutritionally many superiors to other protein sources. It contains as high as 30% protein of quality similar to the protein of eggs or milk. It is also a rich source of vitamins and minerals and richest known source of tocopherol.

11.11.9   Sugar free bread

In sugar free breads, sugar is replaced with enzyme active flour (2.0%).

11.11.10   Low calorie bread

Low calorie breads are made using cereal brans and fat substitutes. Sucrose esters have been found to be a good substitute for fat.

11.11.11   Low salt bread

Salt is an essential ingredient in bakery products. It has stabilizing influence on fermentation and hence production of carbon dioxide and flavour compounds.

It also strengthens the gluten and hence the dough incorporation of salt improves the loaf volume, colour and textural parameters in addition to the taste and flavour of bakery products.

11.11.12   Gluten-free bread

A small segment of population suffers from dietary wheat intolerance, which includes disturbances known as celiac disease. The symptoms may include cramps, diarrhea etc. and the responsible factor has been found to be gliadin. Using wheat starch or any other non-wheat flours and gluten substitutes consisting of pregelatinized starch, guar gum and carboxyl methylcellulose could make gluten free breads.

Brown bread

Brown bread is normally made by mixing maida and whole wheat flour in the ratio of 50:50 or higher whole wheat in maida. The processing conditions are the same as whole wheat bread.

Rye bread

Rye bread is produced in different shapes, sizes and taste to suit the palate of consumers. The main ingredients used are wheat flour rye flour, water, yeast and salt. Other ingredients may include malt, sourdough, colour, etc. Rye flour can be obtained in to three major groups white, medium and dark. They differ in ash content. Rye flour offers more favourable acid environment for yeast than white wheat flour. Its levels of sugars, dextrins are somewhat higher. In addition, rye flour has higher alpha amylase activity and lower active fermentation. Acid flavour in rye bread can be imparted by using sours that contains blends of lactic acid or acetic acid, in various ratios often combined with dried yeast and bacterial cultures with flour as a carrier.

11.12   Bread spoilage and stalling

In commercial bakery units, the shelf life of bread is linked to their profitability and reputation in the market. Bread staling, development of rope and molds are the major problems of bread during storage. Such changes make the bread unfit for human consumption and bakers suffer in reputation and economically too.

This section explains factors influencing shelf life of bread and suitable measures are suggested to extend the period during which bread remained fit for human consumption.

11.12.1   Microbial spoilage of bread

Persistent problem that every baker faces in marketing of his bakery products is that of microbial spoilage caused mainly by mold growth and occasionally by bacterial contamination. Mold growth may occur at any place on the baked product while bacterial spoilage generally occurs in the icings, and fillings. Because of this variability, several materials have been introduced to best cope with these problems and it is advisable to learn what they are and how they can be used to greatest advantage.

Rope development in bread

In baking, rope is a bread disease. The presence of rope in bread results in breaking down of starch and protein, producing a discoloured and sticky condition in the crumb which is accompanied by an odour similar to the odour of a rotten pineapple.

The ropiness generally appears in bread 12–36 hours after baking and mainly during summer months when ambient temperature reaches 35°C or higher. Rope in bread is caused directly by bacteria known as *Bacillus measentericus*.

The rope spores are present in the air and can also be carried by the particle of dust. This, including insanitary conditions in bakery, can infect the bakery raw materials and eventually cause ropiness in the bread.

11.12.2   Mold growth in bread

Bread if kept under warm or moist conditions will develop thread like whiskers, which are generally due to the presence of mold. The colour of these molds can be brown, red, orange, yellow, green blue, pink, white or black. Air, flies or insects can carry mold spores. Mold spores usually settle outside of the bread, therefore, mold growth is noticeable on the exterior surface. Most commonly found molds in bread are of the *Aspergillus, Penicillium, Rhizopus* and *Mucor*. None of which is disease producing microorganisms. The slicer blades, if kept unclean, are source of mold contamination of sliced bread.

11.12.3   Yeast spoilage

Yeast spoilage of bread is rare, but it may occur sometime by wild yeast in long fermentation process. Yeast, like mold does not survive the baking process, it may contaminate bread during cooling and slicing.

Yeast contamination in bread plant occurs mainly through dirty equipments and infected sugars. Fermentative and filamentous yeast are held responsible for bread spoilage. Filamentous yeast is known as 'chalk mold' because they form a white, spreading growth on bread surface which can easily be confused with mold. Use of ethanol spray can prevent yeast growth in bread.

11.13   Bread staling

Bread staling, broadly, includes all changes, except microbial that occur in bread after baking. Changes occur in both crumb and crust of the bread, however, crumb firmness is more correlated with bread staling. Other changes such as loss of flavour, decrease in water absorption capacity, amount of soluble starch, enzyme susceptibility of the starch, increase in starch crystallinity and opacity are also associated with bread staling. All these changes affect to a large extent, the palatability of bread and hence consumer's acceptance. Bread filming is virtually arrested at temperature of $-200°C$ or less, and at $-2°C$ the finning rate is at its maximum. A $60-100°C$, it is possible to reverse most of the stale character of bread due to the heat reversible character of starch retrogradation.

11.13.1   Crust staling

The crust of fresh bread is crisp, brittle and somewhat dry. On staling it becomes soft, tough and leathery. This is caused due to migration of moisture from the moist center crumb to the crust. The hygroscopic crust readily absorbs moisture which diffuses outwards and becomes soft and leathery. The excess humidity, above 80% Relative humidity (RH) in the bread store is also undesirable. Preventing the evaporation of water from crust to the atmosphere can inhibit the staling rate of crust. The bread if packed in the moisture proof wax coated paper or film, which does not permit the moisture from interior of the loaf to pass through crust into the atmosphere, stays fresh for longer time.

11.13.2   Crumb staling

The crumb staling depends on changes in the starch to a certain extent. The proteins take up the water released by starch. The starch cells absorb more water when bread is fresh and have alpha-pattern which on lowering of temperature, changes slowly to the beta pattern, which holds less water. As the crumb of bread stales it becomes drier, less elastic, crumbly and harsh textured. It also looses the fresh flavour and aroma.

11.13.3   Causes of staling

The moisture migration, change in the structure of starch and retrogradation due to various physical factor, are the major causes of the staling. Staling is caused by changes in moisture content and starch structure during storage of bread. The starch, which has been gelatinized during baking holds the moisture contained in bread, and when the loaf comes out of the oven, the gelatinized starch that is distributed throughout the loaf contains the maximum amount of moisture. Some of the moisture does disappear from the loaf by evaporation as the bread cools, but other constituents of the loaf, especially the coagulated gluten, absorb some of the moisture contained in the gelatinized starch. The result of this migration of moisture from the starch to the gluten is claimed to be largely responsible for the development of staleness in bread.

Ethanol reacts with starch and prevents its retrogradation, thereby inhibiting the bread staling process. The ratio of starch to protein in dough is critical in determining the rate of staling.

11.14   Fermentation of cereals

Cereals are grown over 73% of the total world harvested area and contribute over 60% to the world food production providing dietary fiber, proteins, energy, minerals and vitamins required for human health. However, the nutritional quality of cereals and the sensorial properties of their products are sometimes inferior or poor in comparison with milk and dairy products. The reasons behind this are the lower protein content, deficiency of certain essential amino acids, the presence of determined antinutrients (phytic acid, tanins and polyphenols) and the coarse nature.

Fermentation can have multiple effects on the nutritional value of food. Microbial fermentation leads to a decrease in the level of carbohydrates as well as some non-digestible poly- and oligosaccharides. The latter reduces side effects such as abdominal distention and flatulence. Certain amino acids may be synthesized and the availability of B group vitamins may be improved. Fermentation of cereals by lactic acid bacteria has been reported to increase free amino acids and their derivatives by proteolysis and/or by metabolic synthesis. The microbial mass can also supply low molecular mass nitrogenous metabolites by cellular lysis. Fermentation has been shown to improve the nutritional value of grains such as wheat and rice, basically by increasing the content of the essential amino acids lysine, methionine and tryptophan. Fermentation of rice by lactic acid bacteria enhances the flavour, nutritive value and available lysine content.

Natural fermentation of maize increases total soluble solids and non-protein nitrogen and slightly increased protein content. Sixteen hour fermented dough had higher levels of the albumin plus globulin fraction, indicating that natural fermentation of maize results in improvement in the nutritional value of the grain. Also maize protein digestibility is elevated.

Improvement in starch digestibility during fermentation can be related to enzymatic properties of fermenting microflora that brings about the breakdown of starch oligosaccharides. The enzymes bring about cleavage of amylase and amylopectin to maltose and glucose. Reduction in amylase inhibition activity may also be responsible for the starch digestibility. Similarly an improvement in protein digestibility of fermented products is mainly associated with an enhanced proteolytic activity of the fermenting microflora.

Fermentation is such an important process which significantly lowers the content of antinutrients (phytates, tanins, polyphenol) of cereal grains. Fermentation also provides optimum pH conditions for enzymatic degradation of phytate which is present in cereals in the form of complexes with polyvalent cations such as iron, zinc, calcium, magnesium and proteins. Such a reduction in phytate may increase the amount of soluble iron, zinc, calcium several folds.

During fermentation, the reduction of phytic acid content may partly be due to phytase activity as it is known to be possessed by a wide range of microflora. Optimal temperature for phytase activity has been known to range between 35°C and 45°C.

Tanin levels may be reduced as a result of lactic acid fermentation, leading to increased absorption of iron, except in some high tanin cereals, where little or no improvement in iron availability has been observed. Diminishing effect of fermentation on polyphenols may be due to the activity of polyphenol oxidase present in the food grain or microflora.

Traditional fermented foods prepared from most common types of cereals (such as rice, wheat, corn or sorghum) are well known in many parts of the world. Some are utilized as colourants, spices, beverages and breakfast or light meal foods, while a few of them are used as main foods in the diet.

Lactic acid fermentation of cereals is a long established processing method and is being used in Asia and Africa for the production of foods in various forms such as beverages, gruels and porridge. Fermented cereal based foods produced in Africa are presented in Table 11.1.

Although differences exist between regions, the preparation procedure could be generalized. Cereal grains, mainly maize, sorghum, or millet grains are soaked in clean water for 0.5–2 days. Soaking softens the grains and makes them easier to crash or wet-mill into slurry from which hulls, bran particles,

Table 11.1 Fermented cereal-based foods in Africa.

Name of product	Associated microorganisms	Substrate	Area of products
Ogi	*Lactobacillus plantarum, Lactobacillus fermentum, Lactobacillus* spp., *Saccharomyces cerevisiae, Candida mycoderma, Rhodotorula* spp., *Aerobacter cloacae, Candida krusei, Debaromyces hasenu, Klebsiella* spp., *Staphylococcus* spp.	Maize, sorghum or millet	West Africa
Kolo and Kenkey	*Lactobacillus* spp. and yeasts	Maize, sorghum or millet	Ghana
Mahewu (magou)	*Lactobacillus delbrueku* and *Lactobacillus bulgaricus*	Maize, sorghum or millet	South Africa
Uji	*Lactobacillus* spp.	Maize, sorghum or millet	East Africa
Kisra	Information not available	Sorghum	Sudan
Enjara	*Candida guillienmandi*	Sorghum	Ethiopia

and germs can be removed by sieving procedures. During the slurrying or doughing stage, which lasts for 1–3 days, mixed fermentation including lactic acid fermentation takes place. During fermentation, the pH decreases with a simultaneous increase in acidity, as lactic and other organic acids accumulate due to microbial activity. The traditional foods made from cereal grains usually lack flavour and aroma. During cereal fermentations several volatile compounds are formed, which contribute to a complex blend of flavours in products. The presence of aromas represented by diacetyl, acetic acid and butyric acid makes fermented cereal-based products more appetizing. The proteolytic activity of fermentation microorganisms often in combination with malt enzymes may produce precursors of flavour compounds, such as amino acids, which may be deaminated or decarboxylated to aldehydes and these may be oxidized to acids or reduced to alcohols. Knowledge of biochemical pathways leading to flavour production can help in making the right choice of starter culture. However, the end product distribution of lactic acid fermentations depends also on the chemical composition of the substrate (carbohydrate content, presence of electron acceptors, nitrogen availability) and the environmental conditions (pH, temperature, aerobiosis/anaerobiosis), controlling of which would allow specific fermentations to be channelled towards a more desirable product.

11.14.1   Micro-organisms for cereal fermentations

The basic fermentation process involves the enzymatic activities of *lactobacilli*, leuconostoc, pediococci, yeasts and molds. Their metabolic activities result in the production of short chain fatty acids such as lactic, acetic, butyric, formic and propionic acids. The pH of these foods is reduced to values of 4 or less. Acids formed during fermentation process, lower the pH thus inhibiting the growth of spoilage organisms. The pH requirements of a number of pathogens are shown in Table 11.2.

Table 11.2 Approximate pH tolerance of some microorganisms.

Organism	Minimum pH	Maximum pH
Escherichia coli	4.4	9.0
Salmonella typhi	4.5	8.0
Campylobacter jejuni	2.3	
Shigella sp.	4.5	8.0
Streptococcus lactis	4.3–4.8	
Lactobacillus sp.	3.0	7.2
Yeasts	1.5	8.0–8.05
Molds	1.5–2.0	11.0

Lactic acid bacteria are industrially important microbes that are used all over the world in a large variety of industrial food fermentations. The primary activity of the culture in food fermentation is to convert carbohydrates to desired metabolites such as alcohol, acetic acid, lactic acid or CO_2. The cultures used in food fermentation are, however, also contributing by secondary reactions to the formation of flavour and texture. Lactic acid bacteria are considered to have several beneficial physiological effects, such as antimicrobial activity, enhancing of immune potency and prevention of cancer and lower serum cholesterol levels.

The proposed health and nutritional benefits of *Lactobacillus* species are:
- Enzyme (lactase) formation
- Colonization and maintenance of the suitable intestinal microflora
- Competitive exclusion of undesirable microorganisms
- Microbial interference and anti-microbial activities
- Pathogen clearance
- Immuno-stimulation and modulation
- Cholesterol reduction/removal

Lactobacillus spp. play a significant role in most of the fermented cereals. The *Lactobacillus bulgaricus* and *Streptococcus thermophillus* are suitable bacteria for rice fermentation because of their lack of amylase, which is necessary for saccharification of rice starch. It was shown that *Lactobacillus plantarum* and *Pediococcus* spp. dominate the latter stages of maize dough fermentation and may therefore be responsible for the rapid acidification of inoculated dough. Mixed culture of lactic acid bacteria and propionic bacteria appeared to be a potential alternative along with conventional starters in rye sour dough fermentation, providing for an improved mold-free shelf-life of bread.

In case when the cereal grain are used as natural medium for lactic acid fermentation, amylase need to be added before or during fermentation or amylolytic bifidobacteria need to be used because these microorganisms contain enough amylase which is necessary for saccharification of the grain starch. The use of amylolytic lactic acid bacteria as a starter culture offers another alternative by combining both amylase production and acidification in one micro-organism.

Fermentation of cereals with pure yeast cultures has been shown to increase the protein content of the fermented products. The possible function of yeasts in fermented foods and beverages are:

- Fermentation of carbohydrates (formation of alcohols, etc.)
- Production of aroma compounds (esters, alcohols, organic acids, carbonyls, etc.)
- Stimulation of lactic acid bacteria providing essential metabolites
- Inhibition of mycotoxin-producing molds (nutrient competition, toxic compounds, etc.)
- Degradation of mycotoxins
- Degradation of cyanogenic glucosides (linamarase activity)
- Production of tissue-degrading enzymes (cellulases and pectinases)
- Probiotic properties

It has been suggested that the proliferation of yeasts in foods is favoured by the acidic environment created by lactic acid bacteria while the growth of bacteria is stimulated by the presence of yeasts, which may provide growth factors, such as vitamins and soluble nitrogen compounds. Association of lactic acid bacteria and yeasts during fermentation may also contribute metabolites, which could impart taste and flavour to fermented food.

11.14.2   Examples of fermented cereal foods

Tarhana

Popular Turkish fermented wheat–yogurt mixture produced from white wheat flour, yogurt, onions, tomato puree, yeast (baker's yeast), salt, paprika, dill, mint, tarhana otu *(Echinophora sibthorpiana)* and water. The mixture is fermented for several (1–7) days at 30°C. The final pH of tarhana varied in the range 4.3–4.8.

Togwa

Fermented gruel or beverage prepared either from cassava, maize, sorghum, millet or their combinations. Togwa is widely produced in Tanzania for use directly as a weaning food or diluted for use as refreshment.

The fermentation is spontaneous or the lactic acid bacteria *(Lactobacillus brevis, Lactobacillus cellobiosus, Lactobacillus fermentum, Lactobacillus plantarum, Pediococcus pentosaceus)* and yeasts *(Candida pelliculosa, Candida tropicalis, Saccharomyces cerevisiae)* are used at concentration 10^9 CFU ml^{-1} and 10^7 CFU ml^{-1}. Fermentation is carried out at ambient temperature for 9–24 h (to pH $\leq$ 3.8).

Ogi

Sour porridge with a flavour resembling yogurt. It is made by lactic acid fermentation of corn, sorghum or millet. Soyabeans may be added to improve nutritive value. It is considered the most important weaning food for infant in West Africa although it is also consumed by adults. Ogi is a natural fermentation product. Lactic acid bacteria, yeasts and molds are responsible for the fermentation although *Lactobacillus plantarum* is the predominant micro-organism. Other bacteria such as *Corynebacterium* hydrolyse corn starch, and then yeasts of the *Saccharomyces cerevisiae* and *Candida* species also contribute to flavour development. Optimum pH for ogi is 3.6–3.7. The concentration of lactic acid may reach 0.65% and that of acetic acid 0.11% during fermentation.

Yosa

New snack food made from oat bran pudding cooked in water and fermented with lactic acid bacteria and *Bifidobacteria*. After fermentation, the matter is flavoured with sucrose or fructose and fruit jam. It is mainly consumed in Finland and other Scandinavian countries. Oat fiber is a good source of β-glucan, which can lower the cholesterol levels in the consumer blood, which in turn can reduce the risk of heart disease.

Balao balao

Philippine balao is a lactic acid fermented rice/shrimp mixture prepared by mixing boiled rice, whole raw shrimp and solar salt (about 3% w/w), packing in an anaerobic container and allowing the mixture to ferment over several days or weeks. The chitinous shell becomes soft and when the fermented product is cooked, the whole shrimp can be eaten.

Boza

Highly viscous traditional fermented Turkish beverage made from cereals such as maize, rice and wheat flours. The preparation of boza is normally carried out by natural fermentation involving mixed cultures of lactic acid bacteria and yeasts. Alcoholic cereal based fermented beverages can be classified into wines (sake-rice wine consumed in Japan and China, bouza-alcoholic wheat beverage consumed in Egypt and Turkey) and beers (European barley beer, rice beer in the Asia-Pacific countries).

Bouza

Alcoholic wheat beverage is common in Egypt. It is a thick, pale yellow liquid with 3.8–4.0% v/v alcohol and a pH 3.1–4.0. The microorganisms responsible for the fermentation are yeasts and bacteria.

Lactic fermentation of bread dough improves the keeping quality and flavour of the baked products. It also enhances the palatability of bread made from low-grade flours and under-utilized cereals. The basic biochemical changes that occur in the sour-dough bread fermentation are:

1. Acidification of the dough with lactic and acetic acids produced by the *lactobacilli*.
2. Leavening of the dough with CO_2 produced by the yeast and the *lactobacilli*.

Section IV
Fruits and vegetables

Fruit and vegetable biotechnology

12.1   Introduction

Fruits and vegetables contain important vitamins, minerals and fibres and are an important part of health diet, and as are all of great significance in agriculture. However, new varieties with traits achieved through biotechnology processes have played a great role today. Biotechnology of fruit and vegetable production are an aid to conventional breeding and its ability to transfer genes between different organisms. The ability to monitor the absence of such genes in plants is good for plant breeders. In order for the plants to transfer specific genes to one another, it can be added to one's crops by providing a majority of time and effort. Metabolic engineering can allow the genes transfer to fruits and vegetables by adding components to the nutritional values.

The integration of emerging biotechnologies with conventional breeding will greatly facilitate the modification of quality or 'value-added' attributes of fruits and vegetables, (i.e. appearance, organoleptic, nutrition, physiological benefits and safety. Beginning in 1994, the first wave of products from biotechnological applications to vegetables were introduced in pilot test markets. Vine-ripe tomatoes with extended shelf life, processing tomatoes with superior quality and deep red colour, squash with novel virus resistance, and potatoes genetically modified to produce an insect-killing protein are examples of the traits introduced into commercial vegetable varieties with the tools of biotechnology.

Biotechnology, in the simplest and broadest sense, is the utilization of living organisms or their components to provide useful products or processes. This definition encompasses essentially all of agriculture, since agriculture is based on the production of plants and animals to provide food, fiber and other products for human use. It also includes familiar uses of microorganisms, such as the yeast used in brewing or baking or the symbiosis of legume plants with nitrogen-fixing bacteria (such as commercial *Rhizobium* inoculants for peas). Thus, humans have been employing biotechnology for at least 10,000 years since the origin of agriculture. Recently, however, the term 'biotechnology' has become synonymous with 'genetic engineering.' The cloning and movement of genes among life forms as diverse as bacteria, tomatoes or sheep is now possible. In this context, biotechnology refers to a wide range of enabling technologies that allow the alteration of heritable traits outside of the living organism and the subsequent reintroduction of the new trait into an organism for specific purposes. These new varieties are called transgenic because they require the recombination of genetic materials from different organisms to cause the desired change.

Some of these enabling techniques are applied without resulting in a transgenic plant. These advanced technologies are sophisticated enhancements of methods that have long been common practice. For example, potatoes are propagated primarily by planting the buds, or 'eyes,' present on the tubers. This vegetative reproduction ensures that the new plants are identical to their parents. As one tuber can give only a few new plants, one for each 'eye,' several rounds of seed increase are needed to produce commercial quantities of potato 'seed.' Unfortunately, this also allows the transmission of disease-causing bacteria or viruses that may have infected the parent plant.

Micropropagation, the mass production of identical plants from tiny buds of the parent plant, is a biotechnique that can eliminate these pathogens from the progeny plants while retaining the advantages of vegetative reproduction. Similarly, the individual cells of a plant can be separated, multiplied, and regenerated into whole plants through a process known as tissue culture. In this way, thousands of copies of a single plant can be made and certified pest-free. These techniques are valuable for propagating plants and are an essential component of more advanced methods of genetic modification.

While these tissue culture techniques are valuable, they do not, in principle, modify the characteristics that are encoded in the plant's genes.

For many crops, breeders have relied heavily on the introduction of genes from related wild, but interfertile, species to provide new characteristics. In tomatoes and potatoes, for example, resistance to diseases such as *Fusarium* and *Verticillium* and pests such as Root Knot Nematode have been introduced into commercial varieties from related wild relatives discovered in South America. Virtually all of our domesticated crops, including vegetables, are dependent upon genes derived from a wide array of germplasm, or genetically distinct variants, of each species.

12.2   Direct and indirect benefits of biotechnology

The long-term beneficial impact of biotechnology on vegetable production will be realized both directly and indirectly. Many of the products will benefit producers and processors by improving the economic efficiency of production. California vegetable production faces many challenges in maintaining productivity while protecting the environment. Biotechnology will contribute to environmental quality protection by reducing the frequency of agricultural pesticide applications and by allowing more environmentally compatible materials and alternative methods to be employed. For example, plant-based pest resistance in both transgenic and conventionally selected varieties will reduce dependence on broad spectrum pesticides. To enjoy continuing benefits from this approach, producers will need to carefully manage their use of such varieties to prevent or delay the development of resistance in the target organisms. Alternative strategies for reducing herbicide use include introducing colour or leaf morphology modifications into seedlings to enhance the use of precision, robotic cultivation.

Biotechnology is also being applied to develop microorganisms for biological control of pests. For example, an insect-attacking virus (baculovirus) has been modified using rDNA techniques to produce a protein toxin from a gene originally obtained from scorpions. Field trials in California and other states on several vegetable crops have been proposed. These types of products are much slower to be commercialized as they provoke far more public concern than plant-based pest resistance.

On the other hand, DNA-based tools allow the detailed analysis of microbial interactions with plants, with other microbes and with the environment in ways never before possible.

Two major categories of direct consumer benefit from biotechnology are sensory and nutritional quality and food safety.

12.3   Consumer benefits

Biotechnology has been applied to improving the sensory properties and shelf-life of vegetables. Some of the innovations, particularly in texture or flavour enhancement, have not been highly publicized as originating from plants modified by methods included in a broad definition of biotechnology. These include carrot, potato, celery, pepper, and melon varieties improved through applications of tissue culture. Public awareness of developments in agricultural biotechnology is primarily associated with genetic engineering, particularly of tomato.

In 1994, the first genetically engineered food product reached consumer markets. Calgene's Flavr Savr® tomato received widespread publicity during development and commercial introduction.

The targeted benefit, shared by several companies developing new tomato varieties, was to deliver to the market tomatoes with improved flavour and reduced rates of softening or decay.

Although production issues have slowed large-scale commercialization, these enhanced sensory quality tomatoes were generally well received by consumers in test markets and limited retail distribution in United States.

Future products entering the commercial pipeline seek to specifically modify the accumulation and stability of sensory traits by increasing the sweetness and by maintaining the acid balance during maturation and ripening of tomatoes. Both plant and non-plant genes have been introduced into test tomato varieties to increase sucrose accumulation, increase the conversion of sucrose to fructose in the fruit, and sustain organic acids during ripening of tomatoes.

Food safety: Biotechnology is playing a key role in the development of rapid and sensitive diagnostic tools for food borne pathogens, microbial toxins, and other contaminants.

Recent outbreaks of *Salmonella* and *Escherichia coli* (*E. coli*) 0157:H7 in fresh vegetables and processed fruit have heightened awareness of the need for proper sanitation and handling. Biotechnology is contributing new detection methods and information about the sources and persistence of foodborne pathogens in production, processing and distribution to the consumer.

12.3.1   Regulation of biotechnology products

The Federal agencies responsible for food, human, and environmental safety apply the same basic requirements to the products of biotechnology as to conventional vegetable production and consumption.

A coordinated system of review and regulation has evolved and continues to be refined among the United States Department of Agriculture (USDA), Food and Drug Administration (FDA), and the Environmental Protection Agency (EPA). Various resources are available which detail the review process, permitting, and product registration requirements for safety testing and approvals from research through commercialization. In general for products related to vegetable production or consumption, the EPA is the lead agency if the product is a microbe or plant with genetically engineered pesticidal traits.

Current regulations, nonetheless, require special conditions or considerations unique to biotechnology products. A non-transgenic product with the identical product trait will likely have reduced regulatory requirements for approval.

12.3.2   Labelling of biotech products

Essentially, if a foodstuff is produced using genetic engineering, this must be indicated on its label.

Actual labelling practice, however, is far more complicated and must be planned and regulated with issues such as feasibility, legal responsibilities, coherence and standardization in mind.

12.4   Strategies for commercialization

The steps between concept, discovery, reduction to practice and commercialization for biotechnology-based products are very complex. Several branch points in the path are reached at each key phase that affect both technical and business strategic decisions. Access to a broad array of enabling licenses and technologies must be obtained to have 'freedom to operate' in a commercial application.

Very broad patents cover the primary tools of biotechnology and are the ground floor access points. A novel application may result in a new patent, but sale of the resultant product requires negotiation of license rights for those technologies utilized in the construction and introduction of any new gene to a plant or microbial product.

12.5   Biotechnology improves nutritional quality and shelf-life of fruits and vegetables

The importance of fruits and vegetables (F&V) in the diet of mankind cannot be over emphasized. Many reviews have reported the wide range of determinants of desirable quality attributes in fruits and vegetables such as nutritional value, flavour, colour, texture, processing qualities and shelf-life. The understanding of the fundamental processes that influence fruit set, maturation, and ripening are required to manipulate fruits and vegetable yield and quality. Biotechnology has played a significant role in this respect. Typically, biotechnology technique such as genetic modification is used in F&V to enable plants tolerate the biotic and abiotic stresses, and plant resistances to problematic pests and disease, which may provide higher nutritional contents, and extend the shelf-life of the produce.

The broader definition of biotechnology refers to commercial techniques that use living organisms to make or modify a product, including techniques for improving the characteristics traits of plants and animals, and development of microorganisms that act on the environment. Additionally, another view of traditional biotechnology covers well established and widely used technologies in brewing, food fermentation, conventional animal vaccine production, and many others based on the commercial use of living organisms. In recent times, advanced biotechnology techniques involve the use of induced mutations, marker-assisted selection, homologous recombination, genomics, and genetic modifications. The major ones are the tissue or cell culture, cell fusion, embryo transfer, recombinant DNA, and age-old fermentation technique. In the mist of recent global challenges such as increasing population, increasing demand for food, climate change, and water scarcity, plant biotechnology has become a necessity tool for growth and yield performance to meet the food needs of today. The production of quality fruits and vegetables with improved shelf-life is no exception.

12.5.1   Need for biotechnology in fruits and vegetable production

A number of challenges have called for the application of biotechnology in the production of fruits and vegetables. These are population increase, water shortages, climate change, high perishability or postharvest decays, and short shelf-life associated with fruits and vegetables. Fruits and vegetables by their intrinsic properties require more water and in the face of water scarcity throughout the world, biotechnology will be required to develop fruits and vegetables that can withstand water stress and still be able to produce good crop of high quality and yield. For instance, during the last century, world population rose from 1.6 to 6.0 billion creating huge challenges for agriculture. However, new technologies increased crop yields drastically so the predicted catastrophic starvation and resulting conflicts did not occur. There are still serious challenges to be faced. Freshwater, vital for agricultural productivity, is becoming scarce and climate change could increase temperature, drought, and uncertainty. New crop varieties need to be developed quickly to meet these challenges and biotechnology will be needed to enhance existing technologies to achieve this. So far, biotechnology has been successfully used to develop insect and herbicide resistance in a limited number of crops such as corn. In the future, the actual metabolism of crop plants will be altered to produce new varieties or species that are tolerant against environmental stresses. In addition, the nutritional value of crops such as rice will be enhanced. Crops will also be used to harvest the sun for biofuels to replace fossil fuels and reduce the emission of CO_2. Some of these new crop plants are already in field trials and will be available to farmers in the near future.

Postharvest decay of fruits and vegetables are a major challenge throughout the world. The degree of postharvest loss through decay is well documented. In the industrialized countries, it is estimated that about 20–25% of the harvested fruits and vegetables are decayed by pathogens during postharvest handling. The situation is far more exasperating in the developing countries, where postharvest decays are often times over 35%, due to inadequate storage, processing and transportation facilities. The use of synthetic fungicides such as benomyl and iprodione to control postharvest diseases of fruits and vegetables is well known in scientific literature. The health and environmental concerns associated with the continuous use of synthetic fungicides have alarmed legal enforcers and consumers to demand greener technology and quality products from the food industry as well as the scientific community. In the past 40 years, microbial antagonists like yeasts, fungi, and bacteria have been used with limited successes to reduce postharvest decays in fruits and vegetables. For instance, fungal diseases like grey mold, powdery mildew, and downy mildew in grapes do notable only cause losses in yield but also reduce wine quality. However, the advances in biotechnology can be employed to develop fruits and vegetables with improved quality and shelf-life. The ability to maintain the quality of stored F&V during postharvest storage is highly related to the physiological, biochemical, and molecular traits of the plant from which they derive. These traits are genetically determined and can be manipulated using genetic breeding and/or biotechnology.

Published research results have revealed potential genes, which when manipulated can be used to improve postharvest qualities of crop plants. The application of this biotechnological knowledge should not only lead to major improvements in postharvest storage of fresh fruits and vegetables but as well improved human food supply.

12.5.2   Dynamics of ripening and perishability in fruits and vegetables

Fruit ripening and softening are major attributes that contribute to perishability in both climacteric and non-climacteric fruits. Fruits and vegetables such as tomato, banana, mango, avocado, etc. take about a few days after which it is considered inedible due to over-ripening. The spoilage includes excessive softening and changes in taste, aroma and skin colour. This unavoidable process brings significant losses to both farmers and consumers alike. Even though ripening in F&V can be delayed through several external procedures, the physiological and biochemical changes associated with ripening is an irreversible process and once started cannot be stopped. Ethylene has been identified as the major hormone that initiates and controls ripening in fleshy fruits and vegetables. Influencing ethylene biosynthesis during ripening in fleshy commodities has been the foremost attempt for combating post-harvest deterioration.

12.5.3   Biotechnology improves quality of fruits and vegetables

The transfer of genetic material from one organism into the DNA of another called transgenic application has been widely used in fruits and vegetables. Tolerant plants to biotic and abiotic stress, higher nutritional contents and extended shelf-life are some of the advantages of transgenic plants. In addition, once a useful transformant is obtained, vegetative propagation, which is the normal method of multiplying in several fruit plants, provides unlimited production of the desired transgenic lines. Recently, reports indicate that recombinant DNA technology has been used by scientists to delay ripening in fruits and vegetables in order for farmers to have the flexibility in marketing their produce and ensure consumers good quality produce from their farms. Transgenic grapes were developed for modified auxin production, fungal and virus resistance as well as fruit quality and colour modifications. Costantini and others transformed grape cultivar. Thompson Seedless with an ovule-specific auxin-synthesizing (DefH9-iaaM) gene and observed that average number of inflorescence per shoot in transgenic grape lines was doubled as compared to control.

Binnie and McManus identified three ACO genes from apple and showed that all three genes express differently. MdACO1 is restricted to fruit tissues, with optimal expression during fruit ripening, MdACO2 expression occurs more predominantly in younger fruit tissue, with some expression in young leaf tissue, while MdACO3 is expressed predominantly in young and mature leaf tissue. Exposure of these fruit to different concentrations of exogenous ethylene showed that various ripening parameters like pulp softening, biosynthesis of volatile aroma compounds, and starch degradation, had different ethylene sensitivities. Their results suggested that the conversion of starch to sugars (an early ripening event) showed a low dependency on ethylene, but a high sensitivity to low concentrations of ethylene.

On the other hand, late ripening events such as pulp softening and ester volatile production showed a high dependency on ethylene but were less sensitive to low ethylene concentrations. Nora and others constructed a gene having an antisense of apple 1-aminocyclopropane-1-carboxylic acid (ACC) oxidase (pAP4) and transformed melon leaves. The pAP4 gene detected in transformed leaves and fruits showed a low ethylene production.

12.5.4   Effect of biotechnological approaches on nutritional quality of fruits and vegetables

Many reviews have reported the wide range of determinants of desirable quality attributes in fresh fruits and vegetables such as nutritional value, flavour, colour, texture, processing qualities and shelf-life. Studies found that tomato plants transformed with yeast SAMDC gene under the control of E8 promoter showed improvement in tomato lycopene content, better fruit juice quality, and vine life. Fruit colouration and softening were essentially unaffected, and all the seedlings from first generation seed displayed a normal triple response to ethylene. Over-expression of Nr (wild-type) gene, in tomato using constitutive 35S promoter produced plants that were less sensitive to ethylene. As ethylene receptors belong to a multi-gene family, antisense reduction in expression of individual receptors did not show a major effect on ethylene sensitivity possibly due to redundancy except in case of LeETR4. Antisense plants developed using LeETR4 under the control of CaMV35S promoter exhibited a constitutive ethylene response and were severely affected.

In apples, Dandekar and others reported differential regulation of ethylene with respect to fruit quality components. A direct correlation between ethylene and aroma production during apple ripening has been reported.

In a related study with two cultivars of apple, Zhu and others characterized the expression patterns of AAT and ACS gene family members in order to examine the relationship with volatile ester production during on-tree and post harvest ripening. They found that differential expression of AAT genes contributed to phenotypic variation of volatile ester biosynthesis in the apple cultivars.

Nishiyama and others found that there was expressed suppression of the ACO gene of transgenic melon fruit when they examined the cell wall polysaccharide depolymerization and the expression of the wall metabolism-related genes. There was also a complete inhibition of softening in the transgenic melon fruits but were restored by exogenous ethylene treatment. Post harvest application of 1-MCP after the onset of ripening completely suppressed subsequent softening, suggesting that melon fruit softening is ethylene-dependent.

The transgenic melon fruits were 60% smaller in size and recorded increased sucrose and acidity invertase levels, with degraded chloroplast as a result of decreased photosynthetic rate than the control.

In another study involving avaocado fruits, Tateishi and others found that three cloned members of β-galactosidases (PaGAL2, PaGAL3 and PaGAL4) played a significant role in the cell wall metabolism during fruits growth and ripening. The study of expression pattern of the isozymes by the same authors during avocado ripening found that the accumulation pattern of the gene transcripts and the response to ethylene gave a correlation between AV-GAL1 transcript and isozyme AV-GAL III. This could be the reason why post harvest biotechnology of avocado has been strongly limited inspite of the fact that it provided early clues to the ripening mechanism in fleshy fruit. Costantini and others transformed grape cultivar Thompson Seedless with an ovule-specific auxin-synthesizing (DefH9-iaaM) gene and observed that average number of inflorescence per shoot in transgenic grape lines was doubled as compared to the control. In their studies, they reported that auxin enhanced fecundity in grapes, thus resulting in increased yield with lower production costs. Similarly, Symons and others have shown that brassinosteroids (steroidal hormones) might be implicated in ripening of non-climacteric fruits. The study showed that grape ripening was significantly promoted by exogenous application of brassinosteroids (BRs) and ripening could be delayed by brassinozole, an inhibitor of BR biosynthesis.

Recent advances in recombinant DNA technology and genetic engineering have opened up the possibility to manipulate ripening in fast perishable fruits like banana. Towards this, many genes involved in ripening have been cloned and characterized. Ripening in banana is characterized by a biphasic ethylene production with a sharp early peak followed by a post climacteric small peak. During banana fruit ripening, ethylene production triggers a developmental cascade that is accompanied by a huge conversion of starch to sugars, an associated burst of respiratory activity and an increase in protein synthesis. Other changes include fruit softening. Banana fruit softening is attributed to activities of various cell wall hydrolases. Lohani and others reported participation of various cell wall hydrolases in banana softening during ripening.

12.5.5   Effect of biotechnological approaches on the shelf-life of fruits and vegetables

The shelf-life of transgenic tomato fruits was reported to last for at least 60 days at room temperature without significant change in hardiness and colour. After 15–20 days of treatment of the transgenic fruits with ethylene, most of the tomatoes reached the ripe stage. Antisense transgenic lines of tomato have also been raised with anti-ACO gene to alter ethylene biosynthesis. RNAi technique has also been used to produce tomato fruit with delayed ripening using ACO gene. According to Xiong and others transgenic tomato fruits had a prolonged shelf-life of at least 120 days. In another study with apples, Wang and others showed that null mutation in MdACS3 gene leads to longer shelf-life.

12.5.6   Challenges associated with commercialization of biotech fruits and vegetables

Many research findings revealed that even though biotechnological approaches is seen by the scientific community as a panacea to solve recent increased demands for fruits and vegetables, the technology is more of a scientific jargon than a commercially viable entity.

This is because:

- Dilemma and uncertainties remain up to today regarding the consumption of biotechnological fruits and vegetables. The impasse has created challenges with consumption of genetically modified fruits and vegetables in many countries and some continents mainly due to the complexities surrounding its use.

- Although the first biotech crop to be commercialized was a genetically modified tomato for processing as a consumer tomato paste, there have been comparatively few introductions of biotech fruits and vegetables since then. Reported cases with potential benefits for farmers in developing countries include virus resistant papaya in China, now commercially grown, and, more recently, the high profile case of *Bacillus thuringiensis* eggplant, or brinjal, in India. Because of the susceptibility of brinjal to the fruit and shoot borer insect, multiple insecticide applications are required to prevent uneconomic losses of yield in this crop.

Approvals of suitably developed and stewarded high-value vegetables and fruits could carry significant benefits for small farmers, because of the relatively high prices of these crops on the market.

The primary conclusion was that the traits did not reach the market not because of poor performance or lack of grower interest but because of regulatory approval uncertainty and prohibitively high and uneconomic development and regulatory costs.

A review by Smith has argued that opinion and debate on acceptance of transgenic agricultural biotechnology remains polarized both 'for and against' and is often not aligned with rigorous review and balanced, empirically grounded assessment of socio-economic and community benefits, human safety, environmental considerations such as non-target safety, gene flow, biodiversity and associated risks.

At a time when biotech crops have been grown extensively in the Americas and Asia for over 13 years, the precautionary principle prevails in many countries even for the traits embodied in these crops.

In the European Union (EU) for example, genetically modified fruits and vegetables are not allowed on the market and none of the GM plants currently authorized in the EU are intended for direct consumption. For instance, in the case of GM tomatoes, they are lurking in grocery stores in the USA and never received authorization in the EU. The situation is the same for biotech bananas, apples, wine grapes, and papaya production. Recent reports in the

EU member states indicate that whiles countries like Finland, Germany, and Greece have strongly opposed commercialization of GM crops including fruits and vegetables; Spain and UK do not fundamentally oppose cultivating GM crops but have used the precautionary principle. So the question remains that 'is biotechnology in fruit and vegetable plant production a commercial activity or simply a research jargon?' A pragmatic approach proposed by Godfray and others move the debate forward saying 'genetic modification is a potentially valuable technology whose advantages and disadvantages need to be considered rigorously on an evidential, inclusive, case-by-case basis'. Genetic modification should neither be privileged nor automatically dismissed. In addition to governments' policy on regulation, key factors influencing future availability of biotech fruits and vegetables in developing countries are stewardship capability, and liability of technology providers. Excellence through stewardship reveals that stewardship biotechnology of fruits and vegetables includes not just management of biosafety and compliance with regulatory authorities' requirements but also product quality and integrity along the whole product life cycle right from early research ideas to the withdrawal of crop varieties. The need for steward- ship is fully founded. With the expected rise in numbers of commercialized events around the world, including potentially *Bacillus thuringiensis* rice in China, concern is growing about the potential for low-level presence of numerous events in trade channels and the food chain in countries without regulatory authorizations.

International harmonized solutions need to be found. Otherwise the private sector will remain very cautious about supporting technology releases to the public sector to assist smallholder farmers in developing countries, especially for food crops that could cross national borders or enter international trade channels. Additionally, little impact has been realized to date with fruits and vegetables because of development timescales for molecular breeding and development and regulatory costs and political considerations facing biotech crops in many countries.

12.5.7   New paradigm shifts for commercialization of biotech fruits and vegetables

Biotechnological approaches offer much potential to increase the development and introduction of improved varieties and as an enabler for greater genetic diversity, but the full benefits are yet to be established. Constraints to the development and adoption of technology-based solutions to reduce yield gaps need to be overcome.

The new paradigm shift proposed includes: (i) the integration of broader thrust that galvanizes public and private investment for the development and provision of technology with the creation of seed systems and markets supported by agricultural extension and other services for farmers, (ii) a commitment to increase and sustain funding of agricultural R&D, (iii) the need to break barriers at the policy and operational level to enable formation of public–private partnerships for transformational change in research, product development, and the delivery of seed-embodied technology to farmers, (iv) the integration of low risk chemical fungicides, natural antimicrobial substances, and physical means such as hot water treatment, irradiation with ultraviolet light, microwave, and infrared treatment in the postharvest biocontrol process, (v) the enhancement in the expression of crucial recombinant DNA genes and/ or combining genes from different agents in the mass production, formulation and storage, or in response to exposure and contact with parent plant tissue after application, (vi) the use of genetically modified organisms as biocontrol agents to enhance the postharvest quality and shelf-life of fruits and vegetables and (vii) the research towards discovering new DNA genes instead of the ones currently used in practices in that only a small portion of the earth micro flora has been identified and characterized.

12.6   Future prospects

Thus, it is evident that developed biotechnological approaches have the potential to enhance the yield, quality, and shelf-life of fruits and vegetables to meet the demands of the 21st century. However, the developed biotech approaches for fruits and vegetables were more of academic jargon than a commercial reality. To make sure that the current debates and complexities surrounding the registration and the commercialization of genetically modified fruits and vegetables are adequately addressed, various stakeholders in the industry (policy makers), private sectors, agriculturalists, biotechnologists, scientists, extension agents, farmers, and the general public must be engaged in policy formulations, seed embodiments, and products development.

The full benefit of the knowledge can be reaped if there are total commitment by all stakeholders regarding increased and sustained funding, increase agricultural R&D, and less cost and time for registration and commercialization of new traits. Thus, the future strategic application of biotechnology is expected to follow a trend of increasing technological sophistication. During the pre-commercial phase, extensive research is required to develop the basic technologies for gene identification, manipulation of desired traits, understanding of plant developmental regulation of genes and the stable re-introduction into competitive varieties. The progression of products will extend applications from single-gene modifications to multiple-gene introductions and developmentally or environmentally regulated gene expression. The level of participation of growers in the economic consequences of marketing of biotechnology products will be specific to the application. As in the mainstream of the produce industry, strategic alliances between agricultural biotechnology companies, large grower/shippers processors and seed companies is a rapidly developing trend.

Vertical integration to control all components of the production and marketing system is viewed as the best approach to ensure quality, safety, proprietary protection and profitability for both conventional and biotechnology derived products.

Genetically modified fruits

13.1   Introduction

Fruits are major ingredients of human diet and provide several nutritional ingredients including carbohydrates, vitamins and functional food ingredients such as soluble and insoluble fibers, polyphenols and carotenoids. Biochemical changes during fruit ripening make the fruit edible by making them soft, changing the texture through the breakdown of cell wall, converting acids or stored starch into sugars and causing the biosynthesis of pigments and flavour components. Fruits are processed into several products to preserve these qualities. Because of various health benefits associated with the consumption of fruits and various products derived from fruits, these are at the centre stage of human dietary choices in recent days. The selection of trees that produce fruits with ideal edible quality has been a common process throughout human history. Fruits are developmental manifestations of the seed-bearing structures in plants, the ovary. After fertilisation, the hormonal changes induced in the ovary result in the development of the characteristic fruit that may vary in ontogeny, form, structure and quality. Pome fruits such as apple and pear are developed from the development of the thalamus in the flower. Drupe fruits, such as cherry, peach, plum, apricot and so on, are developed from the ovary wall (mesocarp) enclosing a single seed.

Some of the important fruits are discussed below.

13.2   Apples

Apple fruits turn brown quickly after being sliced or bitten. Although this is a natural phenomenon, browning has been considered as an undesirable trait that often discourages consumption and causes unnecessary waste.

Apple browning is caused by a polymer compound of pigment that primarily consists of quinones. Quinones are produced from phenols, which are common in apple fruit cells; through the action of a class of enzymes called polyphenol oxidizes (PPOs). In fact fruit cells, phenols and PPOs are separated in different compartments. When cells are damaged by slicing or biting, phenols and PPOs come into contact and react to produce quinones, which will eventually become the pigmented polymer compound (together with amino acids and proteins).

Therefore, PPOs are the key enzymes responsible for apple browning. To stop or minimize fruit browning, reducing the activities of PPOs has long been considered as an effective approach. The 'non-browning' apple is genetically engineered to keep from going brown after being cut. When apple flesh is cut and exposed to oxygen, it begins to brown but the GM apple or 'Arctic Apple' will not brown for 15–18 days.

'Foreign genes': There are no 'foreign genes' in Arctic Apples. All that is different about them is that certain apple genes are 'turned off.' Most genes in cells of an apple or any other organism are turned off most of the time. The genes to grow roots or leaves are turned off in apple fruits just as our genes to develop the features of an eye are turned off in all the other parts of our body. The scientists that developed the Arctic Apple simply used a natural mechanism to turn off the genes that make the enzymes that turn apples brown when they are cut.

Organic growers: For thousands of years, people have known that if you find a fruit that is good, you need to propagate it from cuttings or buds, not seeds. These are grafted onto rootstocks that have been selected for pest resistance and in most cases for dwarfing so the trees don't get too tall to safely pick. Actually, to get a good crop, growers need to make sure that the bees that visit each tree have typically gotten pollen from some other variety of apples, even crab apples, because otherwise fruit set is less likely.

There is a small chance that a bee will carry pollen from a GMO apple to a few flowers in an organic block. A few of the seeds in those organic apples might be pollinated with the GMO pollen. Only a small part of the seeds (the embryo) in those apples will have that extremely minor change. Apple seeds shouldn't be eaten. They contain compounds that can generate cyanide a rather potent poison. Consumers routinely discard them.

Bees and pollen: Apples are pollinated by bees. The bees will stay very close to their hive when there is enough food (like when an orchard is in bloom) and so the risk of contamination from bees is small.

*Apple seeds: Apple s*eeds do not breed true (they may retain some traits of their parents but the resulting trees are not an exact copy of their parent) so grafting is used to propagate apple trees rather than seeds.

However if apple seeds pollinated with GM pollen germinate they will result in GM apple trees (trees expressing the GM trait or carrying the new gene sequence).

There are many ways that GM apple seeds can spread in our environment such as by humans discarding apple cores, cores in compost piles, seeds scattered by animals, and deliberate plantings.

GM pollen and fate of apples: If an apple tree is pollinated with GM pollen, the genes would be present in the resulting apple seeds, not the apple flesh. If pollen from GM apple trees moves into a non-GE apple orchard, some seeds in apples from the non-GM trees that were pollinated could carry the new gene sequence and could express the new GM trait.

Brand protection: Technology opponents have long known that they have the most leverage if they can threaten an entity with a valuable consumer brand with potential, controversial attention. The apple industry is concerned that even though the Arctic apples would be clearly labelled as such, activists could smear the entire apple industry 'brand' and make consumers afraid that they might be eating something they don't want.

GM apple is not necessary: The GM technology is unnecessary, as there are already techniques that slow browning in apples in our kitchens we use lemon juice, and the food service industry uses ascorbic acid (vitamin C). There are also varieties of naturally slow-browning apples, such as Ambrosia.

13.3   Citrus

Citrus is one of the most important commercial and nutritional fruit crops in the world, hence it needs to be improved to cater to the diverse needs of consumers and crop breeders. Genetic manipulation through conventional techniques in this genus is invariably a difficult task for plant breeders as it poses various biological limitations comprising long juvenile period, high heterozygosity, sexual incompatibility, nucellar polyembryony and large plant size that greatly hinder cultivar improvement. Hence, several attempts were made to improve Citrus sps. by using various *in vitro* techniques. Citrus sps. are widely known for their recalcitrance to transformation and subsequent rooting, but constant research has led to the establishment of improved protocols to ensure the production of uniformly transformed plants, albeit with relatively low efficiency, depending upon the genotype.

Genetic modification through *Agrobacterium*-mediated transformation has emerged as an important tool for introducing agronomically important genes into Citrus sps. Somatic hybridization has been applied to overcome self and cross-incompatibility barriers and generated inter-specific and inter-generic hybrids. Encouraging results have been achieved through transgenics for resistance against viruses and bacteria, thereby augmenting the yield and quality of the fruit. Now, when major transformation and regeneration protocols have sufficiently been standardized for important cultivars, ongoing citrus research focuses mainly on incorporating such genes in citrus genotypes that can combat different biotic and abiotic stresses.

Citrus are subject to many biotic and abiotic stresses and markets are continuously requiring fruits of higher quality. These pose important problems in most citrus growing areas that can only be solved with the establishment of citrus improvement programs to recover new and healthy genotypes to be used as rootstocks and varieties. Success of traditional citrus-breeding strategies is limited by the peculiar genetic and reproductive characteristics of citrus. Biotechnology offers different approaches that can overcome many of these limitations.

Some of the technologies used in Spain are:

1. Shoot tip grafting: It is used worldwide to recover plants free of all known citrus pathogens, and already hundreds of millions of trees originated in this technique have been planted in the field.

2. Embryo rescue: It is used in large programs to recover triploid seedless hybrid varieties from aborted seeds from $2n \times 4n$ and $2n \times 2n$ crosses.

3. Protoplast fusion: It allows the production of allotetraploid hybrids among sexually compatible and incompatible parents to be used as rootstocks or as parentals for interploid breeding. Fusions between diploid and haploid cell lines are also used to produce triploid hybrids.

4. Genetic transformation: Efficient protocols for transformation of several genotypes have been developed and introduction in citrus of genes of potential agronomical interest has been accomplished.

13.4　Cherry

While worldwide production of genetically modified crops has been increasing dramatically since 1996, the use of transformation technologies for Prunus crops has been rather modest when compared to the major commercialized genetically engineered crops. To date, only transgenic plum 'Honey Sweet' with PPV (plum pox virus) resistance has received USDA non-regulated status. The lack of efficient biotechnology platforms is considered to be the 'bottleneck' preventing the improvement of Prunus species through genetic transformation. To overcome this limitation for cherry species, previous efforts have focused on optimizing shoot micro-propagation, plant regeneration, gene delivery, and stable transformation. To date, plant regeneration has been reported for 19 sweet cherry *(Prunus avium L.)* cultivars, 12 sour cherry *(Prunus cerasus L.)* cultivars, and 24 genotypes for other cherry species such as cherry hybrids, black cherries *(Prunus serotina Ehrh.)*, and wild cherries *(Prunus avium L.)*. Stable transgenic plants have been reported for sour cherry 'Montmorency', cherry rootstocks 'Gisela 6' and 'Colt', and *Prunus subhirtella* Miq. 'Rosa', but only reporter genes were used in these successful transformations. Most recently, an interfering RNAs (RNAi) vector targeted to prunus necrotic ringspot virus (PNRSV) has been transformed into the hybrid cherry rootstocks. This RNAi strategy, utilizing transformed rootstocks to achieve virus-resistance in the scion, minimizes concerns about 'transgene-flow' and 'foreign protein production' in commercial scion cultivars, and it is therefore a potential approach for engineering virus resistance for fruit-bearing cherry genotypes without direct transformation of the fruiting genotype itself. As more genomic resources for Prunus species become available, more genes of interest will be identified and isolated.

In the future, success in cherry genetic engineering will still depend on availability of desirable target genes and reliable transformation systems with efficient plant regeneration, efficient gene delivery, and effective selection without using antibiotics or herbicides.

13.5   Guava

Fruits and vegetables have specific healthy attributes. Guava is the most nutritious of all edible fruits. High pectin contents make guava suitable for jelly making. Guava has antidiarrheal, antibacterial, antiamoebic and anti-spasmodic activities. High concentrations of several vitamins, dietary fiber, carotenoids, lectins, saponins, tannins, phenols, triterpenes, and flavonoids altogether make guava therapeutically an important fruit. Ascorbic acid in guava fruit far exceeds that in citrus. Consumption of guava fruit is reported to lower blood cholesterol, triglycerides, hypertension and some cardiac problems. Due to high quercetin contents, guava leaves are used to develop a phytomedicine for diarrhea. Since guava plant is susceptible to frost, genetic improvement to enhance cold hardiness of guava germplasm is necessary for adapting and cultivating it in temperate climates.

Efficient and reliable biotechnology using *in vitro* plant regeneration amenable to genetic transformation will ensure cold hardiness enhancement in guava. Development of biotechnology protocols for morphogenesis and somatic embryogenesis along with production of edible quality guava fruits and their nutritional analysis has been achieved using various tissue explants from mature guava trees growing at Fort Valley State University. For guava micropropagation, researchers have developed and germinated 'synseeds' from somatic embryos. Successful preliminary work on genetic transformation has paved the way for incorporating cold hardy genes into the guava genome. Guava *(Psidium guajava L.)*, also called 'apple of tropics,' is immensely nutraceutical and horticulturally important. Being a tropical plant, it cannot stand temperatures below 25°F and needs frost protection to grow in temperate regions. To adapt in cold climate, cold hardy guava cultivars are needed. The transgenic plants developed using biotechnological approaches of tissue culture and rDNA technology, appear to have great potential. Thus, protocols for *in vitro* propagation of guava were developed via organogenesis and somatic embryogenesis using nodal explants from mature trees and young zygotic embryos, respectively. Nodal explants induced multiple shoots when cultured on MS medium fortified with KIN, BAP and Ad.S. Adding a $(NO_3)_2$ to medium was useful to prevent *in vitro* shoot tip browning of adventitious shoots. Rocker liquid culture greatly increased growth of multiple shoots compared to the agar-based medium. It appears to be a good tool for woody plant tissue culture. Induction of somatic embryos in guava was also achieved on MS medium supplemented with IAA auxin. About 80–90% somatic embryos germinated normally. To achieve Agro-bacterium-mediated gene transfer in guava, on-going co-cultivation of organogenic tissues of guava is to optimize protocols for freeze tolerance gene (CBF1, CBF2, CBF3) transfer.

Plasmid vectors containing selectable markers (nptII gene for antibiotic selection and GUS reporter gene as scorable gene mediated selection), with CaMV 35S promoter gene has been introduced into guava tissues and the resultant plants showed antibiotic resistance.

13.6   Apricots

Apricot *(Prunus armeniaca L.)* is a species originated in China and Central Asia, from Tien Shan to Kashmir. Finally, different studies are being performed to the identification of genomic regions involved in fruit quality traits by joining the phenotypic data with the molecular characterization of different apricot progenies. The related QTL identification linked to quality traits is the first step in order to develop specific molecular markers which will offer the opportunity of optimize apricot breeding programs for fruit quality by introducing an early marker assisted selection (MAS) strategy.

Molecular breeding strategies and genome sequencing: Characterization of genetic variability in apricot species has been traditionally based on morphological traits. However, such traits are not always available for analysis, and do not provide sufficient information to trace the expansion of apricot from the centers of origin; to expose the major genetic events in the evolution of cultivars or to identify and characterize genotypes during the breeding process. Molecular marker technology offers several advantages over the sole use of morphological markers, have proved to be a powerful tool for solving the above problems, and have become an essential tool for the molecular studies.

Isoenzymes were the first molecular markers to be utilized in apricot. Nevertheless, the utilization of these markers is limited due to the small number of *loci* that can be analyzed with staining methods. In this sense, the availability of DNA markers provides a new opportunity to evaluate plant diversity. Restriction fragment length polymorphism (RFLP) markers are based on the differential hybridization of DNA fragments from restriction-enzyme digestion. RFLP markers are codominant and can detect a virtually unlimited number of markers. These markers have been used in the molecular characterization of apricot genotypes and also in the development of the first genetic linkage map in apricot.

More recently, the utilization of PCR-based markers has increased the opportunities for molecular characterization and mapping of populations. Random amplified polymorphic DNA (RAPD) markers are based on the PCR amplification of random locations in the genome. RAPDs are characterized by using arbitrary primers. A single oligonucleotide is utilized for the amplification of genomic DNA. RAPD techniques have been successfully used in apricot for identifying cultivars and also for mapping population. In contrast to isoenzymes and RFLPs, RAPDs are dominant markers. This feature, as well as their variable degree of repeatability and problems in transferring across populations, limits their utilization to map construction. These difficulties can be overcome by converting RAPDs to sequence-characterized amplified regions or SCARs.

On the other hand, amplified restriction fragment length polymorphism (AFLP) technology is a powerful DNA fingerprinting technology based on the selective amplification of a subset of genomic restriction fragments using PCR developed later than RAPD. These markers have been mainly used in apricot for molecular characterization of varieties and genetic linkage mapping. AFLP has a number of advantages over the RAPD: more *loci* analyzed and better reproducibility of banding although presents the inconvenient of the difficulty to the use in routine.

Presently, simple sequence repeat (SSR or microsatellite) markers, also based on the PCR technique but with specific primers, are the best suited markers for the assessment of genetic variability within crop species because of their high polymorphism, abundance, and codominant inheritance. SSRs are extremely abundant and dispersed relatively evenly throughout the genome. DNA flanking SSRs is often well conserved in related species, which allows the cross-species amplification with the same primer pairs in related species. This situation gave the first application of SSR in apricot using primers developed in other *Prunus* species such as peach. More recently, primer sets using apricot sequence information have been developed. These markers have been mainly used in apricot for molecular characterization of apricot varieties, identification of accidental pollination in apricot progenies and genetic mapping.

Genetic linkage analysis: Genetic linkage analysis was initially performed in apricot using the combination of different molecular makers including RFLPs, RAPDs, AFLPs and SSRs. The use of RFLPs provides a virtually unlimited source of high quality markers, located all over the genome. In this sense, the most recent utilization of PCR-based markers has increased the opportunities for mapping and tagging a wide range of traits. New genetic linkage maps in apricot has been developed using a reduced type of markers. AFLPs have been other wide used markers allowing the detection of a higher level of polymorphism in apricot than RFLPs or RAPDs. Finally, SSR markers are currently becoming the markers of choice for genetic mapping in apricot.

The genotype-dependency of apricot transformation could be overcome by transforming meristematic cells and therefore avoiding the bottleneck of most transformation procedures. Such a procedure has been developed for apricot allowing the transformation of four different cultivars.

Although only marker genes have been introduced in apricot commercial cultivars, transgenic apricots plants expressing the *Plum pox virus* (PPV) coat protein gene were successfully developed from seed derived tissue. In this case, the coat protein gene of PPV was used to introduce coat protein mediated resistance. In plum *(Prunus domestica L.)*, PPV resistance has been successfully achieved by post-transcriptional gene silencing.

Future work perspective: A key point for apricot breeding is to maintain production and consumers confidence by assuring acceptable production and quality levels through the introduction of new apricot cultivars on the market. Until now, apricot cultivars have been mainly generated through controlled crosses and open pollination. Additional advantages encouraging the utilization of the new biotechnologies to apricot breeding include high levels of synteny between genomes and a well-established international network of cooperation among researchers. In this sense, future works regarding marker-assisted selection (MAS) of apricot breeding must include the comparative mapping of different progenies. Genomic methodologies including expressed sequences tags (ESTs) cloned gene analogs (CGAs) and single point mutations (single nucleotide polymorphisms, SNPs) may make it possible to discover genes of interest in quality selection in apricot. More recent efforts are being oriented to the elaboration of physical maps, the development of quick gene sequencing and cloning tools, and the complete sequencing of the peach genome to develop efficient molecular markers applicable to assistant selection in peach breeding programs. Finally, the increasing availability of biotechnological techniques such as genetic transformation further complements *in vitro* culture opportunities. In this sense, at this time several apricot genotypes genetically modified are being assayed although to date there is no commercial varieties.

13.7   Papaya

Papaya is cultivated primarily in tropical and sub-tropical regions. It is an important part of the diet in many developing countries. It is the source of the enzyme papain that is used in the pharmaceutical and food industries. For several decades the global production of papaya has been threatened by ringspot disease caused by the papaya ringspot virus (PRSV).

PRSV was identified for the first time in 1945 in the American state of Hawaii. Since then there have been outbreaks of ringspot disease in other areas of the world where papaya is cultivated. Hawaii is responsible for just 0.1% of global papaya production. Researchers from Cornell University (USA) and the University of Hawaii developed a genetically altered variety of papaya that is resistant to PRSV. A remarkably efficient development and approval procedure allowed the GMO papayas to be introduced as soon as in 1998. In less than four years' time papaya production was back to pre-PRSV levels. Hawaiian papaya cultivation had been saved. In the meantime, Hawaii has started exporting its biotech papayas to Canada and Japan. Due to the specificity of the virus and the virus resistance mechanism Hawaiian GMO papayas are resistant to the Hawaiian PRSV in particular.

However, the developments on Hawaii have inspired and stimulated other papaya cultivating countries to use the same methods to develop virus resistant papayas for their local markets. Resistant papaya varieties have now been developed in Brazil, Taiwan, Jamaica, Indonesia, Malaysia, Thailand, Venezuela, Australia and the Philippines.

Role of biotechnology: The development of the virus resistant GMO papayas shows that in certain cases using biotechnology can improve plants more efficiently than traditional plant breeding. By using traditional crossing techniques breeders are limited to the genetic information that is available within a certain plant species. In the case of PRSV, the genetic information required for resistance is not present in papaya *(Caricapapaya)*.

This resistance is known to be present in the *Vasconcellea* papaya. The *Vasconcellea* papaya belongs to the same family as the Carica papaya, but a different genus. Attempts have been made since 1958 to transfer the PRSV resistance in *Vasconcellea* to Carica using intergenetic crossing (between genera). Such hybridizations are exceedingly inefficient and not only do they require the development of special procedure and laboratory techniques, they also rely on a great deal of luck. Over the course of more than 50 years only a small number of infertile offspring have been generated. Filipino and Australian researchers have recently reported the first fertile, classically bred *Carica papaya* plant that is resistant to PRSV. Despite the very significant contribution that classical breeding has made, and will continue to make in the future, the classically bred plant was far too late to save the damaged Hawaiian papaya plantations.

Following evaluation of all the food safety and environmental factors, the import of the Rainbow papaya was approved in Japan on 1 December 2011. Thus, the genetically modified crops are often associated with multinationals and large-scale cultivation. The story of the GMO papaya demonstrates that this does not necessarily always have to be the case. The Hawaiian GMO papaya was developed by the public sector and the intellectual property rights were transferred to the local papaya industry. The GMO papaya has been cultivated since 1998 to deal with the papaya ringspot virus, a pathogen that does not just cause serious losses in yield but can also make commercial papaya cultivation impossible.

Thanks to the use of GMO papayas cultivation has been able to continue on Hawaii. This success story inspired many papaya-producing countries to look for a similar solution to deal with local papaya viruses. Until now virus resistant GMO papayas are only being cultivated in Hawaii and China but more countries are ready to make use of biotech papayas in the fight against the papaya ringspot virus.

13.8   Eggplant

Eggplant is a member of the nightshade family as are tomatoes, peppers, potatoes and sweet potatoes. As such, eggplant is technically a fruit. As with tomatoes and other of nightshade veggies there are a number of varieties of eggplant. The large dark purple pear shaped eggplant is the most common. The smaller version of the larger purple-skinned eggplant is often called Italian or baby eggplant. These have a somewhat more intense flavour and the flesh is much more tender. The straight thin eggplants known as Japanese or Asian eggplant have thin delicate skins like Italian eggplant but the flesh is sweeter. The colour ranges from dark purple to a striped purple as well as a light amethyst.

There is a wide genetic diversity in the cultivated as well as the wild species of eggplant. Cultivated varieties of eggplant are susceptible to a wide array of pests and pathogens as well as to various abiotic stress conditions. In contrast, the majority of wild species is resistant to nearly all known pests and pathogens of eggplant and thereby is a source of desirable traits for crop improvement. Transgenic eggplants for insect resistance, for the production of parthenocarpic fruits and abiotic stress tolerance have been accomplished. However, transgenics of eggplant are yet to be developed for improvement of other agronomic traits, including disease and pest resistance and quality and shelf life of fruits.

FSB-resistant eggplant: FSB-resistant (FSBR) eggplant is an insect resistant eggplant developed with the help of biotechnology. Also called *Bacillus thuringiensis* eggplant or *Bacillus thuringiensis* brinjal, it produces a natural protein that makes it resistant to FSB. Once the FSB caterpillar feed on plant leaves, shoots and fruits, they stop eating and eventually die. The *Bacillus thuringiensis* protein in the biotech eggplant only affects FSB and does not affect humans, farm animals and other non-target organisms. Biotechnology has opened up new vistas for crop improvement. Biotechnological tools like *in vitro* propagation, genetic engineering and molecular biology has helped overcoming constraints of conventional breeding, and identification and introduction of useful genes that confer resistance to pests and diseases, and tolerance to abiotic stresses in eggplant. Although some developments in the field of biotechnological applications have taken place, the full potential is yet to be exploited for improvement of eggplant.

The understanding of specific metabolic pathways directly or indirectly involved in plant morphogenesis has helped in understanding and improving the regeneration potential of eggplant genotypes.

Plant transformation procedures have been well established and are being utilized for producing eggplant transgenics. However, genetic engineering has not yet been utilized to its full potential for eggplant improvement as the existing studies have been confined mainly to the development of insect resistant transgenics and transgenics for parthenocarpic fruit development as well as tolerance to abiotic stresses. Thus, such studies need to be further exploited so as to address the development of transgenic plants tolerant to the wide array of environmental stresses, and fungal and bacterial pathogens.

The prospects of eggplant improvement appear brighter with the advent of biotechnology tools. The areas that need to be strengthened in eggplant research are genetic engineering and molecular biology.

Genetically modified vegetables

14.1   Introduction

Vegetables are very important components of diet. Vegetables are highly perishable, and transportation, storage, and distribution require low-temperature conditions. Vegetables are processed into juice, sauce, and canned under aseptic conditions. This enables long-term storage of processed products. Tomato is the major produce that is processed into juice and sauce. Preservation of nutritional components is compromised during processing. Stability of juice is influenced by the particle size distribution. Vegetables and fruits have many similarities with respect to their composition, harvesting, storage properties, and processing. In the true botanical sense, many vegetables are considered as fruits. Thus, tomatoes, cucumbers, eggplant, peppers, and so on could be considered as fruits, since they develop from ovaries or flower parts and are functionally designed to help the development and maturation of seeds. However, the important distinction between fruits and vegetables is based on their use. In general, most vegetables are immature or partially mature and are consumed with the main course of a meal, whereas fruits are generally eaten alone or as a dessert.

Some of the important vegetables are discussed below.

14.2   Tomatoes

Tomato is a major vegetable crop that has achieved tremendous popularity over the last century and it is grown in almost every country of the world. Tomatoes breeding programs can highly benefit of biotechnological tools, such as gene transfer technology, which allows the introduction of foreign genes into a germplasm, without modifying the genetic background of elite varieties. However, a breeding program associated to biotechnological tools depends upon the development of an efficient *in vitro* plant regeneration system. *In vitro* plant regeneration of tomatoes using protocols for adventitious shoot regeneration from cotyledon segments has been reported. The system is based on three culture steps: a bud induction phase, culturing the explants in medium supplemented with cytokinin; an elongation phase, transferring the shoot buds to medium with a lower concentration of cytokinin, and a rooting phase, using a culture medium supplemented with auxin.

A genetically modified tomato, or transgenic tomato, is a tomato that has had its genes modified, using genetic engineering. The first commercially available genetically modified food was a tomato engineered to have a longer shelf life (the Flavr Savr). Currently there are no genetically modified tomatoes available commercially, but scientists are developing tomatoes with new traits like increased resistance to pests or environmental stresses. Other projects aim to enrich tomatoes with substances that may offer health benefits or be more nutritious. As well as aiming to produce novel crops, scientists produce genetically modified tomatoes to understand the function of genes naturally present in tomatoes.

14.2.1   Genetically engineered traits

At present, tomatoes are the only food that has been marketed with GE delayed-ripening traits. Delaying the ripening process in tomatoes is of interest to producers because it allows more time for shipment of tomatoes from the farmer's fields to the grocer's shelf, and increases the shelf life of the tomatoes for consumers.

Although ripening makes tomatoes edible and flavourful, it also begins the gradual decline towards softening and rot, causing losses for producers and consumers. Tomatoes that are genetically engineered to have delayed ripening can be left longer on the plant to mature, will have longer shelf-life in shipping, and may last longer for consumers.

Some tomatoes have been genetically engineered to alter one particular aspect of tomato ripening: softening. The process of fruit softening is caused in part by the breakdown of pectins—compounds which give support to the walls of tomato cells. Tomatoes have been engineered to have reduced levels of a pectin breakdown enzyme called polygalacturonase. This not only increases shelf life, it makes the tomato products thicker (higher pectin to water ratio), which is of interest to tomato processors. This is the technique used in the well-known 'FlavrSavr' tomato. Tomato plants naturally produce the compound ethylene to trigger the ripening of tomatoes on the plant. Several genetic engineering strategies involve the reduction or prevention of ethylene production to slow or stop the ripening process.

14.2.2   Methods of transformation

There are various methods of transformation available which are followed by different laboratories. To create the transgenic tomato, a gene from *E. coli* (a bacterium which occurs naturally in the mammalian gut) called kan(r) and the FLAVR SAVR gene (from a tomato) were inserted into a plasmid (a circular ring of DNA) and plasmids like these were inserted into a group of tomato cells in a growth medium containing an antibiotic (Engel 77). The Kan(r) gene, when established in the cell, produced a substance called APH (3') II that gave the cell resistance to the antibiotic.

The antibiotic killed cells that did not receive the plasmid. The purpose of the bacterial gene was, therefore, to identify the cells that were genetically transformed. The FLAVR SAVR gene coded for a strand of RNA that was the reverse of a strand of RNA that naturally occurs in the plant. The original RNA strand in the plant is responsible for the production of the enzyme polygalacturonase. Polygalacturonase breaks down pectin in the cell walls of the tomato during the ripening process and causes the entire tomato to become soft. The complementary strand of RNA from the FLAVR SAVR gene binds to the polygalacturonase RNA and the two strands 'cancel each other out,' preventing the production of polygalacturonase and the softening of the tomato.

Tomato transformation and regeneration were analyzed and optimized by Carolina Cortina and coworkers. They infected Cotyledon explants from *Lycopersicon esculentum cv.* UC82B with *Agrobacterium tumefaciens* strain LBA4404 harbouring the neomycin phosphotransferase *(NPTII)* reporter gene. They found that on increasing concentration of thiamine(vitamin) from 0.1 mg l^{-1} in standard medium to 0.4 mg l^{-1} decreased the chlorophyll lost that accompanied the expansion of necrotic areas in cotyledon explants. They observed optimal shoot regeneration rate with a balanced concentration of 0.5 mg l^{-1} auxin indolelacetic acid (IAA) and 0.5 mg l^{-1} cytokinin zeatin riboside. Finally, when the phenolic acetosyringone was present in the co-culture medium at 200 μM, they confirmed transgenic lines reached 50% of antibiotic resistant shoots. The efficiency of transformation reached 12.5% with this protocol.

Chyi Y.S. and coworkers investigated the genetic behaviour of DNA sequences in the backcross progeny of 10 transformed *Lycopersicon esculentum* × *L. pennellii hybrids*. They used Isozyme and restriction fragment length polymorphism (RFLP) markers to test linkage relationships of the insertion in each backcross family. The TDNA inserts in 9 of the 10 transformants were mapped in relation to one or more of these markers, and each mapped to a different chromosomal location. Because only one insertion did not show linkage with the markers employed, it must be located somewhere other than the genomic regions covered by the markers assayed. They concluded that *Agrobacterium*-mediated insertion in the *Lycopersicon* genome appears to be random at the chromosomal level.

Backcross progeny of two nopaline negative transformants showed incomplete correspondence between the T-DNA genotype and the kanamycin resistance phenotype. Two kanamycin resistant progeny plants of one of these two transformants possessed altered T-DNA restriction patterns, indicated genetic instability of the T-DNA in this transformant McCormick S. and others modified leaf disc transformation system in tomato. They used leaf explants and hypocotyls sections to regenerate transformed plants. They found evidences for both single and multicopy insertions of the T-DNA, and have demonstrated inheritance of the T-DNA insert in the expected Mendelian ratios. A reduced efficiency of transformation was observed with binary T-DNA vectors as compared to co-integrate T-DNA vectors.

Table 14.1 shows comparison of ranges of nutrients between transgenic and normal tomatoes.

Table 14.1 Comparison of ranges of nutrients between transgenic and normal tomatoes (per 100 g fruit).

Nutrient	Normal range	Transgenics	Controls
Protein	0.85 g	0.75–1.14 g	0.53–1.05 g
Vitamin A	192–1667 IU	330–1660 IU	420–2200 IU
Thiamin	16–80 µg	38–72 µg	39–64 µg
Riboflavin	20–78 µg	24–36 µg	24–36 µg
Vitamin B6	50–150 µg	86–150 µg	10–140 µg
Vitamin C	8.4–59 mg	15.3–29.2 mg	12.3–29.2 mg
Niacin	0.3–0.85 mg	0.43–0.70 mg	0.43–0.76 mg
Calcium	4.0–21 mg	9–13 mg	10–12 mg
Phosphorus	7.7–53 mg	25–37 mg	29–38 mg
Sodium	1.2–32.7 mg	2–5 mg	2–3 mg

Note: For Table 14.1, the 'normal range' represents values that the researchers looked up in standard references. The 'controls' column represents actual amounts of nutrients found in nontransgenic (traditional) varieties grown by the researchers alongside the transgenic varieties.

Various benefits of transformed plants

The increased consumption of fruits and vegetables is associated with reduced cardiovascular disease. D. Rein and coworkers studied the health effects of wild type tomato (wtTom) and flavonoid-enriched tomato (flTom). Human C-reactive protein transgenic (CRPtg) mice express markers of cardiovascular risk. They analysed markers of general health (bodyweight, food intake, and plasma alanine aminotransferase activities) and of cardio-vascular risk (plasma CRP, fibrinogen, E-selectin, and cholesterol levels). CRPtg mice were fed a diet containing 4 g/kg wtTom, flTom peel, vehicle, or 1 g/kg fenofibrate for 7 weeks which reduced cardiovascular risk. A.L.E. Lopez and coworker showed that transgenic tomato expressing interleukin-12 has a therapeutic effect on progressive pulmonary tuberculosis. They observed that transgenic tomato L-12 administration resulted in a reduction of bacterial loads and tissue damage compared with wild-type tomato (non-TT). In the late infection, a long-term treatment with TT–IL-12 was essential. They successfully demonstrated that TT–IL-12 increases resistance to infection and reduce lung tissue damage during early and late drug-sensitive and drug-resistant mycobacterial infection.

Konijeti R. and coworkers found dietary lycopene combined with other constituents from whole tomatoes have greater chemopreventive effects against prostate cancer as compared to pure lycopene provided in a beadlet formulation in mice. They fed mice with lycopene in form of tomato paste and lycopene beadlets. The incidence of prostate cancer was significantly decreased in the lycopene beadlets LB group relative to the control group up to 95%. Smith and coworkers investigated the expression of coat protein gene of Tomato leaf curl virus (TLCV) into an expression vector and mobilized to *Agrobacterium tumefaciens* through triparental mating.

Cotyledon leaf explants of Pusa Ruby tomato were transformed by co-cultivation with Agrobacterium containing TLCV–CP constructs. Kanamycin-resistant transformants were regenerated and established in glasshouse. They observed that in TI generation transformed plants showed disease tolerance when compared to non-transformed ones.

Thus, the tomato has developed into an excellent model system for the plant molecular biologist. Plants can be easily regenerated from single cells or protoplasts. Selection of single cells has yielded lines genetically resistant to pathogens, herbicides, and other abiotic environs. Tomato protoplasts can be fused with protoplasts of other tomato lines, or species, as well as with protoplasts of other plant genera.

Tomato DNA has been isolated, cloned, and engineered. Successful transformation of tomato has been effected, using a variety of methods to introduce foreign DNA. The critical road block to improvement of tomato varieties through genetic engineering is the discovery and manipulation of useful single genes that can be isolated by molecular geneticists.

14.3   Soyabeans

Soyabean is the oil crop of greatest economic relevance in the world. Its beans contain proportionally more essential amino acids than meat, thus making it one of the most important food crops today. Processed soyabeans are important ingredients in many food products.

To help meet the challenges of increased soyabean demand, biotechnology tools are being used to develop soyabeans with improved nutritional value and greater resistance to disease, herbicides, and drought. Producers are increasingly turning to biotech soyabeans because of the cost and time savings and reasonable yield enhancement these soyabeans offer. Future traits offer the promise of further crop protection benefits, higher yield, and grain value enhancement through oil and protein modification. Despite all the opportunities, biotech soyabeans face numerous challenges. Because of the cost of technology and regulatory clearance, it is challenging for developers to capture an acceptable return on biotechnology investments. In order for the full benefits of biotechnology to be realized by the world's farmers and consumers, global acceptance of biotech crops and grain is critical.

Biotechnology can be defined broadly as a set of tools that allows scientists to genetically characterize or improve living organisms. Several emerging technologies, led by transformation and molecular characterization, are already being used extensively for the purpose of plant improvement. Other emerging sciences, including genomics and proteomics, also are starting to impact plant improvement. Looking forward, biotechnology promises to deliver products with improved nutritional value and yield enhancement through greater resistance to disease, herbicides, and drought. Enhanced productivity is increasingly important, because per-capita soyabean consumption is growing. Soyabean consumption and production are increasing. Soyabean consumption is driven primarily by meat consumption in human diets, as soyabeans are used primarily for animal feed.

14.3.1   Impact of soyabean technology

The soyabean is particularly difficult to transform. Despite the wide-ranging impact and notoriety of biotechnology solutions, there are additional biotechnology tools that are quietly, yet effectively, improving the productivity of soyabeans. One of these is the use of molecular markers, which help scientists track and select key genes during breeding. The story of how Pioneer is tackling a serious soyabean pest is an example of the role this biotechnology tool can play in agriculture.

The soyabean cyst nematode (SCN) is the most destructive soyabean pest in the United States, causing yield losses estimated at more than $1 billion a year. Nematodes attach to roots, causing significant damage, plant stunting and yellowing, and yield loss. While attached, female nematodes are fertilized by male nematodes and produce a large number of eggs. At season's end, the female dies and the eggs remain in her body, which forms a protective shell or cyst. Nematode cysts may remain in infested fields for more than a decade. Gene mapping has been used to identify the location of SCN resistance genes on specific chromosomes.

Another emerging tool of biotechnology is genomics, which refers to the study of the function and structure of genes. The soyabean genome, as with genomes of other species, holds a vast resource of blueprints that determine what this great plant can provide. Genomics is helping researchers understand soyabean DNA structure and function to change traits that affect pest resistance, yield, and grain composition.

Herbicide-tolerant soyabean

Herbicide tolerant soyabean varieties contain a gene that provides resistance to one of two broad spectrum herbicides. This modified soyabean provides better weed control and reduces crop injury. It also improves farm efficiency by optimizing yield, using arable land more efficiently, saving time for the farmer, and increasing the flexibility of crop rotation. It also encourages the adoption of no-till farming-an important part of soil conservation practice.

Insect-resistant soyabean

This biotech soyabean exhibits resistance to lepidopteran pests through the production of Cry1Ac protein. Insect resistant soyabean was developed to reduce or replace high insecticide applications and at the same time maintain soyabean yield potential.

Oleic acid soyabean

This modified soyabean contains high levels of oleic acid, a mono-unsaturated fat. According to health nutritionists, monounsaturated fats are considered 'good' fats compared with saturated fats found in beef, pork, cheese, and other dairy products. Oil processed from these varieties is similar to that of peanut and olive oils. Conventional soyabeans have an oleic acid content of 24%. These new varieties have an oleic acid content that exceeds 80%.

14.3.2   Challenges Ahead

Capturing value for quality-enhancing traits is one of the biggest challenges facing biotech soyabeans. Put simply, to capture value, value must be created. To ensure adoption, this value must be sufficient to reward adequately all participants in the value chain (developer, grain producer, grain handler, and end user). Biotech solutions will likely be used to expand the use of soyabean protein and oil in the food market. This might include improvement of soyabean isolate flavor and functionality and improved oil/protein health characteristics for consumers. Biotechnology also will be used to create traits that lead to soyabeans with increased industrial and energy uses (e.g. biodiesel, lubricants).

However, capturing value also hinges on acceptance of the technology used to create the value. To help ensure acceptance, soyabeans derived through biotechnology must be developed while following stringent safety and regulatory guidelines established by industry and government agencies. These guidelines must be followed from product concept to post-market.

Today, every product is subject to extensive regulatory scrutiny. Approval costs for a product can reach into the millions of dollars. The value of new soyabean traits, however beneficial they may be for producers and consumers, could be derailed by increased regulatory cost. The biotech industry will likely face more and tighter regulations with increased scrutiny of field research and more data requirements. There will be increased pressure on technology globally. If biotech soyabeans are going to provide the hoped-for benefits, the industry and others will need to work to help consumers and regulators understand the impact of additional regulatory proposals and the safety procedures that are already in place.

14.3.3   Soyabeans pros and cons

There are some potential disadvantages to growing GMO soyabeans as well.

Pros of genetically modified soyabeans

Higher yields are a real possibility. Genetic modifications may make it possible for farmers to grow multiple crops of soyabeans on the same land during the year. This would help to make a farm more profitable every year and enhance the food supply that is available to everyone in a region, country, or even the world.

More farmers could plant soyabeans. Although soyabeans are a rather hearty crop, there are places in the world where they just won't grow. Through the use of genetic modifications, the plants could be designed to respond to more difficult environments and this would help farmers anywhere in the world to use their land more efficiently.

They are easy to care for on a daily basis. GMO soyabeans cannot only be engineered to be pest resistant, but also weed resistant. This allows farmers to spend more of their time working on other crops or livestock instead of focusing all of their attention and energy on their soyabeans.

Cons of genetically modified soyabeans

They may increase the rate of food allergies. More people than ever before are allergic to soyabeans. It's become so prevalent that it has become a listed food allergy on grocery products. Because soya protein is extensively used in many food products due to its affordability, the development of a food allergy makes it difficult for people to find foods to eat at times.

The crops are hard on the soil. Many farmers have found that they can grow GMO soyabeans on their land for a limited amount of time before they'll have to do a crop rotation. Unlike other crops, it may take more than one year for land to recover from the growing of genetically modified soyabeans. Sometimes the land isn't usable ever again.

There is no real consumer benefit. Although more soyabeans can potentially be grown, the cost of growing GMO soyabeans is the same. This means that consumers are going to pay the same amount for their food products, whether modifications were made to it or not.

Genetically modified soyabeans may be able to change the world's food supply, but is the cost of that alteration a price that is too high to pay?

14.4   Corn

The large productivity gains in corn production made during the last several decades have come primarily from advanced plant breeding techniques and improved corn management of the crop. GM corn products have increased corn production over a wide range of growing conditions, helping promote yield stability and reduce production risks. Corn products with biotechnology traits have the latest technology integrated with seed genetics to help protect yield potential.

Biotechnology benefits for corn production: There are many options to consider for corn product selection including products which are conventional, herbicide-tolerant, or insect-protected and herbicide-tolerant. Each farmer must determine what product offers the best opportunity to help maximize profitability on his or her farm. Field research trials from a variety of sources show that corn products with genetically modified (GM) traits have a number of benefits over conventional corn to help manage risk under variable yield conditions and protect yield potential.

- GM corn provides more yield than conventional corn.
- GM corn responds to higher plant densities more than conventional corn.

Weed management benefits:

- Reduced plant stress due to weed infestations to protect yield potential and plant health.

- Removes hosts for insects, diseases, and nematodes.
- Facilitates the use of reduced-tillage for soil and water conservation.

Insect management benefits:

- Damage from the multi-pest complex (corn earworm, European corn borer, western bean cutworm, fall armyworm) causes stress and injury to plant tissue, reducing yield potential. Insect damage can allow fungi to infect, proliferate, and produce mycotoxins which have the potential to cause health problems in animals and humans.
- GM corn rootworm protection has been shown to have agronomic benefits in addition to insect management. Improved root growth and activity can allow corn products to export more cytokinins from the roots and utilize nitrogen more effectively after flowering to promote higher kernel weight and yield potential.
- Higher corn plant densities are important for maximizing grain yield potential.

GM products with insect and herbicide tolerance protect corn yield potential and provide other benefits researchers found that GM crops are credited with decreasing pesticide and fuel use, facilitating conservation tillage practices that reduce soil erosion, improving carbon retention, and lowering greenhouse gas emissions.

To sum up farmers planting GM corn products with herbicide resistance and multiple mode of action insect protection traits can realize higher yield potential by using intensive corn management practices to:

- Reduce plant stress from corn borers, ear feeding insects, stalk boring insects, and rootworm root damage.
- Plant corn-intensive crop rotations.
- Maintain higher plant densities to help maximize corn production.
- Harvest better quality grain by preventing insect damage that can lead to stalk and ear rot diseases. Mycotoxins produced by these diseases have the potential to cause health problems in animals and humans.
- Reap the economic benefits of higher yield potential in feedstuffs for cattle.

14.4.1   Pros and cons of genetically modified (GM) corn

Pros

GM corns are genetically engineered to last longer and be in a better condition than regular crops, this can bring down the price of corn.

GM corns are genetically engineered to resist harsh conditions like weed killers (Roundup) and weather, allowing more of the crops to survive.

Since corn are used in so many products as well as feeding livestock (are huge building blocks of the food industry), the extra GM corn and soybeans produced can be used to help the impoverished.

Cons

Farmers are free to use chemical-rich weed killers to kill all other plants except the GM corn leaving these chemicals on the surface of the crops.

GM corn mate just like all other corn and soybeans, with all other corn and soybeans. This causes problems by widely spreading the GM corn and soybean DNA before we know its effects on the environment and human health.

14.5   Carrot

All varieties of carrots contain valuable amounts of antioxidant nutrients. Included here are traditional antioxidants like vitamin C, as well as phyto-nutrient antioxidants like beta-carotene.

Different varieties of carrots contain differing amounts of these antioxidant phytonutrients. Red and purple carrots, for example, are best known for the rich anthocyanin content. Oranges are particularly outstanding in terms of beta-carotene, which accounts for 65% of their total carotenoid content. In yellow carrots, 50% of the total carotenoids come from lutein.

Carrot DNA apparently contains a gene that causes carrots to express orange pigmentation. The orange colour in carrots is caused by beta-carotene, a precursor to vitamin A.

Morris, together with Kendal Hirschi and other Texan colleagues, has found a way to double the calcium content of carrots through genetic modification, making them a rich source of the element that is so vital for bones. Morris and others loaded their super-carrots with a protein called sCAX1, which pumps calcium into the plant's cells. The protein originally hailed from the plant-of-choice for geneticists, *Arabidopsis thaliana,* where it exists in a larger version. Morris's team lopped off a small piece from its tip that stops the protein from funnelling in more calcium once a certain amount has been reached.

In this shortened form, sCAX1 is relentless in its import of calcium and the researchers have found that it can greatly increase the calcium content of several vegetables including tomatoes, potatoes and carrots. These super-charged vegetables could help to reduce the risk of osteoporosis, one of the world's leading nutritional disorders, where a lack of calcium leads to brittle bones.

Benefits and risks: Morris's tinkered carrots are not only more nutritious but also more long lasting and more productive. Again, calcium is the key. Farmers have known this for some time. Many growers soak apples in calcium solutions to keep them firm during shipping and fresh on the shelf. Potato crops are also sprayed with extra calcium, which helps them to tolerate hot conditions and ward off infections. There's good reason to believe that the modified carrots would also enjoy similar benefits.

14.5.1   Biotechnological applications in carrot improvement

Trait mapping

The genetics of two multigenic traits of carrot, carotenoids and restoration of CMS were evaluated in the 1960s–1980s. Several individual genes were and the heritabilities of several carrot traits have been measured. Multigene trait mapping only began with the development of molecular markers and linkage maps. To date, seven monogenic traits have been mapped for carrot: yel, cola, *rs, Mj-1, Y, Y2* and *P1*. Work is underway to localize genes for flavour compounds and pigments, and to localize genes to specific chromosomes using fluorescent *in situ* methods. Quantitative trait loci (QTL) have been mapped for carrot total carotenoids and five component carotenoids: phytoene, α-carotene, Beta-carotene, zeta-carotene, and lycopene; and the majority of the structural genes of the carotenoid pathway is now placed into this map. Marker-assisted selection has been reported for rs gene.

Gene mapping

Bradeen and Simon studied 103 *F2* individuals of the cross B9304 X YC7262, which segregated for core colour. Using bulked segregant analysis combined with *F2* mapping, they identified six AFLP markers linked to and flanking the *Y2* locus. Markers were located between 3.8 cM and 15.8 cM from the gene. Using the same *F2* mapping population, Vivek and Simon subsequently identified a single AFLP marker located 2.2 cM from the *Y2* locus, assigning the locus to one end of linkage group *B. Anthocyanin* accumulation in the carrot phloem is conditioned by the *P1* locus, with purple (P1) dominant to nonpurple (*P1*). Simon studied the inheritance of *P1* and *Y2* in *F2* and BC populations originating from Eastern carrot germplasm and concluded that the two *loci* are unlinked. Consistent with this, Vivek and Simon mapped *P1* to linkage group A, independent of *Y2*. *P1* is flanked by AFLP markers mapping 1.7 and 8.1 cM away from the gene.

The rs allele has recently been found to be a naturally occurring knockout mutant of a carrot invertase isozyme which produces no functional enzyme. Vivek and Simon mapped rs to one end of linkage group C, 8.1 cM away from an AFLP marker. Mapping results are consistent with inheritance data indicating that rs is genetically unlinked to $Y2$ and $P1$.

Quantitative trait loci (QTL) detected

In addition to single genes conditioning important traits, several QTL have been identified in carrot through segregation analysis. To date, QTL conditioning synthesis in carrot roots of provitamin A α-and β-carotenes, the carotene lycopene, and precursors in the carotene pathway have been mapped. Among orange carrots, heritability of 0.40 and approximately 20 major QTL have been reported to control carotenoid content. Most modern carrot breeding effort has exclusively involved intercrosses among orange carrots and the numerous QTL involved in that colour class. A major exception to this generalization has been the use of white wild carrot as a source of CMS. As yellow, red, and even white cultivated carrots become more popular, the major genes and eventually QTL conditioning these colours will be better described.

Smith and others reported that two major interacting loci, Y and $Y2$ on linkage groups 2 and 5, respectively, control much variation for carotenoid accumulation in carrot roots. These two QTLs are associated with carotenoid biosynthetic genes zeaxanthin epoxidase and carotene hydroxylase and carotenoid dioxygenase gene family members as positional candidate genes. Dominant Y allele inhibits carotenoid accumulation. When Y is homozygous recessive, carotenoids that accumulate are either only xanthophylls in $Y2$ plants, or both carotenes and xanthophylls, in $y2y2$ plants. These two genes played a major role in carrot domestication and account for the significant role that modern carrot plays in vitamin A nutrition.

Future directions: Genes conditioning root pigmentation and sugar and terpenoid content are candidates for gene mapping in the near future. MAS may be more widely integrated into carrot breeding programs as converted markers linked to quality traits are generated. Genomics resources such as expressed sequence tag libraries, microarrays, *in situ* hybridization methodologies, and transposon-tagging systems are likely to be developed which would further accelerate the generation and identification of useful variation for carrot quality improvement.

14.6   Potatoes

Potato has always been a close companion to biotechnology. Potato, being vegetatively propagated crop, is highly amenable to asexual clonal propagation techniques *in vitro* and consequently genetic engineering. Besides, potato has the distinct advantage of possessing a commercially viable carbon sink in the form of tuber. Therefore, it has also been looked upon as a potential bioreactor for the production of novel compounds of therapeutic and industrial values.

Potato is the third most important global food crop and the most widely grown noncereal crop. Transformation is highly effective for adding single genes to existing elite potato clones with no, or minimal, disturbances to their genetic background and represents the only effective way to produce isogenic lines of specific genotypes/cultivars. This is virtually impossible via traditional breeding, as due to the high heterozygosity in the tetraploid potato genome, the genetic integrity of potato clones is lost upon sexual reproduction as a result of allele segregation. These genetic attributes have also provided challenges for the development of genetic maps and applications of molecular markers and genomics in potato breeding. Various molecular approaches used to characterize loci, (candidate) genes and alleles in potato, and associating phenotype with genotype.

New molecular biology and plant cell culture tools have enabled scientists to understand better how potato plants reproduce, grow and yield their tubers, how they interact with pests and diseases, and how they cope with environmental stresses. Those advances have unlocked new opportunities for the potato industry by boosting potato yields, improving the tuber's nutritional value, and opening the way to a variety of non-food uses of potato starch, such as the production of plastic polymers.

14.6.1   Producing high-quality propagation material

Unlike other major field crops, potatoes are vegetatively reproduced as clones, ensuring stable, 'true-to-type' propagation. However, tubers taken from diseased plants also transmit the disease to their progenies. To avoid that, potato tuber 'seed' needs to be produced under strict disease control conditions, which adds to the cost of propagation material and therefore limits its availability to farmers in developing countries.

Micropropagation or propagation *in vitro* offers a low-cost solution to the problem of pathogens in seed potato. Plantlets can be multiplied unlimited number of times, by cutting them into single-node pieces and cultivating the cuttings. The plantlets can either be induced to produce small tubers directly within containers or transplanted to the field, where they grow and yield low-cost, disease free tuber 'seed'. This technique is very popular and routinely used commercially in a number of developing and transition countries. For example, in Viet Nam micropropagation directly managed by farmers contributed to the doubling of potato yields in a few years.

Protecting and exploring potato diversity

The potato has the richest genetic diversity of any cultivated plant. Potato genetic resources in the South American Andes include wild relatives, native cultivated species, local farmer-developed varieties, and hybrids of cultivated and wild plants. They contain a wealth of valuable traits, such as resistance to pests and diseases, nutrition value, taste and adaptation to extreme climatic conditions. Continuous efforts are being made to collect, characterize and conserve them in gene banks, and some of their traits have been transferred to commercial potato lines through cross-breeding.

To protect collections of potato varieties and wild and cultivated relatives from possible diseases and pest outbreaks, scientists use a variation of micropropagation techniques to maintain potato samples *in vitro*, under sterile conditions. Accessions are intensively studied using molecular markers, the identifiable DNA sequences found at specific chromosomal locations on the genome and transmitted by the standard laws of inheritance.

Obtaining improved varieties

Potato genetics and inheritance are complex, and developing improved varieties through conventional cross breeding is difficult and time consuming. Molecular-marker based screening and other molecular techniques are now widely used to enhance and expand the traditional approaches to potato in food production. Molecular markers for characteristics of interest help identify desired traits and simplify the selection of improved varieties. Such techniques are currently applied in a number of developing and transition countries.

Through the Potato Genome Sequencing Consortium, significant progress is being made in mapping the complete DNA sequence of the potato genome, which will enhance our knowledge of the plant's genes and proteins, and of their functional traits. Technical advances in the fields of structural and functional potato genomics – and the ability to integrate genes of interest into the potato genome – have expanded the possibility of genetic transformation of the potato using recombinant DNA technologies. Transgenic varieties with resistance to Colorado potato beetle and viral diseases were released for commercial production in the early 1990s in Canada and the USA, and more commercial releases can be expected in the future.

Transgenic potato varieties offer the possibility of increasing potato productivity and production, as well as creating new opportunities for non-food industrial use. However, all biosafety and food safety aspects must be carefully assessed and addressed before their release.

14.6.2   Organic potatoes

Organic potatoes can be grown on a large-scale without commercial pesticides and standard fertilizers. However, production costs for organic potatoes are higher and their yields are lower than for conventionally produced potatoes. Whether prices for organic potatoes can be high enough to offset these costs remains a question.

To demonstrate the production and profitability of commercial organic potatoes, the researchers used two 25- to 30-acre potato fields in Coloma, Wisconsin. The researchers wanted not only to demonstrate commercial organic potato production, but also to determine crop rotation systems best suited to potato production.

In Wisconsin, farmers grow potatoes on light, sandy soils under irrigation and rely heavily on purchased chemical pesticides and fertilizers. Since sandy soils are typically more permeable than heavier soils, many people worry about these chemicals passing through the soil and contaminating the groundwater.

The research team substituted cultural and biological inputs for synthetic chemical inputs. By not putting the chemicals on the land, they hoped to eliminate the negative effects on the environment. It was found that organic potato production removes the environmental concerns associated with pesticides and nutrients.

Superior, Red Norland, and Norkotah Russet varieties were grown in 10-acre blocks according to organic labeling requirements. Researchers closely monitored all crop inputs and operations.

The team used mechanical and cultural practices to control weeds, insects, and potato diseases. Early planting and harvest eliminated or reduced such problems as potato leaf hopper infestation, green peach and potato aphid infestation, and early and late blight.

Also, they planted the potatoes in fields that were planted to alfalfa or corn the year before to reduce nitrogen purchases (alfalfa) and potato diseases (alfalfa and corn) that carry over from year to year.

Section V
Meat, fish and poultry products

Biotechnology of fermented meat

15.1   Introduction

Meat is one of the most valuable and demanding food products. Worldwide meat consumption is growing and the variety of products available as convenience foods is on the increase.

Meat processing in the East European countries, Asia and South America continues on an upward path. These are all factors that contribute to the need for efficient, fully optimized meat processing.

The continuously growing focus on automation along the entire meat processing chain is closely linked to the growing use of information technology. Hygiene and traceability are of crucial importance, requiring full monitoring of all processes along with the continuous documentation of origin and production data, in the interests of consumer protection.

There are a wide variety of meat products that are attractive to consumers because of their characteristic colour, flavour and texture.

The need to improve the processing and quality of these meat products prompted research in the last decades on endogenous enzyme systems that play important roles in these processes, as has been later demonstrated.

It is important to remember that the potential role of a certain enzyme in a specific observed or reported biochemical change can only be established if all the following requirements are met: (i) the enzyme is present in the skeletal muscle or adipose tissue, (ii) the enzyme is able to degrade *in vitro* the natural substance (i.e. a protein in the case of a protease, a triacylglycerol in the case of a lipase, etc.), (iii) the enzyme and substrate are located closely enough in the real meat product for an effective interaction and (iv) the enzyme exhibits enough stability during processing for the changes to be developed.

Muscle enzymes: There are a wide variety of enzymes in the muscle. Most of them have an important role in the *in vivo* muscle functions, but they also serve an important role in biochemical changes such as the proteolysis and lipolysis that occur in postmortem meat and during further processing of meat. Some enzymes are located in the lysosomes, while others are free in the cytosol or linked to membranes. The muscle enzymes with most important roles during meat processing are grouped by families and described below.

Muscle proteases: Proteases are characterised by their ability to degrade proteins and they receive different names depending on respective mode of action. They are endoproteases or proteinases when they are able to hydrolyse internal peptide bonds, but they are exopeptidases when they hydrolyse external peptide bonds either at the amino termini or the carboxy termini.

Lysosomal proteases: Lysosomes contain a variety of proteolytic enzymes including cathepsins. The lysosomal cathepsins are present in all mammalian cell types, with the exception of enucleated red blood cells. The concentration of lysosomes, and hence the proteases, varies in different cells and tissues and is particularly high in liver, spleen, kidney and macrophages.

Proteasome complex: The proteasome is a multicatalytic complex with different functions in living muscle. This enzyme has shown degradation of some myofibrillar proteins.

Exoproteases (peptidases): There are several peptidases in the muscle with the ability to release small peptides of importance for taste. Tripeptidyl-peptidases (TPPs) are enzymes capable of hydrolysing different tripeptides from the amino termini of peptides, while dipeptidyl peptidases (DPPs) are able to hydrolyse different dipeptide sequences.

Exoproteases (aminopeptidases): There are five aminopeptidases, known as leucyl, arginyl, alanyl, pyroglutamyl and methionyl aminopeptioases, based on their respective preference or requirement for a specific N-terminal amino acid. They are able, however, to hydrolyse other amino acids, although at a lower rate.

Lipolytic enzymes: Lipolytic enzymes are characterised by their ability to degrade lipids and they receive different names depending on their mode of action. Lipases and esterases are located either in the skeletal muscle or in the adipose tissue.

Muscle lipases: Lysosomal acid lipase and acid phospholipase are located in the lysosomes. Both have an optimal acid pH (4.5–5.5) and are responsible for the generation of long-chain free fatty acids.

Glycolysis: Glycolysis consists in the hydrolysis of carbohydrates, mainly glucose, either that remaining in the muscle or that formed from glycogen, to give lactic acid as the end product. As lactic acid accumulates in the muscle, pH falls from neutral values to acid values around 5.3–5.8. The contribution of glycolysis to pH drop is restricted to a few hours postmortem, although it is also important in fermented meats where sugar is added for micro-organism growth.

Lipolysis: Lipolysis makes an important contribution to the quality of meat products by the generation of free fatty acids, some of which have a direct influence on flavour and others that, with polyunsaturations, may be oxidised to volatile aromatic compounds, acting as flavour precursors. In addition, the breakdown of triacylglycerols affects the texture of the adipose tissue and an excess of lipolysis/oxidation may contribute to the development of rancid aromas or yellowish colours in fat.

15.2   Fermented meat

Fermented meat is a type of dried meat that has been prepared and dehydrated according to specific techniques that alter some of the chemical make-up of normally perishable foods. Fermentation typically involves the introduction of bacteria or yeast that converts certain meat nutrients into mixtures of carbon dioxide and alcohol. This process of making fermented meat can both increase the flavour intensity and prevent the end product from spoiling for relatively long periods of time. Successful fermentation of meat can only be accomplished under specific environmental conditions in order for the end product to be completely safe to eat. Various culinary traditions often have their own unique types of fermented meat such as salami or chorizo sausage.

Meat must be fermented in an environment without oxygen, which is frequently called an anaerobic condition. Removing the oxygen is essential for the meat to correctly dehydrate without the possibility of harmful bacterial invasion. The fermentation catalyst bacteria are usually introduced in controlled amounts. This type of bacteria is usually one of several varieties derived from lactic acid. Cooks with experience in fermenting frequently use a strain of lactic acid bacteria called *pediococus cerevisiae*. This important step is normally accomplished by sealing the cuts of meat to be dried in an airtight container.

Various flavours of salami are normally made through this basic fermentation process. This kind of dehydrated sausage is often made with a combination of catalyst bacteria and other additives called nitrates. The inclusion of nitrates can often lessen the chances of contamination by allowing the beneficial lactic acid bacteria to grow and spread through the meat at a faster rate. Once the technique of fermenting meat has been perfected, the process is normally not limited to meats such as beef or pork. Some types of seafood can also be successfully fermented, although health professionals sometimes advise that this practice should be approached carefully. Fermented fish can often have high rates of bacterial contamination that can lead to serious food-borne illnesses such as botulism.

Some of the benefits of fermented meat include economic practicality and environmental necessity. People who live in remote geographic areas with harsh winters often ferment large stores of meat due to limited access to other food sources during the worst weather conditions. Fermented meats are generally able to stand up to extreme temperatures at a better rate than other foods.

15.3   Raw material preparation

Various factors are taken into account when producing fermented meats.

15.3.1   Ingredients

Lean meats from pork and beef, in equal amounts, or only pork are generally used. Quality characteristics such as colour, pH, and water-holding capacity are very important. Pork meat with another defect, known as PSE (pale, soft, and exudative), is not recommended because the colour is pale, and the sausage would release water too fast, which could cause casings to wrinkle. Meat from older animals is preferred because of its more intense colour, which is due to the accumulation of myoglobin, a sarcoplasmic protein that is the natural pigment responsible for colour in meat.

15.3.2   Other ingredients and additives

Salt is the oldest additive used in cured meat products since ancient times. Salt, at about 2–4%, serves several functions, including: (i) an initial reduction in a_w, (ii) providing a characteristic salty taste and (iii) contributing to increased solubility of myofibrillar proteins. Nitrite is a typical curing agent used as a preservative against pathogens, especially *Clostridium botulinum*. Nitrite is also responsible for the development of the typical cured meat colour, prevention of oxidation, and contribution to the cured meat flavour.

Carbohydrates like glucose and lactose are used quite often as substrates for microbial growth and development. Disaccharides, and especially polysaccharides, may delay the growth and pH drop rate because they have to be hydrolysed to monosaccharides by microorganisms.

Spices, either in natural form or as extracts, are added to give a characteristic aroma or colour to the fermented sausage. There are a wide variety of spices (pepper, paprika, oregano, rosemary, garlic, onion, etc.), each one giving a particular aroma to the product. Some spices also contain powerful antioxidants.

The most important aromatic volatile compounds may vary depending on the geographical and/or plant origin. For instance, garlic, which gives a pungent and penetrating smell, is typically used in chorizo, and pepper is used in salchichon and salami. Paprika gives a characteristic flavour and colour due to its high content of carotenoids. The presence of manganese in some spices, like red pepper and mustard, is necessary for the activity of several enzymes involved in glycolysis and thus enhances the generation of lactic acid.

15.3.3   Cultures

Cultures for food fermentations are selected primarily on the basis of their stability and their ability to produce desired products or changes efficiently. These cultures may be established ones obtained from other laboratories or may be selected after the testing of numerous strains. Stability is an important characteristic; yields and rates of changes must not be erratic. Some cultures may be improved by breeding, e.g. the sporogenous yeasts, but selection is the most commonly used method for the improvement of strains. Selection of cultures with desirable traits can be made from new strains isolated from the environment, from existing strains or following mutation of strains by various means.

Maintenance of activity of cultures: Once a satisfactory culture has been obtained, it must be kept pure and active. Usually this objective is attained by periodic transfer of the culture into the proper culture medium, incubation until the culture reaches the maximal stationary phase of growth, and then storage at temperatures low enough to prevent further growth. Too frequent transfer of an unstable culture may lead to undesirable changes in its characteristics.

Maintenance of purity of cultures: To ensure the purity of cultures, they should be obtained periodically from a culture laboratory or be checked regularly for purity. Methods for testing a culture for purity vary with the type of culture being tested. Microscopic examination will indicate contamination only if the contaminant differs from the desired organism in appearance and is high in numbers.

Preparation of cultures: Mother Culture is usually prepared daily from a previous mother culture and originally from the stock culture. These mother cultures can be used to inoculate a larger quantity of culture medium to produce the mass or bulk culture to be used in the fermentation process. Often, however, the fermentation is on such a large scale that several intermediate cultures of increasing size must be built up between the mother culture and the final bulk or mass culture. Culture makers attempt to produce and maintain a culture that: (i) contains only the desired micro-organism(s), (ii) is uniform in microbial numbers, proportions (if a mixed culture), and activity from day-to-day, (iii) is active in producing the products desired and (iv) has adequate resistance to unfavourable conditions if necessary, e.g. heat resistance, if it has to take heating in a cheese curd.

Activity of culture: The activity of a culture is judged by its rate of growth and production of products. It should be good if the mother or the intermediate culture is satisfactory and culture medium, incubation time, and temperature are optimal. Deterioration of cultures may result from improper handling and cultivation, frequent transfer over long periods in an inadequate culture medium, selection, variation or mutation, or attack of bacteria by a bacteriophage.

15.3.4   Lactic acid bacteria

Lactic acid bacteria (LAB) are essential agents during meat fermentation improving hygienic and sensory quality of the final product. Its fermentative metabolism prevents the development of spoilage and pathogenic microflora by acidification of the product, also contributing to its colour stabilization and texture improvement. LAB strongly influence the composition of non-volatile and volatile compounds through release/degradation of free amino acids and the oxidation of unsaturated free fatty acids is also prevented. In addition, certain LAB has also been found to produce bacteriocins being used as bioprotective culture to preserve fresh and processed meat and fish.

The typical flavour of dry cured fermented sausages is the result of a careful balance between volatile (alcohol, ketones, aldehydes and furans) and non-volatile compounds (amino acids, peptides, sugars and nucleotides), these coming from raw materials (meat, spices, nitrites and other additives) or generated from biochemical reactions occurring during fermentation and ripening. Non-volatile components provide basic tastes, such as sweet, sour, salty and bitter.

Many of non-volatile compounds such as peptides and amino acids involved in the characteristic flavour development are produced during meat protein hydrolysis. In this regard, meat LAB has a direct or indirect participation on this phenomenon, so its presence determines to a large extent, the sensory characteristics of the final product. Recently, proteomic approaches have been applied to correlate proteolytic profiles with technological parameters in view to detect valuable biomarkers to monitor and predict meat quality. Certain peptidic fractions originated from meat have been identified to be related to sensory attributes.

Trends in meat industry are focalizing in products with high organoleptic standards, long shelf-life and containing specific nutrients to cover special consumer requirements. In parallel, consumers are increasingly demanding pathogen free foods, with minimal processing and few preservatives and additives. Thus, biopreservation has gained increasing attention as means of naturally controlling the shelf-life and safety of meat products. Some microorganisms commonly associated with meat have proved to be antagonistic towards pathogenic and spoilage bacteria.

Based on these considerations, the most significant developments in which LAB are the main protagonists, acting both as starter cultures improving the sensorial quality and as functional starter cultures, are reviewed. Also, a proteomic approach is detailed as example of new means by which LAB are studied to obtain improved strains that will contribute to the quality and safety of meat products.

Role of LAB in meat protein degradation

In general, different authors agree about the importance of muscle proteinases in the initial breakdown of sarcoplasmic and myofibrillar proteins, these enzymes being the main agents of proteolysis during the first days of fermentation. Although the role of microbial enzymes in this event has been accepted with greater reluctance, LAB are endowed with proteolytic activity; mainly intracellular amino, di and tripeptidases. The intracellular enzymes of *Lactobacillus* were reported to be responsible for the generation of small peptides and amino acids which contribute to the process either as direct flavour enhancers or as precursors of other flavour compounds during the ripening of dry fermented sausages. Contrary to what happen for dairy *lactobacilli*, no extracellular proteases have been characterized so far from meat-borne *lactobacilli*.

To elucidate the origin of meat protein degradation during sausage fermentation several studies were carried out using different model systems, such as sausages added with antibiotic or with muscle protease inhibitors, as well as by using liquid meat-based media. In general, data suggest that the addition of protease inhibitors protects myosin, actin and troponin T to a greater extent than antibiotics indicating that in this process endogenous muscle proteases are more important than bacteria.

Proteolytic compounds generated during sausage fermentation

Argentinean dry fermented sausages are produced in different regions using local artisan techniques, these having a strong Spanish and Italian influence. The most common dry fermented sausages are salamis, fuet and milano-type sausages which include beef and/or pork meat, pork fat, salt, and different spices. Even when starter cultures have been applied by local fermented sausage industries, lack of uniformity in the products, inadequate raw materials as well as the need for quick economical incomes, lead to a lower quality products comparing to the European ones.

Even when muscle enzymes are greatly involved in meat protein degradation, bacterial proteolytic activity leads to a richer composition of small peptides and amino acids which contribute to the ripening process either as direct flavour enhancers or as precursors of other flavour compounds. Fermentation temperatures of commercial lactic acid bacteria are shown in Table 15.1.

Table 15.1 Fermentation temperatures of commercial lactic acid bacteria

Name	Temperature range in °C
Lactobacillus curvatus	22–37
Lactobacillus farciminis	22–32
Lactobacillus plantarum	25–35
Lactobacillus sakei	21–32
Pediococcus acidilactici	25–45
Pediococcus pentosaceus	20–37

LAB strategies to cope survive and grow in meat environments

Nowadays, it is accepted that an appropriate starter culture have to be selected from indigenous populations present in a food product in order to be more competitive, well-adapted, and with high metabolic capacities to beneficially affect quality and safety while preserving their typicity. Indeed, research of this technological potential and its adaptive mechanisms to a particular environment has been of outmost interest for the industry of starter cultures. Variations in temperature, osmotic conditions, oxidative or acidic environments are situations to which LAB are routinely subjected during its passage through the gastrointestinal tract or during food processing. Thus, different approaches were used to understand adaptation of bacteria to stressful environments.

Contribution of selected LAB to meat conditioning

It is known that meat is an excellent substrate for bacterial growth and if some methods to restrict their presence are not applied, meat becomes contaminated. A widely used practice in meat marketing is vacuum packaging of primal cuts for distribution and extension of shelf-life. As a result of subsequent practices such as removal of meat from their packages, it can become contaminated with pathogens. The new trends of meat industry involve the use of a new generation starter cultures with industrial or nutritional important functionalities. Functional starter cultures contribute to food safety by producing antimicrobial compounds such as bacteriocins and also provide sensorial, technological, nutritional and/or health advantages.

LAB peptides involved on safety of meat and meat products

The preservative ability of LAB in foods is attributed to their competition for nutrients and to the production of antimicrobial metabolites including organic acids, hydrogen peroxide, enzymes and bacteriocins. Bacteriocins are a heterogeneous group of peptides and proteins ribosomally produced that kill or inhibit the growth of close related microorganisms such as food pathogens (*Listeria monocytogenes* and *Clostridium*) and other meat spoilage bacteria such as *B. thermosphacta*.

Among the many benefits bacteriocins offer, while extending shelf-life of foods, are: (i) an extra protection during temperature abuse conditions, (ii) decrease the risk for transmission of foodborne pathogens through the food chain, (iv) minimization of the economic losses due to food spoilage, (v) reduction of chemical preservatives application, (vi) possibility of less severe heat treatments application, this implying a better preservation of food nutrients and vitamins, and keeping organoleptic properties of foods and (vii) marketing of 'novel' foods (less acidic, with a lower salt content, and with a higher water content).

Bacteriocins can be introduced into food in at least four different ways: (i) as purified or semi-purified bacteriocins added directly to food, (ii) as bacteriocins produced in situ by bacterial cultures that substitute all or part of the starter culture, (iii) as an ingredient based on a fermentate of a bacteriocin-producing strain and (iv) immobilized as constituent of a bioactive food packaging film.

Strategies to improve bioprotective effect in meat processing

The effectiveness of bacteriocins in food systems should consider the adequacy of the environment for bacteriocin stability and/or production. A loss of activity is often detected when bacteriocins are added to meat and meat products, this being a direct consequence of additives (nitrites, NaCl and spices) or meat components, such as proteolytic enzymes, bacteriocin-binding proteins or fat particles.

Thus, *Lactobacilli* play a key role in industrial and artisan food fermentation, including meat products.

They contribute to raw-material preservation due to acidification and to a production of antimicrobial compounds such as bacteriocins, but also because of their capacity to contribute to product characteristics such as flavour and texture.

15.4.4   Micrococcaceae

This group consists of *Staphylococcus* and *Kocuria* (formerly *Micrococcus*), which are major contributors to flavour due to their proteolytic and lipolytic activities.

15.4.5   Yeast

Yeasts are eukaryotic microorganisms classified in the kingdom Fungi, with the 1500 species currently described estimated to be only 1 per cent of all yeast species. Most reproduce asexually by budding, although a few do so by mitosis. Yeasts are unicellular, although some species with yeast forms may become multicellular through the formation of a string of connected budding cells known as pseudohyphae or false hyphae, as seen in most molds. Yeast size can vary greatly depending on the species, typically measuring 3–4 µm in diameter, although some yeast can reach over 40 µm.

The yeast species *Saccharomyces cerevisiae* has been used in baking and in fermenting alcoholic beverages for thousands of years.

15.4.6   Molds

The most used molds are *Penicillium nalgiovense* and *P. chrysogenum*. They contribute to flavour through their proteolytic and lipolytic activity, and to appearance in the form of a white coating on the surface. Inoculation of sausages with natural molds present in the fermentation room is dangerous because toxigenic molds might grow.

15.4.7   Casings

Casings may be natural, semisynthetic, or synthetic, but a common required characteristic is its permeability to water and air. Natural casings are natural portions of the gastrointestinal tract of pigs, sheep, and cattle, and although irregular in shape, they have good elasticity, tensile strength, and permeability. Natural casings are typically used for traditional sausages because they give a homemade aspect to the product.

15.5   Fermented sausages

Fermented sausages are cured sausages and to produce salami of a consistent quality one must strictly obey the rules of sausage making. This field of knowledge has been limited to just a few lucky ones but with today's meat science and starter cultures available to everybody, there is little reason to abstain from making quality salamis at home. It is unlikely that a home sausage maker will measure meat pH (acidity) or a_w (water activity) but he should control temperatures and humidity levels in his drying chamber. Figure 15.1 shows the processing of fermented sausages.

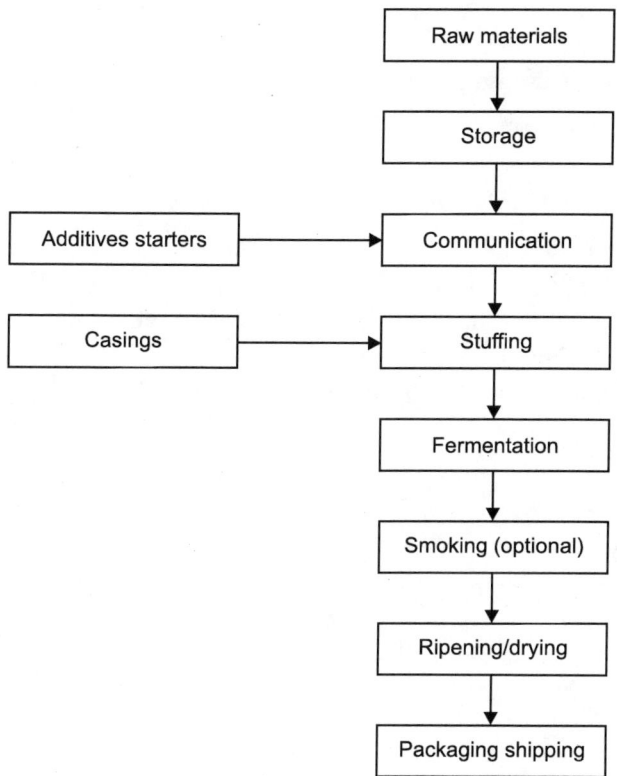

Figure 15.1 Flow diagram showing the processing of fermented sausages

There is a difference in fermented sausage technology between the United States and the European countries. American methods rely on rapid acid production (lowering pH) through a fast fermentation in order to stabilize the sausage against spoilage bacteria. Fast acting starter cultures such as *Lactobacillus plantarum* and *Pediococus acidilactici* are used at high temperatures up to 40°C (104°F). As a result pH drops to 4.6, the sausage is stable but the flavour suffers and the product is sour and tangy. In European countries, the temperatures of 22–26°C (72–78°F) are used and the drying, instead of the acidity (pH) is the main hurdle against spoilage bacteria which favours better flavour development. The final acidity of a traditionally made salami is low (high pH) and the sourly taste is gone.

Some known European sausages are French saucisson, Spanish chorizo, and Italian salami. These are slow-fermented sausages with nitrate addition and moderate drying temperatures. North European sausages such as German or Hungarian salamis are made faster, with nitrite addition and are usually smoked.

Fermented sausages can be divided into two groups:

1. Sliceable raw sausages (Salami, Summer Sausage, Pepperoni)
2. Spreadable raw sausages (Teewurst, Mettwurst)

Depending on the manufacturing method, they can be divided into:

- Fast-fermented
- Medium-fast-fermented
- Slow-fermented. These can be smoked or not, or made with mold or without mold.

Depending on the amount of moisture that they contain, they can be grouped as:

- Moist – 10% weight loss
- Semi-dry – 20% weight loss
- Dry – 30% weight loss

There is also a group of non-fermented cooked salamis that are made in many European countries. This group will cover any sausage that is smoked, cooked and then air-dried for 1–3 weeks at 10–12°C (50–52°F). This reduces a_w (water activity) to about 0.92 which makes the product shelf stable without refrigeration.

15.5.1   Magic behind fermented sausages

Making fermented sausages is a combination of the art of the sausage maker and unseen magic performed by bacteria. The friendly bacteria are working together with a sausage maker, but the dangerous ones are trying to wreak havoc. Using his knowledge the sausage maker monitors temperature and humidity, which allows him to control reactions that take place inside the sausage. This game is played for quite a while and at the end a high quality product is created.

We all know that meat left at room temperature will spoil in time and that is why it is kept in a refrigerator/freezer. Yet fermented or air dried sausages are not cooked and don't have to be stored under refrigeration. What makes them different? Fermented sausages and air dried meats are at an extra risk as in many cases they are not subject to cooking/refrigerating process. In a freshly filled with meat casing, bacteria seem to hold all advantages: temperatures that favour their growth, moisture, food (sugar), oxygen, we have to come up with some radical solutions otherwise we might lose the battle. Fortunately, meat science is on our side and what was a secretive art for many years is being revealed and made accessible to everybody today. Even so, manufacture of fermented products is still a combination of an art and technology.

A meat processing facility develops its own microbiological flora in which bacteria live all over the establishment (walls, ceilings, machinery, tools, etc.). Each place will have its own peculiar flora and some places will contain more bacteria which are needed for making fermented sausages. These bacteria are just waiting to jump on a new piece of meat and start working in. All they need is a bit of food: moisture (meat is 75% water), oxygen (the air) and sugar (meat contains sugar). Sugar has been introduced into sausage recipes for hundreds of years as somehow we have always known that it is needed. Some places in Italy had inside flora which was beneficial to produce a product of a great quality and they suddenly developed a name for making wonderful meat products. They were probably not better sausage makers than their counterparts working in different locations. Most likely they were lucky to have their shop located in their area which was blessed by Mother Nature for making fermented sausages. They did not have much clue to what was happening, this empirical knowledge was passed from father to son but it worked like magic.

Fermented sausages are made by use of 'beneficial or friendly' bacteria that we manipulate to our advantage and they become microscopic labourers performing tasks that we cannot do ourselves. All this bacteria talk should not alarm anybody as we are surrounded with fermented foods: sour bread, wine, yogurt, sourkraut, cheeses, etc. Some of the most dangerous bacteria (*E. coli* and *Listeria monocytongenes*) live in our digestive tract and help us to digest foods; other *(Staph.aureus)* are present in our skin, mouth and nose. The most toxic poison known to men is in soil *(Clostridium botulinum)* and we touch those spores every time when working in the garden, yet we are perfectly fine. Dangerous bacteria are present in meat and we eat them every time when undercooked meat is served and that does not seem to affect us either. This is due to the small number of bacteria present and if their number were higher our immune system would not be able to fight them off.

In regards to sausage making we can divide bacteria as.

Spoilage

These bacteria cause food to deteriorate and develop unpleasant odours, tastes and textures. These one-celled microorganisms can cause fruits and vegetables to get mushy or slimy, or meat to develop a bad odour. Most people would not choose to eat spoiled food. However, if they did, they probably would not get sick.

Dangerous

These are known as pathogenic bacteria and they cause illness. They grow rapidly in the 'Danger Zone' the temperatures between 40°F and 140°F (4–60°C) and do not generally affect the taste, smell, or appearance of food. Food that is left too long at unsafe temperatures could be dangerous to eat, but smell and look just fine. *E. coli* 0157:H7 (Most illness has been associated with eating undercooked, contaminated ground beef), *Campylobacter* (Most cases of illness are associated with handling raw poultry or eating raw or undercooked poultry meat), and *Salmonella* (*Salmonella* is usually transmitted to humans by eating foods contaminated with animal feces).

They are often of animal origin, such as beef, poultry, milk or eggs. Chicken meat is known to contain more *Salmonella* than other meats. *Staph. aureus* is hard to control and to inhibit its growth a_w must be lower than 0.89 and pH below 5.2. *Cl. botulinum,* very toxic and heat resistant, likes moisture but hates oxygen.

Beneficial

These bacteria can be managed to our advantage to produce fermented sausage. They are naturally occurring in meat and are responsible for:

- Converting nitrate to nitrite: *(Micrococcus, Staph.xylosus, Staph. carnosus)*
- Improving flavour: *(Micrococcus)*
- Increasing acidity (lowering pH) by producing lactic acid through sugar metabolism: *(Pediococcus and Lactobacillus)*
- Mold growth: (Penicillium nalgiovense) which is highly desired in some Italian salamis

To produce a quality safe product, it is necessary to:

- Prevent the growth of spoilage bacteria
- Prevent the growth of dangerous bacteria
- Create favourable conditions for the growth of beneficial bacteria

To eliminate the risk of bacteria growth and to prevent meat spoilage we can employ the following steps, also known as 'hurdles':

1. Using meats with a low bacteria count
2. Curing – adding salt and sodium nitrite/nitrate
3. Lowering pH of the meat to 5.2
4. Lowering a_w (water activity) by drying to 0.89
5. Smoking

15.5.2   Meat selection

Meat must be perfectly fresh with the lowest count of bacteria possible. Commercial producers try to keep this number between 100 and 1000 per gram of meat but a home-based sausage maker has to make sure that:

- Meat is very fresh and kept cold.
- Facilities and tools are very clean.
- Working temperatures are as low as possible.

If the above conditions are not met bacteria will multiply and will compete for food with starter cultures inhibiting the growth. As a result, the product will be faulty, especially the slow-fermented salami. It makes little difference to what kind of meat is used and salamis can be made from pork, beef, venison, buffalo, horse and other meats. If chicken is used the thigh (dark meat) will be the better choice than a breast (light meat). Chicken breast being light meat contain little myoglobin that reacts with nitrite to produce curing colour and the finished sausage will have a very light pink colour unless chicken meat will be mixed with darker meats. Keep in mind that chicken meat has higher chances to be infected with *Salmonella* than other types of meat. Also remember that raw pork or venison meat may be infected with trichinosis.

15.5.3   Curing

The application of salt and nitrite is actually our first line of defence against the growth of spoilage bacteria as in many cases, there is very little we can do about a selected meat's bacteria count except making sure that the meat is fresh. As the sausage slowly dries out it loses moisture but not the original amount of salt which remains inside. As a result, in time the sausage becomes much saltier to bacteria. In about 3–6 days, a_w drops to about 0.95 and the sausage is microbiologically more stable as some pathogenic bacteria (for example *Salmonella*) stop multiplying now.

Using a combination of different hurdles is more effective that relaying on one method only. For example the first hurdle is an application of salt and sodium nitrite which eliminates some of the microbiological spoilage. This will not be enough to produce a stable sausage if we don't follow up with additional hurdles such as lowering pH (increasing acidity) and then lowering water activity a_w (eliminating moisture by drying).

Most bacteria can tolerate water activity levels (a_w) up to 0.92. For example *Clostridium botulinum* (food poisoning) bacteria are active all the way down to 0.93 a_w. To bring water activity level down to 0.93 level about 10% salt solution is needed. We will have to add 100 g of salt to 1 kg of meat to be sure that *Clostridium* growth will be inhibited. But any salt level above 3% will make meat unpalatable to most people so salting alone will not cut it. Additional hurdles such as lowering water activity and lowering of pH (increasing acidity) will have to be implemented. The more salt applied to meat, the stronger fence is created against bacteria and some compromise has to be made as the salt plays a very important role in preventing bacteria growth, especially in the first stages of a process.

15.5.4   pH – acidity

Foods with a low pH value (high acidity) develop resistance against microbiological spoilage. Pickles, sourkraut, eggs, pig feet – anything submerged in vinegar will have a long shelf-life. Even ordinary meat jelly (headcheese) will last longer if some vinegar is added and this type of headcheese is known as 'souse'. Bacteria hate acidic foods and this fact plays an important role in the production and stabilization of fermented sausages. Ideally the pH value of meat to be used for making fermented products should be below 5.8.

Sugar, glucono-lactone (GDL) and citric acid

Sugar, Glucono-lactone (GDL) and citric acid are important additives in the manufacturing of fast and medium-fermented salamis as in these sausages pH reduction (increasing acidity) is the main hurdle against bacteria growth. In slow-fermented sausages which are dried for a much longer time, lowering moisture (a_w) is the main hurdle employed to inhibit bacteria growth.

Glucono-delta-lactone is manufactured by microbial fermentation of pure glucose to gluconic acid but is also produced by the fermentation of glucose derived from rice. It is soluble in water and is non-toxic and completely metabolized in our bodies. It can be found in honey, fruit juices, wine and many fermented products. It is a natural food acid (it has roughly a third of the sourness of citric acid) and it contributes to the tangy flavour of various foods. Since it lowers the pH, it also helps preserve the food from deterioration by enzymes and microorganisms. It is metabolized to glucose; one gram of GDL is equivalent to one gram of sugar. Glucono-Delta-Lactone is often used to make cottage cheese, Tofu, bakery products and fermented sausages.

About 1 g (0.1%) of GDL per 1 kg of meat lowers pH of meat by 0.1 pH. It shall be noted that the addition of sugar already lowers the pH of the meat and adding GDL will lower the pH even more. As it is a natural acid, adding more than 10 g may cause a bitter and sour flavour.

Citric acid is a weak organic acid found in citrus fruits. It is a natural preservative and is used to add an acidic (sour) taste to foods, soft drinks and wine. In lemons and limes it can account for as much as 8% of the dry weight of the fruit. Citric acid is more for its informational value in lowering pH than by its practical usefulness in making fermented sausages. It acts about three times faster than GDL (1 g of citric acid added to 1 kg of meat lowers pH of meat by about 0.3 units) and in higher doses it will contribute to a sour taste. Its usefulness is therefore strictly limited.

Sugar is mainly added to provide food for starter cultures. The pH drop in sausage depends on the type and amount of sugar utilized. Introduction of more sugar generally leads to lower pH and stronger acidification. What is notable is that lactic bacteria process different sugars differently. Only dextrose (glucose) can be fermented directly into lactic acid and by all lactic bacteria. Other sugars molecular structure must be broken down until monosaccharides are produced and this takes time and some lactic bacteria are more effective than others. Sugar introduction also helps to offset the sourly and tangy flavour of fast and medium-fermented sausages and acts as a minor hurdle in lowering water activity. The types of sugar which may be used in making fermented sausages are listed in order of their importance on producing lactic acid by lactic acid bacteria:

- *Glucose:* 'Dextrose' is glucose sugar refined from corn starch which is approximately 70% as sweet as sucrose but it has an advantage of being directly fermented into lactic acid and is the fastest acting sugar for lowering pH. As lowering pH is the main hurdle against bacteria growth in fast-fermented sausages, dextrose is obviously the sugar of choice. It can be easily obtained from all sausage equipment and supplies companies.

- *Sucrose:* Common sugar (also called saccharose) is made from sugar cane and sugar beets but also appears in fruit, honey, sugar maple and in many other sources. It is the second fastest acting sugar. It can be used with GDL in medium-fermented sausages. In slow-fermented sausages common sugar should be chosen as it has been used for hundreds of years. There is no need to lower fast pH and sugar contributes better to a strong curing colour and better flavour.

- *Maltose:* Malt sugar is made from germinating cereals such as barley, is an important part of the brewing process. It's added mainly to offset sour flavour and to lower water activity.
- *Actose:* Actose also referred to as milk sugar is found most notably in milk. Lactose makes up around 2–8% of milk (by weight). Maltose and lactose are less important as primary fermenting sugars but may be used in combinations with common sugar to bring extra flavour.

About 1 g (0.1%) of dextrose per 1 kg of meat lowers pH of meat by 0.1 pH. This means that 10 g of dextrose added to meat with initial pH value of 5.9 will lower pH by one full unit to 4.9. Sugar levels of 0.5–0.7% are usually added for reducing pH levels to just under 5.0.

When using acidification as a main safety hurdle, salami is microbiologically stable when pH is 5.2 or lower and this normally requires about 48 hrs fermentation time for fast-fermented product and 72–96 hours for medium-fermented type. In slow-fermented salami pH does not drop lower than 5.5 but the sausage is microbiologically stable due to its low moisture level (prolonged drying).

15.5.5   Water activity (a_w)

Water activity is an indication of how tightly water is 'bound' inside of a product. It does not say how much water is there, but how much is available to support the growth of bacteria, yeasts or molds (fungi). Adding salt or sugar 'binds' some of this free water inside of the product and lowers the amount of available water to bacteria which compete very poorly with salt. Molds are very good competitors for free water. We could make a_w lower by lowering the temperature of the product but that is not practical as the temperatures for making fermented sausages are well defined. A much better solution is to lower water activity by drying.

- Air drying is the process employed in lowering water activity (moisture removal) and has to be properly controlled otherwise it may lead to a number of defects including a total loss of product.
- During the long drying process of salami, the original hurdles lose some of their original strength as the nitrite is depleted and the number of lactic-acid bacteria decreases and the pH increases. This is offset by drying which lowers water activity by removing moisture and the sausage becomes more stable in time.
- When using drying as a main safety hurdle, salami is microbiologically stable when a_w is 0.89 or lower.
- The drying chamber should not be overloaded as a uniform air draft is needed for proper drying and mold prevention.

15.5.6   Smoking

Smoking may or may not be utilized in a production of fermented sausages. It has been used in countries in Northern Europe where due to colder climate and shorter seasons, the drying conditions were less favourable than in Spain or Italy. Smoking imparts a different flavour, has some effect of fighting bacteria, especially on the surface of the product and thus prevents growth of molds on fermented sausages. Mold is desired on some traditionally made Italian salamis and obviously smoking is not deployed. Most consumers prefer sausages without mold on the surface, and smoking is an old method to prevent it. As mold can already grow in the first days of fermentation, it is recommended to smoke sausages at the early stage of production. In order not to unbalance ongoing fermentation, the temperature of the smoke should approximate the temperature of the fermentation chamber which in raw slow fermented sausages falls into 18–22°C, (66–72°F) range. For the same reason, cold smoking will be applied to semi-dry sausages which will be fermented but not cooked.

Cold smoking is performed with a thin smoke, 70–80% humidity, and good air ventilation to remove excess moisture. Cold smoking is drying with smoke and should be interrupted by drying periods without smoke. Often recipes call for 3–4 days of cold smoking, but that does not mean that the smoking is continuous. Heavy continuous smoke application for such a long period may impart a bitter taste to the product. All raw fermented sausages which are not subject to heat treatment and which are smoked must be smoked with cold smoke.

Warm smoking (25–50°C, 77–122°F) will be applied to semi-dry sausages which will be cooked. If sausages are fully cooked (68–71°C, 154–160°F), the hot smoke (60–80°C, 140–176°F) may be administered. Smoke may be applied late during the fermentation stage (preferably after) when the surface of the sausage is dry. If mold appears before smoking was performed, or shortly after, the sausage should be rinsed and wiped off and then smoke may be applied again. Applying smoke for about 4 hours after fermentation prevents mold growth but only for some time.

Fermented sausages which will be cooked may be smoked with warmer smoke, which will increase the core temperature of the sausage and will shorten the cooking process. If mold reappears later, the sausage is wiped off and the smoke is reapplied again. Applying heavy smoke early during the fermentation stage is not the best idea as smoke contains many ingredients (phenols, carbonyls, acids, etc.) which may impede reactions between meat and beneficial bacteria, especially in the surface area. After hot smoking/ cooking, shower sausages with hot water (removes grease and soot), then with cold water and then transfer them to storage. At home, the sausages are normally smoked/cooked outside and they are cold showered only. This prevents them from shrivelling and shortens the meat's exposure to high temperatures.

Most semi-dry sausages are smoked, many with cold smoke. Traditionally made slow fermented sausages (Hungarian salami) are cold smoked. In the past cold smoking was firstly a preservation step with a benefit of better flavour. Today, cold smoking is seldom performed as a preservation step due to the widespread use of refrigeration. Some foods, notably cold smoked salmon (lox) are still made with cold smoke but the majority of regular sausages are hot smoked today. Think of cold smoke as a part of the drying/fermentation cycle and not as the flavouring step. If the temperature of the smoke is close to the fermentation temperature, there is very little difference between the two. The sausage will still ferment and the drying will continue and the extra benefit is the prevention of mold that would normally accumulate on the surface. Cold smoking is performed with a dry, thin smoke. Applying heavy smoke for a long time would definitely inhibit the growth of colour and flavour-forming bacteria which are so important for the development of flavour in slow-fermented sausages. As drying continues for a long time and cold smoking is a part of it, it makes little difference whether cold smoke is interrupted and then re-applied again. In traditional smokehouses the fire was started in the morning and burning logs produced smoke until late night hours till the fire died out. Then it was re-ignited again and the smoke continued. The drying temperature falls into 15–18°C (59–64°F) range and cold smoke (< 22°C, 72°F) fits nicely into this range. To sum it up, the length of cold smoking is loosely defined, but the upper temperature should remain below 22°C (72°F). Unfortunately, this rule puts some restraints on making slow-fermented sausages in hot climates for most of the year, when using an outside smokehouse. You can't produce cooler smoke than the ambient temperature around the smokehouse, unless some cooling methods are devised. By the same token, people living in cooler climates can make those sausages for most of the year. Semi-dry sausages, which are of fast-fermented type, are fermented at higher temperatures. These sausages can be smoked with warmer smoke as they are subsequently cooked.

15.5.7   Manufacturing technology

The first manufacturing steps such as meat selection, grinding, mixing and stuffing are common to all sausages whether fresh, smoked or fermented types. The main difference is that no water should be added to meat during processing as water is the necessary nutrient for bacteria. The technology of making dry sausages relies on removal of water and not on bringing water in. After being stuffed with meat the fermented sausages are submitted to:

- Conditioning (optional)
- Fermenting
- Drying
- Storing

Conditioning: Conditioning is an optional step for a home sausage maker as he has to exercise his own judgment. In commercial plants the process of grinding, mixing and stuffing salami is undertaken at a low temperature (0°C, 32°F) and as the cold sausage is placed in a warmer (fermenting/drying) room, not needed condensation will appear on the surface of the sausage. The sausage must remain there for 1–6 hours (depending on its diameter) at low humidity (no air draft) until the moisture evaporates. Then we can start the fermentation process.

If the sausage casing is dry, there is no need for conditioning. It should also be very carefully monitored (or even eliminated) in small diameter casings which can dry out too quickly on the surface. This will eliminate moisture (food) for lactic bacteria and they will not reduce pH within the outer layers. As a result the sliced sausage will have a different colour in its outer layer.

Fermentation: Fermentation refers to the production of lactic acid and to produce consistent quality product parameters such as temperature, humidity and air flow should be carefully monitored. The humidity in a drying room is increased to about 92–95 % and the temperature is increased to 18°–26°C, (66–78°F). The temperature range depends on the type of the sausage produced (fast, medium or slow-fermented) and the type of the starter culture used. The air flow is kept about 0.8 m/s. Commercial plants monitor a_w (water activity) of the sausage and readjust the correspondingly humidity level of the drying chamber. There is normally a difference of less than 5% between moisture level of the sausage and relative humidity of the room, the latter figure being lower. This means if the a_w of the sausage is 0.95, the humidity is set at 90%. Then when a_w drops to 0.90, the humidity drops to 85% and so on.

When the fermentation starts the main hurdles against microbiological spoilage of the sausage are the low bacteria count of the meat, the presence of nitrite and salt. Keep in mind that in time the sausage will be losing more and more moisture but the salt remains inside and the percentage of salt in a finished sausage will be higher. In about 48 hours lactic bacteria metabolize enough sugar to produce a sufficient amount of lactic acid to drop pH (increase acidity) of the sausage and this stabilizes the sausage making it more resistant to spoilage.

Lactic acid producing bacteria widely used in starter cultures are:

- *Lactobacillus: Lb. sakei, Lb. plantarum, Lb. farcimis, Lb. curvatus*
- *Pediococcus: Pediococcus pentosaceus, Pediococcus acidilactici*

They all have different recommended growth temperatures and can be optimized for making fast or slow-fermented sausages. In the USA fast fermented sausages dominate the market and *Pediococcus acidilactici* is widely used as it allows fermentation at temperatures as high as 45°C (114°F).

15.5.8   Drying

Drying is a very important process especially in the initial stages of production. One may say why not to dry a sausage very fast which will remove moisture and be done with all this pH stuff and bacteria. Well, there are basically two reasons:

- The outside layer of the sausage must not be hardened as it may prevent removal of the remaining moisture. It may affect the curing of the outside layer which will become visible when slicing the sausage.
- Naturally existing in meat, bacteria and/or introduced starter cultures need moisture and some time before they can metabolize sugar and produce lactic acid which lowers pH. They are not going to multiply in one second and start heavy production of acid. Similar to yeasts used to ferment wine, these bacteria need some time to accommodate themselves in this new environment, and keep on eating sugar.

Even if we could rapidly dry out the sausage without hardening its surface this will inhibit beneficial bacteria from doing their work by removing moisture which they need. The only possibility will be to lower pH using chemical reactions such as adding GDL or citric acid. This method does not depend on bacteria but unfortunately it will add so much acidity that the product will not be edible. Moisture removal during fermentation (it is a part of drying) must proceed slowly.

Water activity (a_w) can be lowered faster in a sausage which contains more fat than a leaner sausage. Fat contains only about 10% water and a fatter sausage having proportionally less meat also contains less water. It will dry out faster.

Drying basically starts already in the fermentation stage and the humidity is kept at a high level of about 92%. Air flow is quite fast (0.8 m/s) to permit fast moisture removal but the high humidity level moisturizes the surface of the casing preventing it from hardening. After about 48 hours the fermentation stage ends but the drying continues to remove more moisture from the sausage.

Maintaining previous fast air flow may harden the surface of the casing so the air speed is decreased to about 0.5 m/s (1.8 miles/per hour slow walk). The temperature is lowered to create less favourable conditions for the growth of bacteria. At this time the medium-fermented sausage will be finished.

Slow-fermented sausages require additional drying time and the humidity is lowered again just to be a few per cent lower than the moisture content of a sausage and that falls into 75–80% range. Air flow is decreased again to about 1 ft/s. The temperature is lowered to 15°C (60°F) to create less favourable conditions for the growth of bacteria. At those conditions the sausage will remain in a drying chamber for an additional 4–8 weeks, depending on the diameter of the casing.

Sausage is microbiologically stable and can remain at the above settings for a very long time. It should be kept in a dark room which will prevent colour change and fat rancidity. There is very little need for the air flow now and it can be kept to the minimum. Some air flow is welcome as it inhibits formation of mold. The temperature is set to about 10–15°C (50–60°F). The humidity should remain at about 75% as lower humidity will increasing drying and the sausage will lose more of its weight. Much higher humidity levels may create favourable conditions for development of mold. If any mold develops it can be easily wiped off with a solution of water and vinegar. The sausage can also be cold smoked for a few hours which will inhibit the growth of a new mold. At these temperatures and humidity levels, the sausage has an almost indefinite shelf life. Depending on the method of manufacture (drying time), diameter of a casing and the content of fat in a sausage mass, fermented sausages lose from 5–40% of its original weight.

Air speed is a factor that helps remove moisture and stale air, and of course it influences drying. Sausages will dry faster at higher temperatures, but in order to prevent the growth of bacteria, drying must be performed at lower levels, generally between 15°C and 12°C (59–53°F). The speed of drying does not remain constant, but changes throughout the process: it is the fastest during the beginning of fermentation, and then slows down to a trickle. At the beginning of fermentation humidity is very high due to the high moisture content of the sausage. When starter cultures are used the temperature is at the highest during fermentation which speeds up moisture escape from the sausage. The surface of the sausage contains a lot of moisture and it must be constantly removed otherwise slime might appear.

If the sausages are soaking wet during fermentation, the humidity should be lowered. At the beginning of fermentation the fastest air speed is applied, about 0.8–1.0 m/s. The speed of 3.6 km/h (2.2 mile/h) corresponds to the speed of 1 m/s. Ideally, the amount of removed moisture should equal the amount of moisture moving to the surface.

Fermentation is performed at high humidity (92–95%) to prevent case hardening. If the humidity were low and the air speed fast, the moisture would evaporate from the surface so fast, that the moisture from the inside of the sausage would not make it to the surface in time. The surface of the casing will harden, creating a barrier to the subsequent drying process. In slow fermented sausages this will create a big problem as the inside of the sausage may never dry out and the product will spoil. As the sausage enters the drying stage, less moisture remains inside and the humidity and air speed are lowered. After about a week the air speed is only about 0.5 m/s and after another week it drops to 0.1 m/s (4 in./s). It will stay below this value for the duration of the drying.

Fast moisture removal is not beneficial in fast-fermented sausages, either. Lactic acid bacteria need water to grow and if we suddenly removed this moisture, they would stop producing lactic acid which would affect fermentation and safety of the product. The technology of making fast fermented sausages relies on pH and not on drying and the air speed control is less crucial as there is little drying.

Spreadable sausages: Course ground spreadable sausages are fermented at 95% humidity and have an air speed of about 0.8 m/s dropping down about 0.1 m/s every 2 days.

Finely ground spreadable sausages are fermented at about 90% humidity but with a slower air speed. As they contain more fat (it helps with spreadability) there is less water to remove. It is much harder for the moisture to manoeuvre among fine meat particles on its way to the surface and the distance is longer, too. As a result, less moisture gets to the surface and the air speed of about 0.1 m/s generally suffices. A typical medium-fermented salami process is shown in Table 15.2.

Table 15.2 A typical medium-fermented salami process

Process	Temp		Humidity %	a_w	pH	Air speed	Time
	°C	°F				(m/s)	
Conditioning	20–25	68–77	< 60	0.96–0.97	5.8	0	< 6 hours
Fermenting	18–25	66–77	98–92	0.94–0.96	5.6–5.2	0.8	2–4 days
Drying	18–22	66–70	85–90	0.95–0.90	5.2–4.8	0.5	5–10 days

Note: The speed of 3.6 km/h (2.2 mile/hour) corresponds to the speed of 1 m/s. Based on parameters in the table above, a medium-fermented salami will lose about 1.0–1.5% of its mass daily.

15.5.9   Use of spices in fermented sausages

Throughout history spices were known to possess antibacterial properties and cinnamon, cumin, and thyme were used in the mummification of bodies in ancient Egypt. It is hard to imagine anything that is being cooked in India without curry powder (coriander, turmeric, cumin, fenugreek and other spices). Latest research establishes that spices such as mustard, cinnamon, and cloves are helpful in slowing the growth of molds, yeast, and bacteria. Garlic and clove are effective against some common strains of *E. coli*. Spices alone cannot be used as a hurdle against meat spoilage as the average amount added to meat is only about 1% (1 g/1 kg). To inhibit bacteria the amounts of spices will have to be very large and that will alter the taste of the sausage. Rosemary and sage have antioxidant properties that can delay the rancidity of fat. Marjoram is a proxidant and will speed up the rancidity of fats.

15.6   Finished product

Once the product is finished, it is packaged and distributed. Fermented sausages can be sold as either entire or as thin slices. The developed colour, texture, and flavour depend on the processing and type of product.

15.6.1   Meat colour

The first impression consumers have of any meat product is its colour and thus colour is of utmost importance. The colour of meat may vary from the deep purplish-red of freshly cut beef to the light gray of faded cured pork. Fortunately, the colour of meat can be controlled if the many factors that influence it are understood.

Fresh and cured meat colour both depend on myoglobin, but are considerably different from each other in terms of how they are formed and their overall stability.

Myoglobin is a water-soluble protein that stores oxygen for aerobic metabolism in the muscle. It consists of a protein portion and a nonprotein porphyrin ring with a central iron atom. The iron atom is an important player in meat colour. The defining factors of meat colour are the oxidation (chemical) state of the iron the compounds of which (oxygen, water or nitric oxide) are attached to the iron portion of the molecule. Because muscles differ greatly in activity, their oxygen demand varies. Consequently different myoglobin concentrations are found in the various muscles of the animal. Also, as the animal gets older there is more myoglobin. A greater myoglobin concentration yields a more intense colour. Muscle pigment concentration also differs among animal species. For example, beef has considerably more myoglobin than pork or lamb, thus giving it a more intense colour.

Meat colour reaction

Immediately after cutting, meat colour is quite dark-beef would be a deep purplish-red. As oxygen from the air comes into contact with the exposed meat surfaces it is absorbed and binds to the iron. The surface of the meat blooms as myoglobin is oxygenated. This pigment, called oxymyoglobin, gives beef its bright cherry red colour. It is the colour that consumers associate with freshness.

Myoglobin and oxymyoglobin have the capacity to lose an electron (called oxidation) which turns the pigment to a brown colour and yields metmyoglobin. Thus, myoglobin can change from a dark purple colour to a bright red colour simply from oxygenation or to a brown colour by losing electrons. The pigments myoglobin, oxymyoglobin and metmyoglobin can be changed from one to the other, depending on the conditions at which the meat is stored.

After cooking, a brown pigment called denatured metmyoglobin is formed, which normally cannot be changed to form another pigment.

Oxymyoglobin, commonly known as the fresh meat colour, is the most desirable colour for fresh meats. Maintaining this colour requires that the meat surface be free from any contamination which would cause a chemical reaction resulting in the formation of the brown pigment metmyoglobin. Also, oxygen must be available at a sufficient concentration in order to combine with the myoglobin to form oxymyoglobin. This reaction is reversible and dependents on the availability of oxygen, active enzymes and reducing compounds in the muscle.

Vacuum packaged fresh meat has a dark, purplish red colour because the oxygen has been removed from the package and reducing enzymes have converted the meat pigment back to myoglobin. Once the meat is taken out of the vacuum package it will recover its bright red colour, albeit for a shorter time period than its unvacuum packaged counterpart. The change from myoglobin to oxymyoglobin (Fig. 15.2) and vice versa usually occurs quite readily. Similarly, the reaction that produces the brown meat metmyoglobin occurs quite easily, but the reverse of this is more difficult. In raw meat there is a dynamic cycle such that in the presence of oxygen the three pigments myoglobin, oxymyoglobin and metmyoglobin are constantly interconverted, all three forms are in equilibrium with one another.

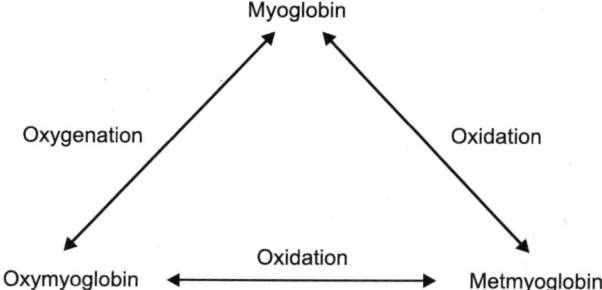

Interconversion of meat pigments. Myoglobin when oxygenated
is bright red in colour and called oxymyoglobin. Both myoglobin and
oxymyoglobin can lose an electron (oxidize) to form metmyoglobin

Figure 15.2 Dynamic cycle showing the presence of oxygen and the three pigments myoglobin, oxymyoglobin and metmyoglobin which are constantly interconverted.

Metmyoglobin is associated with chilled meat that has been stored too long (i.e. reducing enzyme activity available to reduce metmyoglobin to myoglobin has been exhausted), but also appears when oxygen partial pressure is low, such as when meat pieces are stacked one on top of another. Oxygen partial pressure can also be reduced when aerobic bacteria use up the oxygen and it is unavailable to react with the myoglobin.

Effect of meat pH

The rate and extent that muscle pH declines postmortem are both variable and have a great impact on the colour of meat and meat products. The normal pH decline in muscles is from approximately 7.0–7.2 down to near pH 5.5–5.7 over about 24 hours. With this pH decline, whole tissue is the characteristic colour of the species. If the pH declines to the normal pH of 5.5–5.7 within 45 min or less, the muscle will appear very pale and soft (PSE). A very low ultimate pH (<5.4) will also result in a paler colour. If the pH does not drop much postmortem, the meat will be dark with a dull, dry surface, dark firm and dry (DFD). As the ultimate pH increases, the meat gradually becomes darker. This darkening of colour becomes noticeable when the muscle pH exceeds 5.7. The colour changes observed with Pale, soft and exudative (PSE) and DFD meat are mostly due to structural changes in muscle. Some of the changes are due to the rate of myoglobin oxygenation. The changes in pH affect the charge on the proteins making up the muscle. These changes alter the spacing between the fibres of the meat, and the change in structure affects how light is reflected and absorbed, and thus affects the visual appearance.

Colour stability of fresh meat

High ultimate pH can affect the colour stability of fresh meat because it affects enzyme activity and the rate of oxygenation. Reducing enzymes are necessary to convert metmyoglobin back to oxymyoglobin. The high ultimate pH has a dry surface and this inhibits the penetration of oxygen into the meat and thus slows down the oxygenation process.

The length of time the meat has been stored postmortem affects the colour stability of the meat or meat product. Increased time from slaughter results in reduced colour stability because co-factors necessary for the reduction of metmyoglobin are depleted as postmortem time increases. A similar problem is seen if frozen meat is used. Products made with frozen meat will be darker initially and will not maintain the fresh colour for as long as products made from meat which has never been frozen. Both storage time and temperature have a great effect on colour stability. Colour acceptability decreases as storage time increases; however, the length of time the colour is acceptable is greatly affected by storage temperature. Fresh meat and meat products should be stored at temperatures of $-1.5°C$ ($29.3°F$) to give maximum colour shelf-life and safety of products.

Particle size reduction and mixing

Air is incorporated into meat products during the grinding process or mixing in of ingredients. The more air that is incorporated, the more stress is put upon the natural reducing systems of meat which help maintain oxymyoglobin stability and keep metmyoglobin formation in check. Longer mixing times and smaller meat particles result in shorter colour shelf-life for meat products. Use of vacuum mixers helps to improve colour stability, but will not completely bring the stability back to what would be expected in whole muscle products.

Cooked meat pigment

During the cooking process, myoglobin is denatured. All of the pigment is not affected at the same time or to the same extent and this is why you get reddish colour at different end point temperatures when cooking. The cooked pigment is denatured metmyoglobin. It is brown and is easily recognized in cooked meat products. Certain meat conditions can result in protection of the myoglobin. The ultimate pH of the muscle is one of these conditions. The ultimate pH of meat or meat products will affect how the meat colour changes during cooking. If the meat has a high pH, it will have to be cooked to higher end-point temperatures to get the same visual degree of doneness as one with normal pH. Frequently, complaints of this hard to cook defect are associated with a high pH of the meat or meat product. This meat appears raw in colour, dark red to purple, long after appropriate cooking temperatures have been reached.

Cured meat colour

Curing of meat has been done for centuries for the preservation of meat products. Today cured products are made for their unique flavour and texture. Cured products have a pink pigment that is relatively stable. To form this pigment, sodium nitrite is either rubbed onto the surface or injected into the meat with a needle injector. The nitrite, when added to water forms nitrous acid and nitric oxide, and which penetrate the meat and combine with the myoglobin to form nitric oxide myoglobin. This pigment is not stable until after cooking when the final cured pigment, nitrosylhemochrome, is formed. The cooked pigment is more stable, but is still sensitive to oxygen presence, temperature and light. This is why most cured products are vacuum packaged in special UV protective films.

Many problems can occur in the curing of meat products that can result in the development of strange colours. One of the most common is the oxidation of the pigment to form a green or grey colour. This is usually caused by metal contamination from molds or smoke sticks. This sometimes happens if the molds are old or they have been welded improperly with solder that contains copper or iron.

Pinking of uncured cooked products

Sometimes a pink colour can form in uncured cooked meat products. This can be caused by many factors and should not be confused with the hard to cook phenomena caused by high meat pH.

Cooked beef or other meats can become contaminated with the curing salt nitrite from contact with a cured product, incomplete cleaning of utensils used with cured products or water contaminated with nitrites. It takes only small amounts of nitrite to develop cured meat colour. Although 50 parts per million are necessary to maintain the pink colour in cooked ground beef, pink colour can show up with levels as low as 5 parts per million – that is 5 milligrams per kilogram of beef. The cured meat colour can be on the surface if the contamination occurs during the cooking process but will be throughout the meat product if it occurred earlier.

A pink colour can also be formed in slow cooked meat products that have not been contaminated with nitrite. It is caused by specific conditions that promote interaction of natural meat pigments and nitrogen containing constituents of meat. This colour is actually desired in products such as Texas barbecue.

Surface pinking, also termed 'pink ring' can occur if gas ovens or barbecue grills are used to cook meat products. Incomplete burning of the gas or contaminates in the gas result in the formation of nitrogen dioxide and nitric oxide. Nitric oxide is the active form of nitrite that yields the pink colour. In roast beef this is generally considered a serious colour defect because consumers might associate it with undercooking.

Naturally occurring nitrates found in water and vegetables can be converted to nitrites during thermal processing and can cause a pink colour to form. Carrots and cabbage are examples of vegetables that have naturally occurring nitrates that can contribute to pinking of cooked meats. This occurrence is rare because nitrates take longer periods of time to be converted to nitrites which in turn yield nitric oxide that forms the colour pigment necessary for the pink colour. This colour is usually seen in slow cooked soups or stews.

Another cause of pink colour is the presence of carbon monoxide (CO). Carbon monoxide combines with natural pigments in meat to produce a dark red colour in raw and cooked meat. Minute amounts of the gas may come from dry ice, or carbon dioxide freezer tunnels used in hamburger production and it will affect colour. Carbon monoxide, which has a greater affinity for myoglobin than oxygen, binds almost irreversibly to the raw colour pigment. However, this occurrence is relatively rare. Researchers have shown that if CO is used in modified atmosphere packages a dark red colour develops and it remains after cooking.

Iridescence in processed meat products

Iridescence is a common problem in sliced roast beef and ham products. The dominant colour is frequently green and consumers sometimes confuse this with green myoglobin pigments associated with microbial growth. The iridescence of meat products is produced by a combination of the angle of incidence of the light on the muscle fibers and the wetness of the surface. If the fibers are pulled slightly out of alignment during slicing, the light strikes the fiber at an angle scattering light which appears as the rainbow or greenish colour on the surface of the meat. Addition of phosphate seems to exacerbate the problem by increasing the amount of water that is retained by the product.

15.6.2   Texture

The consistency of fermented meats is initiated with the salt addition and pH reduction. The water-binding capacity of myofibrillar proteins decreases as pH approaches its isoelectric point and releases water. The solubility of myofibrillar proteins is also reduced, with a trend toward aggregation and coagulation, forming a gel. The consistency of this gel increases with water loss during drying. The meat: fat ratio may affect some of these textural characteristics, but, in general, the final texture of the sausage will mainly depend on the extent of drying.

15.6.3   Aroma flavour

In order to have a cooked meat product with its desirable aroma flavours as expectation of consumer, it is important to understand how aroma flavours are derived, the mechanisms by which flavour components are generated, and the factors affect formation of flavour compounds then determine the final aroma flavour characteristics of cooked meat. Regarding the effects of factors on aroma flavours of cooked meat and to minimize the detrimental effects it is suggested if increasing the polyunsaturated fatty acids (e.g. C18:3n-3, DHA, EPA) to increase nutritional benefits to the consumer by using fat-supplemented diets however the undesirable flavours may result. Because the breakdown products of these fatty acids have a shorter chain length therefore are more volatile and they affect meat flavours by interacting with the Maillard reaction results in reducing levels of meaty aroma compounds such as sulphur-substituted thiophenes. Therefore, diets, feeding regimes, welfare and management of animals should be taken into account. Cooking conditions such as temperature, holding time and cooking methods play an important role in determining the formation volatile flavour compounds. In general, it has been demonstrated that cooking meat at high temperature (by roasting, grilling) will produce better aroma flavour characteristics due to which the important Maillard products are formed. In addition to the cooking effect, it is suggested that a slow cooking and longer hold time can allow the undesirable volatile flavour compounds to dissipate, thus reduce warm-off flavour.

15.6.4   Taste

Sour taste, mainly resulting from lactic acid generation through microbial glycolysis, is the most relevant taste in fermented meats. Sourness is also correlated with other microbial metabolites such as acetic acid. Salty taste is usually perceived as a direct taste from salt addition. Other taste contributors are those compounds resulting from protein hydrolysis.

Genetically engineered fish

16.1 Introduction

Biotechnology, the use of biological systems or living organisms in production process has a wide range of useful applications in fisheries and aquaculture. The potential of biotechnology to contribute to increasing agricultural, food and feed production, improving human and animal health, abating pollution and protecting the environment, has been acknowledged. Biotechnology makes it possible to achieve increased growth rate in farmed species, boost the nutritional value of aqua feeds, improve fish health, help restore and protect environment, extend the range of aquatic species and improve the management and conservation of wild stocks.

Some biotechnologies are simple with a long history of application: e.g. fertilization of ponds to increase feed availability. Others are more advanced and take advantage of increasing knowledge of molecular biology and genetics, e.g. genetic engineering and DNA disease diagnosis. The field of genetic biotechnology similarly ranges from simple techniques such as hybridization, to more complex processes such as the transfer of specific genes between species to create Genetically Modified Organisms (GMOs).

16.2 Fish feed biotechnology

Biotechnology is helping to answer some of the technical and environmental concerns of fish farming. Presently, the most common protein source for many fish diets is fishmeal. Fishmeal a by-product of fish processing is used because of its high quality and high protein content. However, it has some disadvantages. One disadvantage for fish producers is that it is expensive. So any cheaper alternative protein source would be welcomed. Another concern regarding fishmeal is the stability of supply. Fishmeal comes from by-products of wild fish, but world fish stocks are declining. The use of fishmeal in aquaculture causes environmental problems. It contains levels of phosphorous far above the requirement for optimal growth in fish. The excess phosphorus goes into the water, causing problems such as eutrophication or excess algal growth.

As a result of these concerns with fishmeal, researchers are using biotechnology to produce alternative plant-based protein source. Plant protein has the potential to address the problem of phosphorus pollution. For Prairie crops to be used as the main protein source for fish, they must be processed into a concentrate. Biotechnology is often used in this processing; plant protein also requires processing because plants contain what are called anti-nutritional compounds that serve as a defense mechanism. These compounds must be destroyed during processing or they could harm the fish or interfere with the fish's ability to utilize the feed.

Researchers are also trying to deal with these anti-nutritional factors by producing feed enzymes to counter them. Phytase is one example. This enzyme can help fish make the best use of the phosphorus available in a plant protein based feed.

Bio-remediation: Farmed aquatic animals are much more sensitive to their immediate environment than land animals. The water, in which they depend for oxygen and a range of other important chemicals also takes up their waste products and may carry pollution from the nearby environment. The process of disease in aquaculture species is thus much more strongly connected to environmental factors than would be the case say, with cattle. A further biotechnology field that has developed in aquaculture, because of the nature of this relationship, is that of bio-remediation. This refers to the use of friendly bacteria or 'probiotics' to treat water or feeds and by natural processes, discourages the development of 'unfriendly' bacteria that potentially would cause disease.

16.3   Transgenic fish

Researchers are seeking to improve the genetic traits of the fish used in aquaculture by using different transgenic techniques. Researchers are trying to develop fish which are larger and grow faster, more efficient in converting their feed into muscle, resistant to disease, tolerant of low oxygen levels in the water and tolerant to freezing temperatures.

The exploitation of Tilapia in small-scale aquaculture in developing countries is constrained by the performance characteristics of fish currently in use. There have been significant advances in the genetic improvement of Tilapia used in aquaculture in recent years; for example, through the use of selective breeding and monosex techniques. However, limited growth rate and excessive reproduction resulting in fish that are small and variable in size, still pose a considerable constraint to the exploitation of Tilapia in developing countries. Transgenic techniques offer the means of producing immediate large quantum changes in performance, for example in growth rate, that far exceed those attainable with other approaches.

16.4   Genetically engineered fish and seafood: Environmental concerns

Farmers and scientists have a history of modifying animals to maximize desirable traits. Genetic modification is one of the current approaches for modifying animals to increase their beneficial traits. In the broadest sense, genetic modification refers to changes in an organism's genetic makeup not occurring in nature, including the production of conventional hybrids. With the advent of modern biotechnology (e.g. genetic engineering or bioengineering), it is now possible to take the gene (or genes) for a specific trait either from an organism of the same species or from an entirely different one and transfer it to create an organism having a unique genetic code. This technique can add both speed and efficiency to the development of new foods and products. Genetically engineered plant varieties, such as herbicide-resistant corn and soyabeans, have already been widely adopted by the US farmers, and some advocate using similar techniques to produce genetically engineered fish or seafood for the aquaculture industry.

A number of environmental concerns have been raised related to the development of genetically modified (GM) fish, including the potential for detrimental competition with wild fish, and possible interbreeding with wild fish so as to allow the modified genetic material to escape into the wild fish population. Sterilization and bioconfinement have been proposed as means of isolating GM fish to minimize the potential for harming wild fish populations. In the process of congressional oversight of executive agency regulatory action, concerns have been raised about the adequacy of the US Food and Drug Administration's review of applications for approval of GM animals, with respect to the potential for environmental harm.

Genetic engineering and fish: Genetically engineered (also called transgenic) fish are those that carry and transmit one or more copies of a recombinant DNA sequence (i.e. a DNA sequence produced in a laboratory using *in vitro* techniques). Because genetic engineering is defined by the technology that is used to create and transfer the DNA sequence, and not the source species of the donor DNA, even fish that are engineered with DNA derived entirely from fish species are considered to be genetically engineered. Currently, no genetically engineered fish has been approved for food production in the United States. To date only one company, AquaBounty, has publicly announced that it has requested FDA approval to market a genetically engineered food animal, a growth enhanced Atlantic salmon that is capable of growing 4–6 times faster (but not larger) than standard salmon grown under the same conditions.

Science-based concerns associated with genetically engineered fish:
The greatest science-based concerns associated with genetically engineered fish are those related to their inadvertent release or escape. Concerns range from interbreeding with native fish populations to ecosystem effects resulting from heightened competition for food and prey species. There is, in principle, no difference between the types of concerns associated with the escape of genetically engineered fish and those related to the escape of fish that differ from native populations in some other way, such as captively bred populations. Ecological risk assessment requires an evaluation of the fitness of the genetically engineered fish relative to non-genetically engineered fish in the receiving population in order to determine the probability that the transgene will spread into the native population. Ecological impacts are the result of the characteristics of the organism, regardless of whether the organism acquired those characteristics through natural selection, artificial selection, or genetic engineering. The presence of genetically engineered fish does not a *priori* have a negative effect on native populations. If genetically engineered fish are ill-suited to an environment or are physically unable to survive outside of containment, they may pose little risk to the native ecosystems. Regulators apply a scientifically derived, risk-based framework to assess the ecological risks involved with each transgene, species, and receiving ecosystem combination on a case-by-case basis. Risks will be quite specific to the gene, species, and site in question, and simple generalizations concerning the risks (and benefits) of genetically engineered fish are not scientifically meaningful.

Can containment be used to isolate genetically engineered fish:
Commercialization of genetically engineered fish will likely depend on the development of effective containment strategies. If genetically engineered fish are adequately contained, they pose little risk to native populations. The NRC recommended the simultaneous use of multiple containment strategies for genetically engineered fish (National Research Council). Physical containment is an obvious first line of defense to prevent the escape of genetically engineered fish. Examples of such measures may include building facilities on land or in locations removed from native populations, or ensuring that water chemistry (temperature, pH, salinity, and concentrations of certain chemicals) is lethal to one or more life stages of the genetically engineered fish, such as treating effluent water to prevent the release of viable gametes or fry. Biological containment or bioconfinement approaches such as sterilization are also being developed. To ensuring food safety, the FDA also evaluates environmental risks posed by genetically engineered animals as directed by the National Environmental Policy Act (NEPA). Under NEPA,

Federal agencies are obligated to cooperate with other involved federal agencies. In the case of the AquaBounty genetically engineered salmon, this cooperation includes involvement of the U.S. Fish and Wildlife Service and the National Marine Fisheries Service in the development of a scientifically based environmental risk assessment.

Will consumers accept genetically engineered fish: Ultimately, it is the marketplace, and not science, that decides the fate of new technologies and acceptability of certain risks. Food retailers and even farmers may be unwilling to stock genetically engineered fish and risk having their market become the target of an organized antibiotech campaign. Such a scenario occurred in Europe, where activist campaigns targeted retailers stocking labelled genetically engineered food products, and attempts to differentiate brands resulted in the removal of these products from supermarket shelves altogether. Despite strong public support for medical applications of genetic engineering, there is less public support for agricultural biotechnology. However, market response and consumer behaviour may differ markedly between affluent Western countries and developing countries. Even if genetically engineered fish are approved by the FDA in the United States, it will likely be activist, food retailer, and consumer response in the market place that will ultimately decide whether genetically engineered food fish will sink or swim.

16.5   Environmental concerns and control options

In addition to its responsibility for assuring food safety, FDA is charged with assessing the potential environmental impacts of newly engineered plants and animals. To fully assess these potential impacts, FDA consults with the Fish and Wildlife Service and the National Marine Fisheries Service (NOAA Fisheries). Despite this consultation, critics question whether FDA has the mandate and sufficient expertise to identify and protect against all potential ecological damage that might result from the widespread use of transgenic fish. The possible impacts from the escape of GM organisms from aquaculture facilities are of great concern to some scientists and environmental groups. A National Research Council report stated that transgenic fish pose the 'greatest science-based concerns associated with animal biotechnology, in large part due to the uncertainty inherent in identifying environmental problems early on and the difficulty of remediation once a problem has been identified.'

16.6   Interbreeding with wild fish

Critics and scientists argue that GM fish could breed with wild populations of the same species and potentially spread undesirable genes. One study postulated a 'Trojan gene hypothesis' after observing that GM Japanese medaka, a fish commonly used as an experimental model, were able to out-compete nonaltered fish for mates in a laboratory environment. However, the resulting offspring of this mating between GM fish and wild fish were less fit, lacking certain physical or behavioural attributes that resulted in the eventual demise of the modified population. The ecological risks of stocking GM shellfish in the wild have not yet been thoroughly examined, since confining and isolating these organisms is more difficult than confinement of many fish species, due to their methods of reproduction and dispersal.

16.7   Competition with wild fish

Even if fast-growing GM fish do not spread their genes among their wild counterparts, critics fear GM fish could disrupt the ecology of streams by competing with native fish for scarce resources. Escaped transgenic fish could harm wild fish through increased competition or predation. In addition, some argue that transgenic fish, especially if modified to improve their ability to withstand wider ranges of salinity or temperature, could be difficult or impossible to eradicate, similar to an invasive species. The consequences of such competition would depend on many factors, including the health of the wild population, the number and specific genetic strain of the escaped fish, and local environmental conditions. Critics maintain that an indication of the magnitude of this potential problem may be noted where non-GM Atlantic salmon from near shore net pens in the northwest United States and British Columbia have escaped and entered streams, in some cases outnumbering their wild Pacific salmon counterparts.

However, it is not known whether GM fish could survive in the wild in sufficient numbers to inflict permanent population damage. One study indicated that, when food supplies were low, GM fish might have the ability to harm a wild population, although the researchers have caution that laboratory experiments may not reflect what would happen in the wild. Biotechnology proponents argue that GM fish, if they escape, would be less likely to survive in the wild, especially when they are reared in protected artificial habitats and have not learned to avoid predators.

16.8   Potential control options

A number of potential safeguards to address these environmental concerns exist and could be required.

16.8.1   Sterilization

FDA could require that only sterile GM fish be approved for culture. Fertilized fish eggs that are subjected to a heat or pressure shock retain an extra set of chromosomes. The resulting triploid fish do not develop normal sexual characteristics and, in general, the degree of sterility in triploid females is greater than males. Thus, all-female lines of triploid fish are considered to be one of the best current methods to insure nonbreeding populations of GM fish. Nonetheless, there are batch-to-batch variations, and it is uncertain whether this method could be effective for all species; it has not been successful for shrimp, for example. Also, critics question whether escaped triploid fish, which in some species have sufficient sex hormone levels to enable normal courtship behaviour, could mate with wild individuals, lowering reproductive success of the wild population.

Other sterilization methods are currently under study, and it is likely that research in this area will increase options. Critics of GM fish counter that the risks to native fish populations, however small, may outweigh the potential benefits of this technology, especially where native fish populations are already threatened or endangered.

16.9   Possible benefits and disadvantages of genetically engineered fish and seafood

16.9.1   Potential benefits

Biotechnology proponents maintain that genetic modification techniques have many advantages over traditional breeding methods, including faster and more specific selection of beneficial traits. Because scientists are able to directly select traits they wish to create or amplify, the desired change can be achieved in very few generations, making it faster and lower in cost than traditional methods, which may require many generations of selective breeding. Genetic modification techniques allow scientists to precisely select traits for improvement, enabling them to create an organism that is not just larger and faster-growing, but potentially improved, for example, by increasing nutritional content. Proponents claim that faster-growing fish could make fish farming more productive, increasing yields while reducing the amount of feed needed, which in turn could reduce waste. With intense exploitation of wild fish stocks, GM fish and seafood could be important means to meet increasing human nutrition needs and address food security concerns.

Shellfish and finfish, genetically modified to improve disease resistance, could reduce the use of antibiotics. Increased cold resistance in fish could lead to the ability to grow seafood in previously inhospitable environments, allowing aquaculture to expand into previously unsuitable areas. Research efforts are also under way to improve human health by genetically modifying fish to produce human drugs like a blood clotting factor and to create shellfish that will not provoke allergic reactions. Biotechnology proponents claim that these advantages could translate into a number of potential benefits, such as reduced costs to producers, lower prices for consumers for edible fish and pharmaceuticals, and environmental benefits, such as reduced water pollution from wastes. Food scientists and the aquaculture industry may support the introduction of genetic engineering, provided that issues of product safety, environmental concerns, ethics, and information are satisfactorily addressed.

16.9.2   Disadvantages

On the other hand, while the majority of consumers in the United States appear to have generally accepted GM food and feed crops, it is uncertain whether consumers will be as accepting of GM fish. Although such fish may taste the same and are expected, like their traditionally bred counterparts, to be less expensive than wild-caught fish, ethical concerns over the appropriate use of animals, in addition to environmental concerns, may affect public acceptance of GM fish as food.

In addition, the commercial fishing industry says that it has successfully educated the public to discriminate among fish from different sources, such wild and farmed salmon. It is possible that a publicized escape of GM fish could lead to reduced public acceptance of their wild product. Environmental and consumer groups are asking that genetically engineered products be specially labelled. However, industry groups are concerned that such labelling might lead consumers to believe that their products are unsafe for consumption.

16.10   Future applications of biotechnology

In the future we can expect the application of biotechnology in aquaculture to increase as new methods to improve fish performance are developed and perfected. In the area of chromosome manipulation gynogenesis will become a practical first step in the generation of monosex female stocks when combined with endocrine masculinization. Androgenesis, although technically more difficult, may see eventual use as a practical means of producing supermales in male heterogametic species thus facilitating the production of all male stocks. Current research on meiotic and mitotic gynogenesis may possibly lead to the feasibility of producing clones of identical high performance fish. Triploidy induction may be used increasing as a means of sterilizing genetically altered fish. Tetraploid fish will be further developed as a means of producing diploid gametes and thus allotriploids.

In the field of controlled sex differentiation non-aromatizable androgens will be used as masculinizing agents and non-steroidal aromatase inhibitors may also be utilized Y-specific-DNA probes will be developed and applied in additional fish species and many species where one sex is more desirable for culture or more valuable in the marketplace will be grown in monosex culture.

Acceptable means of accelerating growth will be applied which utilize somatotropin, placental lactogen, related peptides or their analogues applied by injection or implantation in slow release formulations or devices or via the diet. Biotechnology will have a major impact on fish nutrition. Practical diets will be developed where suitably modified plant protein sources totally replace fish meal. Astaxanthin from the yeast *Phaffia rhodozyma* will provide a natural source of pigment for salmonid diets. Dietary optimization of omega-3 polyunsaturated fatty acid levels may be used to produce designer seafood for health conscious consumers.

The techniques of molecular biology will be utilized in many different ways to produce high performance fish. In the field of fish health further diagnostic tests based on the polymerase chain reaction, dot blot and *in situ* hybridization, etc. will facilitate disease and parasite detection. Increasing knowledge of the relationship between stress and the immune response and the utilization of glucans to stimulate non-specific defense mechanisms will all contribute to minimizing the effects of stress on performance.

The utilization of molecular biology to develop new or more effective vaccines will further contribute to the elimination of disease outbreaks in aquacultured fish. In the area of genetics the introduction of marker assisted selection into broodstock development should assist in the genetic improvement of aquacultured stocks. The engineering of transgenic fish in which specific characteristics are enhanced or suppressed may provide a quantum leap in the development of high performance fish.

While still in this infancy this technology may provide means of improving growth and food conversion efficiency, accelerating smoltification, modifying nutritional requirements, increasing disease resistance, inhibiting or redirecting sexual development and facilitating adaptation to new environments. Early results show considerable promise for acceleration of growth and smoltification and potentially improvement of food conversion efficiency in transgenic salmonids, however, selection over several generations will be required to generate true breeding transgenic stocks.

17.1　　Introduction

Poultry products are nutritious and add variety to the human diet. Most of the products are from chickens, but ducks and turkeys also are important sources. In developing countries, rural production systems for maintaining poultry are not far removed from conditions encountered by wild chickens and ducks. Where reliable refrigeration is not available, consumers purchase small amounts of poultry meat and eggs on a regular basis. In developed countries, commercial production systems provide large supplies of meat and eggs. The food processing industry provides meat in the form of whole carcass, cut-up parts, or further processed meat. Eggs are available as whole (shell) eggs or liquid or component parts. By-products from food processing are recycled, usually by inclusion in animal feed. Domesticated birds are an important source of foods for people in many parts of the world. Much of that food is in the form of meat, but eggs also are important to add nutrition and variety to human diets. Chickens now provide more meat and eggs than all other poultry species combined, so additional food is derived from turkeys, ducks and geese (waterfowl), and pheasants and quail (game birds).

Species of birds that were domesticated were present on earth prior to humans. Each species was wild and developed in a part of the world that permitted it to survive and maintain its numbers. When humans began to grow their food instead of collect it from the wild, some species of wild animals had characteristics that made domestication possible.

These animals accompanied humans as civilizations progressed. In some cultures, birds were a source of food. In other cultures, birds were kept as pets, for entertainment or for sacrifice during religious rituals.

Poultry, or domestic birds, are raised for their meat and eggs and are an important source of edible animal protein. Poultry meat and eggs are highly nutritious. The meat is rich in proteins and is a good source of phosphorus and other minerals, and of B-complex vitamins. Poultry meat contains less fat than most cuts of beef and pork. Poultry liver is especially rich in vitamin A. It has a higher proportion of unsaturated fatty acids than saturated fatty acids. This fatty acid ratio suggests that poultry may be a more healthful alternative to red meat.

17.2   Different varieties of poultry meat

Chickens and turkeys are the most common sources of poultry meat (87% and 6.7% of total poultry production, respectively). However, other commercially available poultry meats include meat from ducks (4% of total poultry production) and from geese, pigeons, quails, pheasants, ostriches and emus (combined about 2.7% of total poultry production). In the United States and Canada, turkey meat is the second most important poultry meat consumed after chicken meat; however, in other countries turkey meat is less important.

17.2.1   Chicken

Archaeological evidence indicates that chickens originated in Asia. There are four species of junglefowl from the area that may have contributed to present day chickens. The red junglefowl (*Gallus gallus*) is considered the most likely ancestor. Junglefowl are members of the pheasant family, with a comb and wattles that distinguish them from other pheasants. Red junglefowl are considered the wild type from which other colour mutations have occurred. This species is found in the wild in Pakistan and in adjacent areas of India and China. Males weigh from 800 grams to 1360 grams (1.75–3.00 pounds), but females are only about 60% as large. The number of eggs that a hen lays for incubation is 4–8, with an incubation time of 19–21 days.

Cornish stock was imported into North America, where they provided genetic material for the meaty broiler chickens. Continued selection for broad breasts has increased the percentage of the body that is white meat, the part of the chicken that is prized most by North Americans. The story of the development of the chicken meat industry shows how chickens have moved from one continent to another and how humans have changed the function of the chicken in the process.

There is a different story for chickens that produce most of the white-shelled eggs. Domesticated chickens were taken from Asia to areas around the Mediterranean Sea. In that area, selection must have been to produce more and larger eggs. Instead of producing 10 or 12 eggs per year, hens that were imported to North America from the Mediterranean area produced dozens of eggs per year. This class of chickens provided the genetic material that was used for egg production breeds. These chickens are somewhat heavier than the junglefowl but not nearly as heavy as chickens that are used for meat production.

An important reason for the large number of chicken breeds is the number of mutations that is known. In the wild, mutants often do not survive or cannot be identified. In captivity, breeding only a small number of closely related chickens resulted in the visual expression of mutations. For example, the red junglefowl has a single comb. Mutants such as rose, pea, walnut and buttercup comb are known. The male red junglefowl has mostly red or brown feathers to go with its black breast plumage. The wild-type colour is due to at least 5 different gene combinations. Mutations in any of these genes can produce variations in colour and pattern. Therefore, chickens may be red, black, white, or many combinations of these colours.

Mutations have occurred in the legs, so that chickens have four to six toes, yellow, white or black scales, and clean or feathered legs. Mutations in head parts have produced crests, muffs and tufts. Mutations in feather structure and location have produced naked necks, frizzles and silkies. In addition, for most standard sized chickens, there is a bantam counterpart that weighs about 450 grams (one pound). All of these mutations presented the possibility for different combinations of mutations to form breeds. Poultry fanciers seized the opportunity to make numerous combinations in the nineteenth and early twentieth centuries when many breeds of chickens were developed. An individual or group of individuals combined breeds of chickens to develop a new breed. The new breed was displayed at poultry shows to advertise its presence. When the breed had been improved to the point that it would produce offspring essentially like itself, the breed was registered with the appropriate poultry organization. A number of the breeds that were developed at that time are no longer available, but poultry fanciers continue to breed and show purebreds.

17.2.2   Turkeys

There is much more certainty about the origin and domestication of the turkey than of the chicken. Turkeys are native to North America. When Europeans arrived in North America, turkeys were present in much of what is now the United States. They were also present in much of Mexico and ranged into parts of Canada. Their habitat was woods interspersed with open fields. Domesticated turkeys descended from the wild turkey, *Meleagris gallopavo*. There are seven sub-species that had different geographical locations. One of these sub-species (Mexican) had been domesticated by the Native Americans when the Spanish arrived in Mexico. Records indicate that turkeys were introduced in Spain no later than 1511. The first record of turkeys in England is in 1541, so they spread very quickly through the western world.

In North America, domesticated turkeys were brought from Europe during the first half of the 1600s. The eastern wild turkey, one sub-species of *M. gallopavo*, was native to the area and was hunted as a source of food.

Although it was not domesticated, the eastern wild turkey was important in the development of the commercial turkey. The domesticated turkey that the Europeans brought to North America was black in appearance and was smaller than the eastern turkey. Wild eastern males took the initiative to mate with domesticated Mexican females and the hybrid vigor of the offspring became obvious. The offspring had a colour pattern similar to the eastern turkey, which was called bronze.

None of these birds, however, had the large amount of breast muscle that we now associate with the turkey. Years of selection for that characteristic resulted in the Broad Breasted Bronze, which generally was the turkey of choice for decades. Eventually, the commercial industry chose a turkey with white feathers, drawing on feather colour mutations known when the Spaniards first encountered the turkey in Mexico.

17.2.3   Ducks

The consensus is that domesticated ducks came from mallards (*Anas platyrhynchos*). These ducks have adapted to a broad range of habitat in the northern hemisphere. There are seven sub-species of mallard, but the most adaptable is the common mallard. In the wild, they were found in Europe and Asia. It is thought that domesticated ducks originated from this sub-species with the colourful male.

A number of breeds of ducks have been developed over the years. Some of them retain the colour pattern of the common mallard, but most do not. Size ranges from the call ducks, which are smaller than mallards, to the heavy breeds, which are several times larger than mallards. The breed that is most popular for meat is the White Pekin. It was developed in China and was imported into several countries in the 1840s. Commercial companies are changing this breed to meet the desires of consumers.

Several breeds of ducks are also excellent egg layers. Runners, ducks that have the posture of penguins, may surpass the best chickens for egg production. An egg a day for at least 300 days has been recorded. Other breeds of ducks almost equal Runners for egg production.

17.3   Broilers for meat production

Broilers are the main type of chicken produced by modern integrated poultry raising facilities due to their high feed-meat conversion ratio. Broilers are generally grown for a specific number of days and until they reach a specific weight. In North America, 7-week-old chickens are classified as *broilers* or *fryers* and 14-week-old chickens are classified as *roasters*.

17.3.1   Production process in the broiler industry

The average growth cycle is about 6 weeks for a broiler. The length of the cycle is influenced by the degree to which the feeding diet is balanced and considers the cost of feed per 1 kg of meat produced, the feed-to-meat conversion ratio and the sale price of boiler meat. Feed quality, heat regulation, veterinary/ sanitary control and animal density within breeding houses (on average 10 animals/m²) are the most important factors affecting growth. After 6 weeks, broilers reach an average weight of 2.5 kg. They are then gathered into cages and sold to processors for slaughtering. In broiler meat production, cages are used only for transportation purposes and not for containing broilers during their growth.

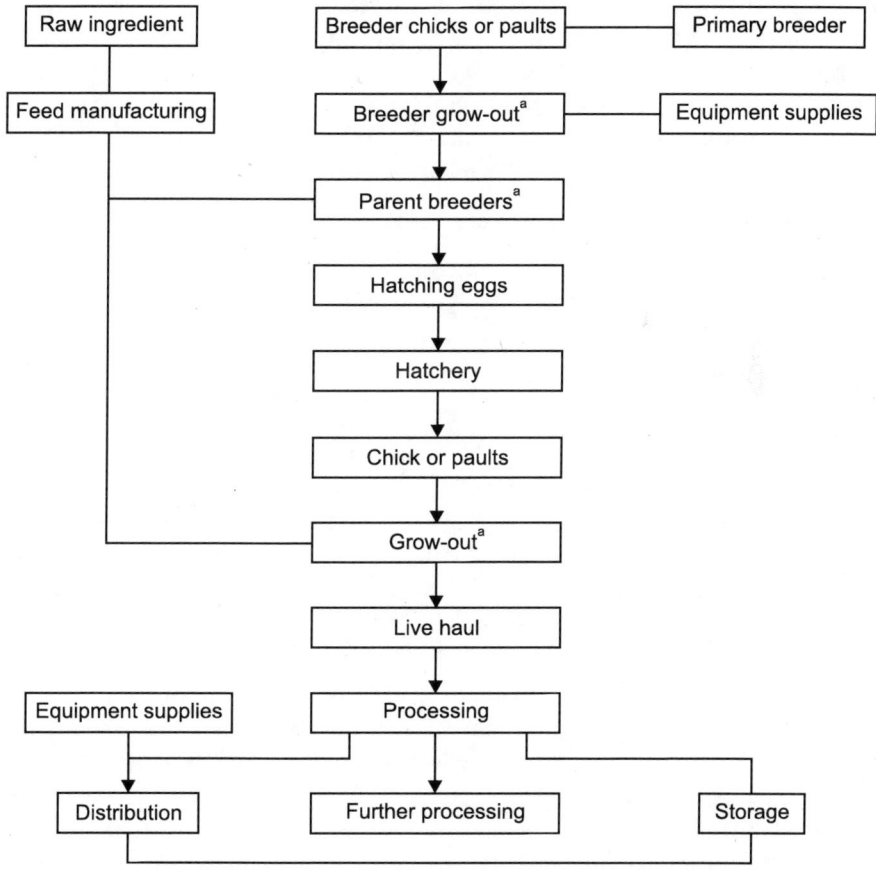

Note: [a] Normally produced under contact arrangement with grower

Figure 17.1 A typical integrated broiler operation

Figure 17.1 shows the following major elements of a modern broiler production chain:

- The hatching–the egg farm provides fertile eggs to the hatchery.
- The hatchery where fertile eggs incubate and day-old chicks (DOCs) hatch from the eggs. All healthy DOCs are sent to broiler grow-out facilities. The DOCs are raised for about 42 days, and after that time they are ready for slaughter or to be sold live.
- The processing plant where birds are slaughtered and either prepared as ready-to-cook whole chicken or cut-up chicken parts, or further processed into deboned chicken, polony and viennas.

There are various arrangements for the raising of broilers, depending on the level of integration of the companies in the broiler industry. In some countries (the United States, Canada, the United Kingdom), integrated companies own and control the hatcheries, feed mills, processing plants and distribution facilities but contract up to 90% of the broiler grow-out with private farms. In the Russian Federation and Ukraine, where contract farming is not developed, all major broiler meat producers also own and control the broiler grow-out facilities.

Advantages of farming broilers
The advantages of farming broilers over other animals are:

- Broilers have a growth cycle of 6–7 weeks, which allows for repeated production throughout the year.
- More broilers than layers can be placed in a shed.
- Broilers have high feed conversion efficiency (FCE) in comparison with other birds or livestock. FCE of broilers is 2, of turkeys is 2.5 and of ducks is 2.5–3. FCE depends on many factors, including the age of birds, feed quality, duration of lighting, etc. Cattle need more than 8 kg of feed per head to put on 1 kg of live weight per head, a ruminant will convert about 7 kg of feed to 1 kg of meat and a pig will convert about 3.5 kg of feed to 1 kg of meat.
- Return from the investment in broilers is fast. Different growth cycles have a direct impact on net cash flow. For example, eight generations of birds can be raised per year compared with from two to two-and-one-half generations of pigs and less than one generation (0.8) of cattle.
- High consumer demand and preference for broilers.

Today's poultry industry is characterized by vertical integration whereby a single company owns the facilities and controls the breeding, hatching and/or processing of broilers but contracts with private poultry farmers for the raising of the chicks. Although an integrated company, referred to as an integrator, owns the birds and provides feed, medication and other supplies, the grower owns the chicken house and provides litter, labour and utilities.

17.3.2   Breeding and feeding

There are two systems of poultry raising: extensive and intensive production.

Extensive (or pasture-based) production system

From a production standpoint, birds are pastured to obtain nutrients from the pasture, improve land fertility and improve bird health. Poultry obtain nutrients from young, vegetative forage plants but because they cannot digest cellulose as ruminants do, they do not make use of the vast energy stored in the plant fiber.

In addition to foraging for plants, poultry in pasture also forage for seeds and live protein such as worms and insects. Extensive breeding is largely practiced in developing countries where natural native genetic lines are used. Small farmers raise poultry in 'free-range' or pasture-based systems that are part of a diversified farm.

There is a niche market for 'free-range' poultry, which is rare in the modern industry. Most producers would rather not expose birds to predators and wildlife to avoid diseases. However, some farmers manage this type of production successfully and obtain premium prices for their poultry (for instance, in France, up to 30% of poultry is produced on range under the Label Rouge certification programme). However, the price premium for 'free-range' chickens in the EU, North America and other developed markets should be viewed against consumer disposable income and willingness to pay for this type of product.

Intensive (or cereal-oilseed protein-based) production system

Under the intensive system, all the nutrients required by the birds must be provided in the feed, usually in the form of a balanced and mixed feed. The following main features distinguish an intensive poultry production system from an extensive poultry production system:

- Intensive range poultry production can be a stand-alone enterprise and requires only a small amount of land. However, careful manure management is needed to prevent environmental damage.
- Extensive range poultry production requires much more land and is usually part of a diversified operation with ruminants. Mixed husbandry can be very important in range poultry production.

17.3.3   Chicken slaughtering and processing

The most common product produced in poultry slaughterhouses is the whole bird. However, poultry meat can be further processed into various products based on the type of poultry meat desired (e.g. from simple cuts to ready-to-eat meals). In fact, during recent years there has been a shift from fresh, whole-bird sales to sales of cut-up bird parts and convenience products because these products have higher value. Whole chickens are available fresh, frozen, bone-in, boneless, uncooked, fully cooked and seasoned.

Chicken parts are available as drumsticks, thighs, wings and breasts. They are also available as legs (drumstick and thigh attached), leg quarters, breast quarters, breast halves and poultry halves. Wingettes and drummettes made from the wing are available. Chicken products taken from the breast and wing are considered white meat and the products taken from the drumstick and thigh are considered dark meat. Cutting up and further processing chickens add value to the product and increase convenience to consumers.

17.3.4   Poultry processing operations

Because poultry meat is a perishable product (chilled poultry meat must be sold to consumers within 72 hours after processing), the conception/ organization of the slaughterhouse must be closely adapted to feed cost and availability, meat supply conditions and consumer markets (average weight and age of animals, available quantity of animals per week and seasonality of supplies).

Preprocessing

The birds are usually transported by truck to the poultry slaughterhouse. Upon arrival, the birds are held in the reception area in the transport crates, pending veterinarian inspection. In most countries, the official veterinarian then inspects each transport crate of live birds to approve them for human consumption. Sick birds are killed and disposed of. After inspection, the birds are removed from the crates in the reception area and put on the killing line. The birds are hung upside down by their feet by shackles on a conveyor, which moves them to the stunning area. Once the birds are shackled, stunning is carried out using one of three possible methods that include: (i) an electrically charged water bath, (ii) gas inhalation and (iii) a blow to the head using a blunt object.

Slaughtering, bleeding and scalding

Slaughtering can be performed manually or by using an automatic circular knife system. The birds should bleed for at least 2 minutes to ensure a total bleed-out. The blood is collected in a tank and handled as an animal by-product for further processing. After bleeding, the birds are exposed to either steam or hot water as part of the scalding procedure. Scalding loosens the feathers and facilitates plucking.

Further processing and evisceration

Feathers are removed in a specially designed plucking machine or by hand. Feathers are collected and treated as an animal by-product. The birds are showered with water during the automated plucking operation and the feathers are collected in a trough under the plucking machine. Following scalding and plucking activities, the head and feet are removed. Inedible organs, including the intestinal tract and lungs, are removed and treated as animal by-products. The eviscerated carcass should be rinsed internally and externally with potable water before further processing. Depending on country-specific regulations, companies may use disinfectants (chlorine or tri-sodium phosphate solutions) to reduce the bacteriological contamination on meat surfaces.

Storing and packaging

After rinsing, the carcass should be cooled as quickly as possible to or below 4°C. Several methods are used for chilling, including: air chilling, which takes place in either a chill room or by continuous air blast; spray chilling whereby water aerosols are added to the air; and immersion chilling, which involves moving carcasses through a counter-flow current in a water bath.

Birds are weighed individually and sorted according to their weight. After weighing, the birds are inspected visually and categorized. Whole birds are typically packed in plastic bags or in containers wrapped in film. Birds are stored before sale at or below 4°C. Birds intended to be sold as quick frozen poultry are frozen in a blast freezer or similar equipment that enables rapid freezing.

Cleaning of the processing plant

Cleaning of the plant is one of the most important tasks in a poultry processing plant. Some rinsing and cleaning should occur during working hours. After working hours, a total cleaning and disinfection of the plant is carried out, normally on a daily basis. Cleaning involves the following major steps, including disassembling of machinery and equipment, as necessary; physical removal of solid material; cycles of rinsing and washing; disinfection; drying; and application of lubricants.

Rendering

Rendering is the heat treatment of animal by-products to eliminate the risk of spreading disease to animals and humans and to produce usable products such as proteins and fat. Rendering involves evaporative processes that may generate a foul odour.

17.3.5   Poultry by-products

Low-risk material

Low-risk by-products are by-products obtained from poultry that have been approved as fit for human consumption (e.g. blood, heads and feet).

Blood is collected in a separate tank. Depending on the storage time before further processing, the need for cooling and chemicals that can prevent coagulation should be considered. Blood is filtered and spray-dried to produce blood meal. Blood meal can be used for feeding fish, pets and other animals.

Feathers are collected in a separate container. Before transfer to the container, water from the scalding process has to be pressed out of the feathers. Because the plucking process can remove portions of the heads as well, some head bits may be present among the feathers. Feathers can be burned to produce heat or processed with heat to hydrolyze the proteins. The low-value proteins from feathers can be used in pet food or animal feed.

Heads and *feet* that are not destined for human consumption are collected in a separate container. When these by-products are to be used for human consumption, they should be approved during the inspection process. Typically, feet used for human consumption are heat treated in order to remove skin and nails before packing.

Heads are normally not used for human consumption, although duck tongues are consumed in some countries.

High-risk material

High-risk by-products include birds that have died for reasons other than slaughtering, condemned birds, and condemned parts of birds, as well as all other by-products not intended for human consumption. Solid organic material that is captured in the wastewater treatment system screens and has a particle size of 6 mm or greater should also be treated as a high-risk by-product and sent for rendering. Grids used in the slaughterhouse and prefiltering of waste streams should be designed so that these kinds of animal by-products can be recovered and sent for rendering.

Processing of by-products

By-products should be collected in separate containers, which are isolated in such a way that food safety is not jeopardized. The containers should be covered to prevent wild birds and animals from coming into contact with the material they contain. At the rendering plant, the materials are chopped up and then heated under pressure (e.g. in the conventional batch dry rendering method) to kill microorganisms and remove moisture.

The liquefied fat and solid protein are separated by centrifugation or pressing. The solid product can then be ground into various animal protein powders for animal feed or pet food.

The effectiveness of the heat process used for rendering depends on various factors, including the holding time, the core temperature and the particle size of the products treated. The rendering process should produce final products that are free from salmonella and clostridium and contain only a limited number of enterobacteriaceae.

Environmental aspects

- *Waste and wastewater*: Water contamination has become a major issue confronting industrial poultry operators.
- *Chicken manure*: Almost 18 billion birds are raised each year in the world and produce more than 22 million tons of manure. Poultry manure is rich in nitrogen and phosphorous and contaminates groundwater and surface waterways such as rivers and bays. Ammonia gas must be ventilated from the chicken houses and can contaminate soil and water. Arsenic, an additive to chicken feed, contaminates litter or waste generated each year by the broiler chicken industry and also contaminates the communities in which it is generated or disposed.
- *Chicken processing*: Chicken guts, heads, feathers, blood and wastewater that remain from the processing are rendered down to their essence before being hauled as sludge to fertilize farmlands in the area. Treated wastewater is released into nearby streams or sprayed on farmlands in the area.
- *Water use*: Three thousand five-hundred liters of water are used in the production of 1 kg of meat. Modern broiler houses (e.g. typically 500 ft long) require almost 38 liters of water per minute.

17.4   Laying hens for egg production

It is almost impossible that any companies rear layers in exactly the same way. All companies use a slight variation of the typical rearing programme described in this section. Management differences in rearing layers may be accounted for by economics (breed selection, vaccination package and the decision as to when to allow moulting), producer preference (breed and strain selection) and/or geography (breed selection and vaccination package).

Hatching and placement: Egg producers purchase their layer stock (i.e. day old Leghorn or other genetic lines of chicks) from a hatchery. At the hatchery, chicks are vaccinated according to the producer's specifications. Hatcheries deliver chicks to the producer within 1–2 days of hatching. Upon arrival, chicks are placed in typical layer pens or are reared in a pullet house.

Lighting and temperature: From chick placement through to approximately 16 weeks of life, the pullets are fed according to body-weight gain and/or age. The goal is to raise a strong and healthy bird that can support egg production. An increase in daily light exposure triggers hens to begin laying eggs. If the laying hen has not reached proper body weight (usually 1 kg) by week 18, egg production will cease very quickly.

Feeding: Feed is usually offered to birds via the chain system. The chain system transports feed into the metal feeder at precise times during the day. In addition to monitoring dietary protein, producers must closely examine other feed ingredients: lysine, methionine, calcium and phosphorus. Before being slaughtered, laying hens are given appropriate feed to reconstitute muscle volume.

Egg production: Hens start laying regularly at around 18–20 weeks of age and in commercial systems, they typically lay for about a year before being sent for slaughter. Producers begin to photo stimulate (regulate the light and its intensity) and adjust the diet around 18 weeks of age in order to support egg production. The calcium levels in the diet are approximately 5–7 times greater than phosphorus levels. When a flock first enters egg production, the rate of egg lay will be around 10–20%. This means that 10–20% of the hens are laying eggs at 18–22 weeks of age. The flock quickly reaches peak egg production (>90%) at 30–32 weeks of age. After about 50 weeks, when the laying curve decreases greatly, hens are gathered a few weeks out of the regulated light cycle in order to perform the so-called moulting (change of feathers). A second laying cycle can then begin. Post-peak egg production (after 30–32 weeks of age) continually decreases to approximately 50% of the hens around 60–70 weeks of age. At this point an economic decision must be made by the producer; production by 50% of the hens is near the 'break-even' point for egg producers (e.g. feed cost = market price of eggs).

The time from ovulation to laying is about 25 hours. About 30 minutes after laying, the hen will begin to make another egg. Commercial hens have been bred to produce a very high yield of around 300 eggs per year. Feed quality, heat regulation, sanitary control and animal density within hen houses are the key factors that affect chick mortality rates.

Egg collection: Hens lay eggs onto an angled wire floor and the eggs roll towards the front of the cage and onto a belt. The belt transports eggs out of the house either to the egg packaging facility or to a storage cooler. Once the eggs enter the egg-packaging centre, they are washed (detergent solution near 40°C and pH 11 that removes soil) within minutes or no later than 12–14 hours post-lay.

Bacteria can be on the outside of an eggshell because the egg exits the hen's body through the same passageway as faeces; thus, the reason for washing and sanitizing eggs at the processing plant. Bacteria can be inside an uncracked, whole egg. Contamination of eggs may be due to bacteria such as *Salmonella Enteritidis* in the hen's ovary or oviduct before the shell forms around the yolk and white. *Salmonella Enteritidis* does not make the hen sick. It is also possible for eggs to become infected by *Salmonella Enteritidis* faecal contamination via the pores of the shells after they have been laid. After washing, eggs are visually inspected (checked for eggshell problems, cracks and blood spots), and then graded for packaging. The weight of a hen's egg may vary from 50 g to 70 g, but the average weight is 55 g. Following packaging, eggs are moved to a cool room (4–7°C), where they await shipment to retail outlets.

17.4.1   Composition and nutritional value of eggs

Eggs are one of nature's highest quality sources of protein and indeed contain many important nutrients. They are a good source of high biological-value protein and they are easily digested. Therefore, they are valuable food for people who are recovering from illness.

Eggs are composed of three main parts:

1. Shell
2. Egg white
3. Egg yolk

88.5% of an egg is edible.

Shell: The shell of an egg is porous to allow the developing chick to obtain oxygen. Other than oxygen, bacteria and odours can enter the egg. The pores also allow water and carbon dioxide to escape. The membrane that lines the inside of a shell acts as a filter to protect against bacteria. At one end of the egg, the membrane separates into an air space to supply the chick with oxygen. The shell is generally strong but the older birds tend to produce weaker shells. The shell's colour varies according to the breed of the bird.

Egg white: There are two layers of egg white:

1. The thick white layer (nearest to the yolk)
2. The thin white layer (nearest to the shell)

Yolk: The colour of the egg yolk is related to the diet of the hen and is due to the presence of carotenes and colourings added to a hen's feed. The nutritional value of the egg is not affected by the colour of the yolk.

The nutritional content of egg white is:

- 10.5% protein
- 88.5% water
- Riboflavin and other B vitamins
- A trace of fat

The nutritional content of egg yolk is:

- 16.5% protein
- 33% fat
- 50% water
- Fat-soluble vitamins A, D, E and K
- Mineral elements, including iron
- Lecithin (an emulsifier)

17.4.2   Egg products

The term 'egg products' refers to eggs that are removed from their shells for processing. The processing of egg products includes breaking eggs, filtering, mixing, stabilizing, blending, pasteurizing, cooling, freezing or drying, and packaging. Liquid, frozen, and dried egg products are used widely by the food-service industry and the commercial-food industry. They are scrambled or made into omelettes, or used as ingredients in egg dishes or other foods such as mayonnaise or ice cream. Pasteurized egg products are products that were rapidly heated and held at a minimum required temperature for a specified time. This destroys *Salmonella*, but it does not cook the eggs or affect their colour, flavour, nutritional value or use.

Dried whites are pasteurized by heating them in the dried form again for a specified time and at a minimum required temperature.

Egg white powder is dried egg white (pure albumen). It can be reconstituted by mixing the powder with water. The reconstituted powder whips like fresh egg white and, because it is pasteurized, can be used safely without cooking or baking it. The product is usually sold along with supplies for cake baking and decorating.

Eggs are commonly used in food preparation for:

- *Thickening:* Eggs are used to thicken custards, sauces, soups, etc. because of the coagulation of the egg proteins.

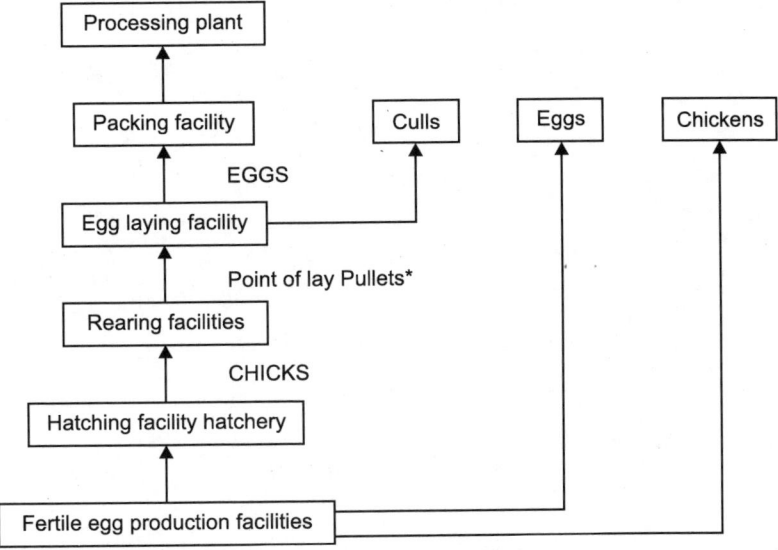

* A young domestic hen, usually one that is less than one-year old.

Figure 17.2 Egg production process.

- *Emulsifying:* Egg yolk contains lecithin, which is an emulsifier and enables oil and water to be mixed without separating. It is used in making mayonnaise and cakes.
- *Binding:* Ingredients for rissoles, croquettes, and meat or fish cakes can be bound together with egg, which when heated will coagulate and hold the ingredients together.
- *Coating:* Eggs are used as a coating for fried foods because they form a protective layer on the outside of the food that sets and holds the food together and prevents it from being overcooked.
- *Glazing:* Egg yolk, egg white or whole egg is brushed over pastries and bread to produce a golden brown shiny glaze during baking.

Impact of biotechnology on poultry nutrition

18.1 Introduction

Poultry farming is the raising of domesticated birds such as chickens, ducks, turkeys and geese for the purpose of farming meat or eggs for food. Poultry are farmed in great numbers. Chickens raised for eggs are usually called layers while chickens raised for meat are often called broilers. The increase of productivity in the poultry industry has been accompanied by various impacts, including emergence of a large variety of pathogens and bacterial resistance. These impacts are in part due to the indiscriminate use of chemotherapeutic agents as a result of management practices in rearing cycles. This chapter summarizes the use of probiotics for prevention of bacterial diseases in poultry, as well as demonstrating the potential role of probiotics in the growth performance and immune response of poultry, safety and whole-someness of dressed poultry meat evidencing consumer's protection.

The poultry industry has become an important economic activity in many countries. In large-scale rearing facilities, where poultry are exposed to stressful conditions, problems related to diseases and deterioration of environmental conditions often occur and result in serious economic losses. Prevention and control of diseases have led during recent decades to a substantial increase in the use of veterinary medicines. However, the utility of antimicrobial agents as a preventive measure has been questioned, given extensive documentation of the evolution of antimicrobial resistance among pathogenic bacteria. So, the possibility of antibiotics ceasing to be used as growth stimulants for poultry and the concern about the side effects of their use as therapeutic agents has produced a climate in which both consumer and manufacturer are looking for alternatives. Probiotics are being considered to fill this gap and already some farmers are using them in preference to antibiotics.

18.2 Impact of biotechnology on poultry nutrition

The impact of biotechnology in poultry nutrition is of significant importance. Biotechnology plays a vital role in the poultry feed industry. Nutritionists are continually putting their efforts into producing better and more economical feed. Good feed alone will not serve the purpose but its better utilization is also essential. Dietary changes as well as lack of a healthy diet can influence the balance of the microflora in the gut thus predisposing to digestion upsets. A well-balanced ration sufficient in energy and nutrients is also of great importance in maintaining a healthy gut. A great deal of attention has recently been received from nutritionists and veterinary experts for proper utilization of nutrients and the use of probiotics for growth promotion of poultry.

In broiler nutrition, probiotic species belonging to *Lactobacillus, Streptococcus, Bacillus, Bifidobacterium, Enterococcus, Aspergillus, Candida,* and *Saccharomyces* have a beneficial effect on broiler performance, modulation of intestinal microflora and pathogen inhibition, intestinal histological changes, immunomodulation, certain haemato-biochemical parameters, improving sensory characteristics of dressed broiler meat and promoting microbiological meat quality of broilers.

Probiotics are 'live microorganisms which when administered in adequate amounts confer a health benefit on the host'. More precisely, probiotics are live microorganisms of nonpathogenic and nontoxic in nature, which when administered through the digestive route, are favourable to the host's health.

It is believed by most investigators that there is an unsteady balance of beneficial and non-beneficial bacteria in the tract of normal, healthy, non-stressed poultry. When a balance exists, the bird performs to its maximum efficiency, but if stress is imposed, the beneficial flora, especially *lactobacilli,* have a tendency to decrease in numbers and an overgrowth of the non-beneficial ones seems to occur. This occurrence may predispose frank disease, i.e. diarrhea, or be subclinical and reduce production parameters of growth, feed efficiency, etc. The protective flora which establishes itself in the gut is very stable, but it can be influenced by some dietary and environmental factors. The three most important are excessive hygiene, antibiotic therapy and stress. In the wild, the chicken would receive a complete gut flora from its mother's faeces and would consequently be protected against infection (Fig. 18.1). However, commercially reared chickens are hatched in incubators which are clean and do not usually contain organisms commonly found in the chicken gut. There is an effect of shell microbiological contamination which may influence gut microflora characteristics. Moreover, also HCl gastric secretion, which starts at 18 days of incubation, has a deep impact on microflora selection. Therefore, an immediate use of probiotics supplementation at birth is more important and useful in avian species than in other animals. The chicken is an extreme example of a young animal which is deprived of contact with its mother or other adults and which is, therefore, likely to benefit from supplementation with microbial preparations designed to restore the protective gut microflora.

The species currently being used in probiotic preparations are varied and many. These are mostly *Lactobacillus bulgaricus, Lactobacillus acidophilus, Lactobacillus casei, Lactobacillus helveticus, Lactobacillus lactis, Lactobacillus salivarius*, *Lactobacillus plantarum*, *Streptococcus thermophilus*, *Enterococcus faecium*, *Enterococcus faecalis*, *Bifidobacterium* spp. and *Escherichia coli*. With two exceptions, these are all intestinal strains. The two exceptions, *Lactobacillus bulgaricus* and *Streptococcus thermophilus*, are yoghurt starter organisms. Some other probiotics are microscopic fungi such as strains of yeasts belonging to *Saccharomyces cerevisiae* species.

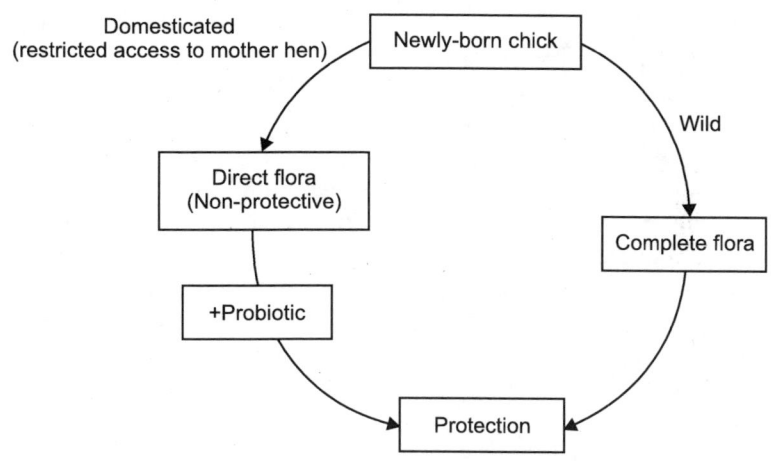

Figure 18.1 Schematic representation of the concept of probiotics

18.3   Mechanisms of action

Enhancement of colonization resistance and/or direct inhibitory effects against pathogens are important factors where probiotics have reduced the incidence and duration of diseases. Probiotic strains have been shown to inhibit pathogenic bacteria both *in vitro* and *in vivo* through several different mechanisms.

The mode of action of probiotics in poultry includes: (i) maintaining normal intestinal microflora by competitive exclusion and antagonism, (ii) altering metabolism by increasing digestive enzyme activity and decreasing bacterial enzyme activity and ammonia production, (iii) improving feed intake and digestion and (iv) stimulating the immune system.

Probiotic and competitive exclusion approaches have been used as one method to control endemic and zoonotic agents in poultry. In traditional terms, competitive exclusion in poultry has implied the use of naturally occurring intestinal microorganisms in chicks and poults that were ready to be placed in brooder house. Nurmi and Rantala first applied the concept when they attempted to control a severe outbreak of *S. infantis* in Finnish broiler flocks. In their studies, it was determined that very low challenge doses of

Salmonella (1–10 cells into the crop) were sufficient to initiate *salmonellosis* in chickens. Additionally, they determined that it was during the 1st week post-hatch that the chick was most susceptible to *Salmonella* infections. Use of a *Lactobacillus* strain did not produce protection, and this forced them to evaluate an unmanipulated population of intestinal bacteria from adult chickens that were resistant to *S. infantis*. On oral administration of this undefined mixed culture, adult-type resistance to *Salmonella* was achieved. This procedure later became known as the Nurmi or competitive exclusion concept. The competitive exclusion approach of inoculating day-old chicks with an adult microflora successfully demonstrates the impact of the intestinal microbiota on intestinal function and disease resistance. Although competitive exclusion fits the definition of probiotics, the competitive exclusion approach instantaneously provides the chick with an adult intestinal microbiota instead of adding one or a few bacterial species to an established microbial population. Inoculating day-old chicks with competitive exclusion cultures or more classical probiotics serves as a nice model for determining the modes of action and efficacy of these microorganisms. Because of the susceptibility of day-old chicks to infection, this practice is also of commercial importance. By using this model, a number of probiotics have been shown to reduce colonization and shedding of *Salmonella* and *Campylobacter*. Competitive exclusion is a very effective measure to protect newly hatched chicks, turkey poults, quails and pheasants and possibly other game birds, too, against *Salmonella* and other entero-pathogens.

Upon consumption, probiotics deliver many lactic acid bacteria into the gastrointestinal tract. These microorganisms have been reputed to modify the intestinal milieu and to deliver enzymes and other beneficial substances into the intestines. Supplementation of *L. acidophilus* or a mixture of *Lactobacillus* cultures to chickens significantly increased (P<0.05) the levels of amylase after 40 days of feeding. This result is similar to the finding of Collington and others, who reported that inclusion of a probiotic (a mixture of multiple strains of *Lactobacillus* spp. and *Streptococcus faecium*) resulted in significantly higher carbohydrase enzyme activities in the small intestine of piglets. The *lactobacilli* colonizing the intestine may secrete the enzyme, thus increasing the intestinal amylase activity. It is well established that probiotics alter gastrointestinal pH and flora to favour an increased activity of intestinal enzymes and digestibility of nutrients.

The effect of *Aspergillus oryzae* on macronutrients metabolism in laying hens was observed, of which findings might be of practical relevance. They postulated that active amylolytic and proteolytic enzymes residing in *Aspergillus oryzae* may influence the digested nutrients. Similarly, it was reported that an increase in the digestibility of dry matter was closely related to the enzymes released by yeast. In addition, probiotics may contribute to the improvement of health status of birds by reducing ammonia production in the intestines.

Probiotic is a generic term, and products can contain yeast cells, bacterial cultures, or both that stimulate microorganisms capable of modifying the gastrointestinal environment to favour health status and improve feed efficiency. Mechanisms by which probiotics improve feed conversion efficiency include alteration in intestinal flora, enhancement of growth of nonpathogenic facultative anaerobic and gram positive bacteria forming lactic acid and hydrogen peroxide, suppression of growth of intestinal pathogens, and enhancement of digestion and utilization of nutrients. Therefore, the major outcomes from using probiotics include improvement in growth, reduction in mortality, and improvement in feed conversion efficiency. These results are consistent with previous experiment of Tortuero and Fernandez, who observed improved feed conversion efficiency with the supplementation of probiotic to the diet.

The manipulation of gut microbiota via the administration of probiotics influences the development of the immune response. The exact mechanisms that mediate the immunomodulatory activities of probiotics are not clear. However, it has been shown that probiotics stimulate different subsets of immune system cells to produce cytokines, which in turn play a role in the induction and regulation of the immune response. Stimulation of human peripheral blood mononuclear cells with *Lactobacillus* rhamnosus strain GG *in vitro* resulted in the production of interleukin 4 (IL-4), IL-6, IL-10, tumor necrosis factor alpha, and gamma interferon. Other studies have provided confirmatory evidence that Th2 cytokines, such as IL-4 and IL-10, are induced by *lactobacilli*. The outcome of the production of Th2 cytokines is the development of B cells and the immunoglobulin isotype switching required for the production of antibodies. The production of the mucosal IgA response is dependent on other cytokines, such as transforming growth factor β. Importantly, various species and strains of *lactobacilli* are able to induce the production of transforming growth factor β, albeit to various degrees. Probiotics, especially *lactobacilli*, could modulate the systemic antibody response to antigens in chickens.

18.4   Criteria for selection of probiotics in the poultry industry

The perceived desirable traits for selection of functional probiotics are many. The probiotic bacteria must fulfill the following conditions: it must be a normal inhabitant of the gut, and it must be able to adhere to the intestinal epithelium to overcome potential hurdles, such as the low pH of the stomach, the presence of bile acids in the intestines, and the competition against other microorganisms in the gastro-intestinal tract. The tentative ways for selection of probiotics as biocontrol agents in the poultry industry are illustrated in Fig. 18.2. Many *in vitro* assays have been developed for the pre-selection of probiotic strains. The competitiveness of the most promising strains selected by *in vitro* assays was evaluated *in vivo* for monitoring of their persistence in chickens. In addition, potential probiotics must exert its beneficial effects (e.g. enhanced nutrition and increased immune response) in the host. Finally, the probiotic must be viable under normal storage conditions and technologically suitable for industrial processes (e.g. lyophilized).

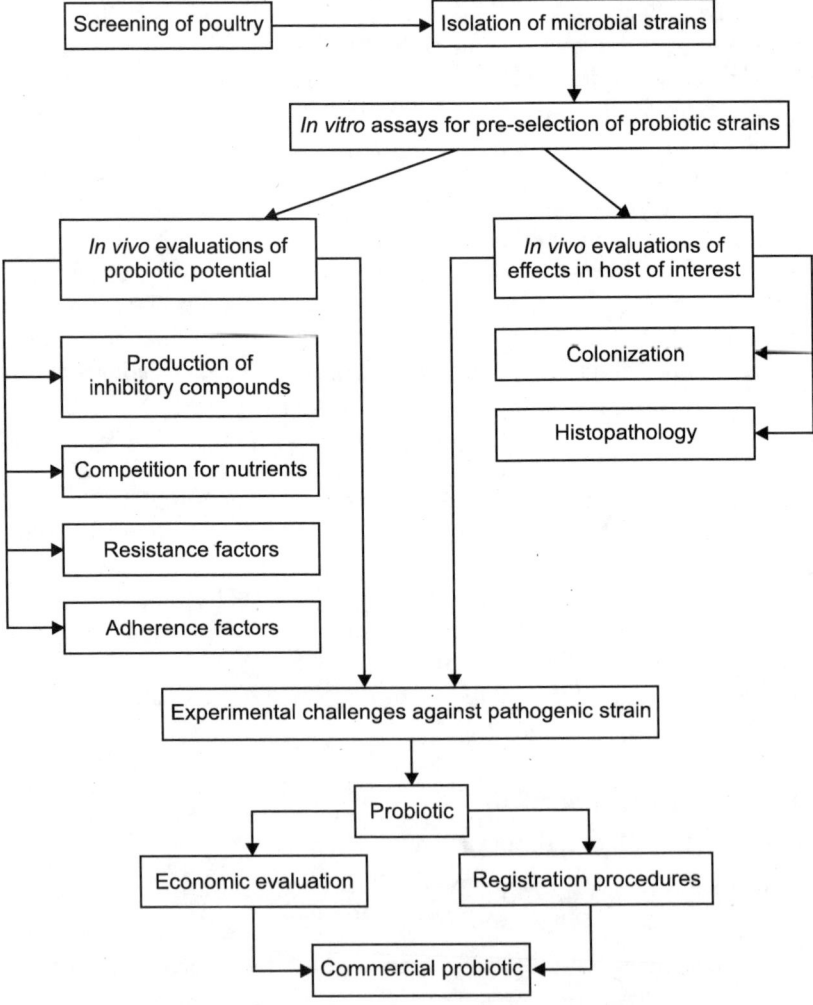

Figure 18.2 Diagram for selection of probiotics in the poultry industry

18.5   Evaluating probiotic effects on growth performance

Studies on the beneficial impact on poultry performance have indicated that probiotic supplementation can have positive effects. It is clearly evident from the result of Kabir and other that the live weight gains were significantly (P<0.01) higher in experimental birds as compared to control ones at all levels during the period of 2nd, 4th, 5th and 6th weeks of age, both in vaccinated and non-vaccinated birds. This result is in agreement with many investigators who demonstrated increased live weight gain in probiotic fed birds. On the other hand, Lan and others found higher (P<0.01) weight gains in broilers subjected to two probiotic species. Huang and others demonstrated that inactivated probiotics, disrupted by a high-pressure homogenizer, have positive effects on the production performance of broiler chickens when used at certain concentrations.

Kabir and others reported the occurrence of a significantly (P<0.01) higher carcass yield in broiler chicks fed with the probiotics on the 2nd, 4th and 6th week of age both in vaccinated and non-vaccinated birds. Although Mahajan and others recorded in their study that mean values of giblets, hot dress weight, cold dress weight and dressing percentage were significantly (P<0.05) higher for probiotic (Lacto-Sacc) fed broilers. On the other hand, Mutus and others investigated the effects of a dietary supplemental probiotic on morphometric parameters and yield stress of the tibia and they found that tibiotarsi weight, length, and weight/length index, robusticity index, diaphysis diameter, modulus of elasticity, yield stress parameters, and percentage Ca content were not affected by the dietary supplementation of probiotic, whereas thickness of the medial and lateral wall of the tibia, tibiotarsal index, percentage ash, and P content were significantly improved by the probiotic.

18.6   Evaluating probiotic effects on the intestinal microbiota and intestinal morphology

Kabir and others attempted to evaluate the effect of probiotics with regard to clearing bacterial infections and regulating intestinal flora by determining the total viable count (TVC) and total *lactobacillus* count (TLC) of the crop and cecum samples of probiotics and conventional fed groups at the 2nd, 4th and 6th week of age. Their result revealed competitive antagonism. The result of their study also evidenced that probiotic organisms inhibited some non-beneficial pathogens by occupying intestinal wall space. They also demonstrated that broilers fed with probiotics had a tendency to display pronounced intestinal histological changes such as active impetus in cell mitosis and increased nuclear size of cells, than the controls.

18.7   Evaluating probiotic effects on immune response

Kabir and others evaluated the dynamics of probiotics on immune response of broilers and they reported significantly higher antibody production (P<0.01) in experimental birds as compared to control ones. They also demonstrated that the differences in the weight of spleen and bursa of probiotics and conventional fed broilers could be attributed to different level of antibody production in response to SRBC. They also suggested a positive impact of the probiotic in stimulating some of the early immune responses against *E. acervulina*, as characterized by early IFN-γ and IL-2 secretions, resulting in improved local immune defenses against coccidiosis. Brisbin and others investigated spatial and temporal expression of immune system genes in chicken cecal tonsil and spleen mononuclear cells in response to structural constituents of *L. acidophilus* and they found that cecal tonsil cells responded more rapidly than spleen cells to the bacterial stimuli, with the most potent stimulus for cecal tonsil cells being DNA and for splenocytes being the bacterial cell wall components.

18.8   Evaluating probiotic effects on meat quality

Kabir and others evaluated the effects of probiotics on the sensory characteristics and microbiological quality of dressed broiler meat and reported that supplementation of probiotics in broiler ration improved the meat quality both at pre-freezing and post-freezing storage. Mahajan and others stated that the scores for the sensory attributes of the meat balls appearance, texture, juiciness and overall acceptability were significantly (p60.001) higher and those for flavour were lower in the probiotic (Lacto-Sacc) fed group. Simultaneously, Mahajan and others reported that meat from probiotic (Lacto-Sacc) fed birds showed lower total viable count as compared to the meat obtained from control birds. On the other hand, Loddi and others reported that neither probiotic nor antibiotic affected sensory characteristics (intensity of aroma, strange aroma, flavour, strange flavour, tenderness, juiciness, acceptability, characteristic colour and overall aspects) of breast and leg meats. On the other hand, Zhang and others conducted an experiment with 240, day-old, male broilers to investigate the effects of *Saccharomyces cerevisiae* (SC) cell components on the meat quality and they reported that meat tenderness could be improved by the whole yeast (WY) or *Saccharomyces cerevisiae* extract (YE).

Thus, the concept of probiotics constitutes an important aspect of applied biotechnological research and therefore as opposed to antibiotics and chemotherapeutic agents can be employed for growth promotion in poultry. In past years, men considered all bacteria as harmful, forgetting about the use of the organisms in food preparation and preservation, thus making probiotic concept somewhat difficult to accept. Scientists now are triggering effort to establish the delicate symbiotic relationship of poultry with their bacteria, especially in the digestive tract, where they are very important to the well being of man and poultry. Since probiotics do not result in the development and spread of microbial resistance, they offer immense potential to become an alternative to antibiotics.

Section VI
Beverages

Beverages

19.1 Introduction

A beverage is any type of drink. It's something we might offer a guest in our house; it's also the favourite moniker of companies that manufacture both soda and juice—they call themselves beverage companies. Three groups of beverages are commonly consumed. These include: (i) carbonated non-alcoholic beverages or soft drinks, (ii) non-carbonated non-alcoholic stimulating beverages such as coffee and tea and (iii) alcoholic beverages.

19.2 Carbonated non-alcoholic beverages

The most popular soft drinks include those based on cola (extract from the cola tree), orange, ginger, lemon and lime. Soft drinks may be classified into three types as carbonated, fruit flavoured and sparkling soda water. The carbonated beverages may contain artificial flavour or natural fruit juice. The major ingredients of soft drinks include water, sugar and/or sweeteners, flavour emulsions and cloudifiers, colouring agents, preservatives and acids and carbon dioxide. Water constitutes the largest component to the extent of 92–93%. It must be free from suspended matter, colouring matter, objectionable odours and minerals which may interfere with the flavour and colour of the soft drink. Sucrose is the most commonly used sweetener.

19.3 Stimulating beverages

19.3.1 Coffee

The process of bringing the harvested coffee fruits to consumers as a beverage involves a series of steps. Greater control of each step of this process improves the ability to produce a good quality coffee. After harvesting, fruits undergo primary processing to separate the seeds. Secondary processing such as decaffeination and steam treatment are performed before roasting. After roasting, the coffee is ground and packed or further processed to produce instant coffee.

Good harvesting methods are important to produce good-quality coffee. Although the awareness of quality is important throughout the entire agricultural process, the degree of coffee fruit maturation and avoiding mold contamination during harvesting, drying, and storage of the seeds are especially critical. Mold contamination affects not only the aroma and flavour of the final beverage, but also its bioactivity, since undesirable mycotoxins and biogenic amines can affect human health.

Coffee fruits are typically harvested in one of three ways: picking, stripping, or mechanical harvest. After harvest, coffee fruits undergo pulp extraction to produce green coffee seeds.

Coffee processing: The processing step involves removal of the skin, pulp, parchment and silver skin. Either of the two methods, viz. dry method or wet method is employed. In the dry method of curing, the cherries are hulled and air-dried in the sun or artificially. In the wet or washed-coffee method, the cherries after removal of the outer skin are soaked in water and passed through pulping machines to separate the pulp from the rest of the bean. Sometimes an acid fermentation by lactic acid bacteria may follow the natural fermentation. After the pulp and its residues have been washed away, the beans are dried to reduce the moisture content to about 10–12%. Drying must be uniform because during the process of drying, the colour and flavour attributes are modified. Over drying or fluctuations in temperature will affect the quality of coffee bean. The dried beans are hulled to remove the thin layer of parchment yielding bluish-green coffee bean (green coffee).

Roasting: Roasting is typically carried out at about 260°C for about 5 minutes when the bean temperature reaches about 200°C. Physical and chemical changes occur during roasting. The beans swell in size, become brittle and change colour to deep brown. Moisture and a small percentage of volatiles are lost and carbon dioxide is produced in relatively large amounts, part of it escaping while the rest is absorbed by the roasted bean. Carbohydrates decompose, caramelize and in combination with other constituents contribute to the flavour and aroma. Fatty substances and proteins are degraded. However, caffeine remains unaffected during roasting.

Freshly roasted coffee has the best flavour and aroma, which deteriorate on standing. Similarly, coffee exposed to air undergoes oxidative changes and becomes stale. The staling of coffee is prevented by the presence of carbon dioxide in the roasted bean.

Grinding: The roasted beans are cooled and ground to a size depending upon its intended end use such as home brewing, vacuum extraction, percolator or vending machine. In each case the particle size and size distribution affect the brewing time, turbidity in the extract and other characteristics. Ground coffee loses its flavour and aroma quickly and hence freshly ground coffee is used for brewing or the ground coffee may be packed in hermetically sealed cans or jars under vacuum or inert gas.

Brewing: Freshly ground coffee is contacted with hot water in suitable vessels to extract the soluble caffeine and flavouring materials. Vacuum coffee is made in a two-part container. The upper compartment has the ground coffee and contains an open tube that extends to the bottom compartment containing water. The water in the lower compartment, on heating, exerts sufficient pressure to force the hot water through the open tube into the upper compartment and contacts the ground coffee. When the heat is reduced in the lower compartment the coffee extract is sucked into the bottom compartment. Drip coffee is made in coffee filter—a two-compartment container, the perforated upper compartment holding the ground coffee. Boiling water is poured which percolates down through a bed of ground coffee and the brew is collected in the lower compartment. The strength and flavour of the beverage depends on ratio of coffee to water, particle size, temperature of water, time and any mixing action.

Coffee contains a number of compounds that contribute to the flavour and bioactivity of the brew. Complex reactions take place during roasting at high temperatures and modify considerably coffee's chemical composition, with some beneficial compounds degraded and some created. A small amount of harmful compounds is also created during roasting; however, the beneficial compounds appear to predominate. To obtain a functional, healthy coffee, it is important to consider every aspect of coffee production, starting with high-quality seeds roasted to light-medium to dark-medium colour degree, preferably at low to medium temperatures.

Medium-roast coffees contain relatively high amounts of antioxidant compounds compared with other food products, a considerable amount of niacin, low acrylamide content, and typically no PAHs. Decaffeinated coffee is indicated for individuals sensitive to caffeine's effects and those who wish to use coffee as an additional tool to reduce the risk of type 2 diabetes.

19.3.2   Tea

All real teas are made from the tea plant, *Camellia sinensis*, a caffeine-producing bush. On an average, a cup of tea has 40–50 mg of caffeine, which is approximately half the caffeine content in a cup of coffee.

Caffeine content is not related to the level of fermentation (oxidation), so white, green, oolong, and black teas made from the same variety of *Camellia sinensis* will have essentially the same level of caffeine in the dry leaf.

The processes used to produce most black teas may play a factor in the caffeine extraction rate due to the crushing of the leaf cells during manufacture. The caffeine extraction from unprocessed leaf (white tea) is theoretically slower than for the highly processed (crushed cell) leaf. Also, *china* varietals (including those grown in Japan) have less caffeine than *assamica* varietals, further explaining why green and white teas tend to show lower caffeine levels when evaluated scientifically.

Black tea is the most popular variety. Processing steps include withering, rolling, fermentation, drying or firing, grading and packing. The plucked leaves and buds are spread thin and withered in trays or drying racks in drying rooms or drum dried. During this withering step, which lasts over 4 to 18 hours, partial drying of leaves occurs reducing the moisture content of the fresh leaves from about 75% to about 60–65% so that the leaves become flaccid and are ready for the next stage of processing. During withering, proteins are hydrolyzed to amino acids, a part of which is transaminated to keto acids. These acids are the precursors for aroma substances.

The next step after withering is rolling. In the rolling step the withered leaves and buds are conditioned in rollers to break open the cells and release the juices and enzymes. During this conditioning, a uniform distribution of polyphenol oxidase in leaves is achieved and the substrate and enzymes are brought together. A variety of rolling techniques are used which determine to some extent the flavour characteristics of the different tea varieties. In the final rolling step, the tea leaves are macerated by crank rollers under pressure. The rolling step may be regarded as the first stage in fermentation. After rolling the leaves are spread out thinly in layers of 5–7 cm thick and fermentation occurs over about 1–4 hours at 35–40°C. During fermentation the enzymes, particularly polyphenol oxidase, bring about the oxidation of the polyphenolic compounds present in the juices, resulting in a change of colour from green to bright coppery red and development of an odour similar to that of sour apples. The formation of pigments and aroma substances occurs during this step.

In addition, oxidation of amino acids, carotenoids, and unsaturated fatty acids also occurs. The enzymatic oxidation of flavanols via the corresponding *o*-quinones provides condensation reactions leading to formation of theaflavins, epitheaflavic acids, benzotropolone derivatives and thearubigins, a group of compounds responsible for the characteristic reddish-yellow colour and astringent taste of black tea extract.

Important tea aroma constituents such as dihydroactinidiolide, theaspirone, linalool and others such as β-ionone and its derivatives are formed. In addition, biosynthesis of aroma compounds such as nerolidol also occurs. Firing or heating the fermented leaves in ovens at 85–95°C for about 20 minutes stops the fermentation process.

The firing step reduces the moisture content to 3%, fixes the tea aroma and changes the colour to black (hence black tea).

During the firing step, the enzyme activity initially rises and about 15% of the theaflavins are formed during initial stages and finally the enzymes are inactivated. The tea becomes black due to the conversion of chlorophyll to pbeophytin at high temperatures in acidic environment. The undesired brown colour develops at higher pH values. The astringent character of tea decreases due to the formation of complexes between phenolic compounds and proteins. The firing step also influences the balance of volatiles, with a loss of some of the volatile compounds while simultaneous build-up of other aroma constituents such as β-ionone, dihydroactinidiolide and theaspirone occurs. In addition, pyrazines, pyridines and quinolines are formed due to sugar-amino acid interactions. The various constituents in black tea include oxidized polyphenolic compounds (~25%), proteins (15%), caffeine (4%), fiber (25%), other carbohydrate (7%), lipids (7%), amino acids (4%) and minerals (5%).

The dried tea is cleaned and graded into various commercial varieties according to the quality. The desirable quality of black tea is related to the polyphenol and enzyme content of the first two leaves and buds used in processing, which are maximum at about 27–28%. The 'leaf grade' tea is of highest quality followed by 'broken grade' which contains smaller and cut leaves. The leaf and broken grades are further categorized into orange pekoe, pekoe and souchang. The first two categories refer to the size of the leaf only; the orange pekoe having the largest size is the best quality tea. Souchang tea quality is the lowest. Apart from these, lesser grades comprising of waste products as fannings and dustings are also marketed. Tea is generally blended and packed into aluminium foil lined wooden boxes or paper cartons of smaller sizes.

19.4   Alcoholic beverages

19.4.1   Beer

Beer and ale are the principal malt beverages. Beer is made from barley malt to which hops (dried flowers of the hop plant) and cereal or malt adjuncts (starch or sugars obtained from corn, rice, wheat, soyabean, potato, cassava or barley) are added and fermented.

Beer is manufactured by the brewing process. The steps involved are: (i) malting, (ii) mashing to prepare wort (unfermented beer), (iii) boiling the wort with hops, (iv) fermentation, (v) maturation or ageing and (vi) finishing.

Beer is up to 97% water. It only makes sense then that the quality of the water would have a significant impact on the quality of the beer. In fact, because of the chemical processes involved, the mineral content of the brewing water is important to the brewing process. Certain historical beer styles evolved in the places they did in large part because of the water.

Ingredients of beer

Grains: Malted grains are the meat and potatoes of beer. They provide the sugars that are fermented by yeast to produce alcohol and CO_2. They also provide the essential nutrients yeast need to reproduce. They are the primary source of colour and body in beer. Malted grains are a major contributor to the flavour and aroma profiles of beer. The malting process is simply a controlled sprouting and kilning of the grain. The sprouting begins to break down starches contained in the grain making them accessible to the brewer and the kilning provides colour and flavour. Grains kilned at higher temperatures or longer periods of time are darker. Malted barley is by far the most widely used grain in beer making, but it is not the only one. Other malted grains commonly used in brewing include wheat, rye and oats.

Hops: Hops are the spice of beer. They provide bitterness to balance the sweetness of the malt, as well as flavours and aromas ranging from citrus and pine to earthy and spicy. Hops are the cone-like flower of a rapidly growing vine (a *bine* actually) in the *cannabis* family. Waxy yellow *lupulin glands* hidden within the leaves of the flower contain the acids and essential oils that give hops their character. Bitterness comes from *alpha acids* that must be chemically altered through boiling in order to be utilized. Hops more than any other brewing ingredient are subject to the phenomenon of *terroir*, as different growing regions produce hops with different flavour and aroma characteristics. The chief hop growing regions are the Northwestern US, Southern England, Germany, Czech Republic, and China.

Yeast: It is said that brewers make *wort* (the word for unfermented beer) and yeast makes beer. Yeast metabolizes the sugars from the grains and produces alcohol and CO_2. Yeast also produces an assortment of other fermentation by-products such as *phenols* and *esters* that add significant flavour and aroma character to beer ranging from delectable fruitiness to peppery spice. There are two main types of yeast, typically called *ale* and *lager*. *Ale* yeast ferment at higher temperatures between 65°F and 75°F.

Higher fermentation temperatures encourage the production of greater amounts of *esters* and *phenols*, resulting in beers with more yeast-derived flavour and aroma. *Lager* yeasts ferment colder, between 45°F and 55°F. Colder temperatures inhibit the production of the various fermentation byproducts resulting in beers with a 'clean' yeast profile. Certain beer styles are fermented with an assortment of wild yeasts and bacteria that produce a range of funky flavours from sour to barnyard.

Other common beer ingredients

Unmalted grains: In addition to the malted grains listed above there are a number of unmalted *adjunct grains* that are commonly used in beer. These grains are used to add flavour, improve mouth feel, lighten body and colour, and to make gluten free beers. The adjunct grains include corn, rice, wheat, rye, oats, and some lesser-used grains like sorghum. Perhaps the most common example of the use of these grains is the International Lager style that typically contains up to 30% rice or corn. Others include Oatmeal Stout (oats) and Belgian Witbier (wheat).

Other sugars: The use of non-grain sugars in beer is probably as old as brewing itself. While yeast cannot ferment all sugars derived from grain, these other sugars are often fully fermentable. For this reason they can improve drinkability in very strong beers by lightening the body without sacrificing the desired alcohol content. This is a common practice in Strong Belgian Ales and English Barley wines which can become thick and cloying when made entirely from grain sugars. Non-grain sugars are also used to enhance beer flavour and colour.

Common brewing sugars include brown sugar, cane sugar, molasses, honey, and the caramelized sugar syrups employed by Belgian brewers. Some American craft brewers are experimenting with more exotic sugars like Thai palm sugar.

Spices: The use of spices in beer goes back to the days before hops. As mentioned above, before hops became the preferred bittering agent for beer, brewers used a mix of herbs and spices called *gruit*. Using herbs and spices to add flavour and aromatic qualities continues to be popular today. Possibly every spice in existence has found its way into beer at some point. Common spices used in brewing include cinnamon, nutmeg, allspice, coriander, grains of paradise, ginger, orange peel, heather and anise, just to name a few. Use of spices is common in holiday ales and winter warmers. Belgian brewers are known for their subtle use of spice in beers such as Saison and Belgian Witbier which is flavoured with coriander and bitter orange peel.

Fruits and vegetables: Fruits and vegetables are used in beer to provide flavour, aroma, colour, and fermentable sugars. Fruits like berries, apricots, and cherries are most commonly used in wheat beers and Belgian sour Lambic beers. However they are sometimes also used in other styles including porters and stouts. American craft brewers have been experimenting with a wide variety of vegetables in beer, including pumpkins, sweet potatoes, beets, and chili peppers.

Brewing process

Beer is an industrial product. A brewery is literally a beer factory in which the brewer takes advantage of and manipulates natural processes to create the perfect growth medium for yeast. On the surface the brewing process is simple. But if we look a little deeper we will find that there is a complex set of chemical reactions at work in the creation of beer. An overview of the brewing process is shown in Fig. 19.1.

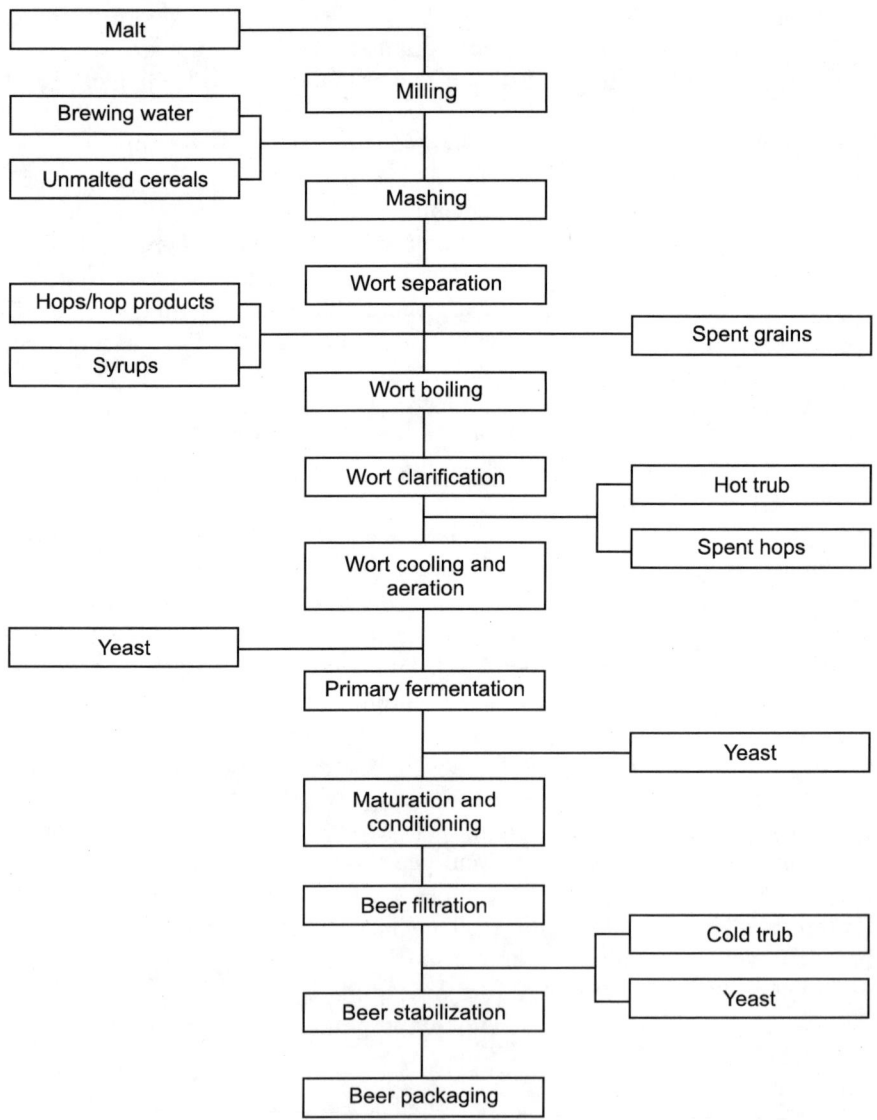

Figure 19.1 Schematic overview of the brewing process (input flows are indicated on the left side and output flows on the right side).

Milling: The first step of the process is crushing the malt. This breaks apart the grains, exposing the starchy ball inside and making it accessible to the brewer. The grains are only lightly crushed, leaving the hulls intact to serve as a filter bed for the *lautering* process later on.

Mashing: Mashing is the process by which the brewer extracts fermentable sugars from the grain. Basically it consists of steeping the grains in water at temperatures between 140°F and 160°F for a period of 60–90 minutes at a thickness similar to porridge. This activates naturally occurring enzymes in the grain that convert the grain starches into sugars, like maltose, that yeast can metabolize. This process occurs in a vessel called a *mash tun*.

Lautering: During *lautering* the fermentable sugars are rinsed from the grains. As the sugary liquor from the mash, now called *wort*, is slowly drained from the bottom of the *mash tun*, heated water is pumped in from the top at the same slow rate. As the water flows through the grains, it raises the temperature to about 170°F, making the sugars more soluble and easier to remove. As mentioned above, the intact grain hulls form a natural filter, removing bits of grain and proteins from the *wort*. In some brewing systems lautering occurs in the *mash tun*. In others the entire mash is pumped to a special *lauter tun*.

Boiling: From the *lautering* stage, the wort is pumped to the *kettle* where it is boiled. A vigorous boil is maintained for 60 minutes or more. During the boil the *wort* is sterilized and concentrated to the proper sugar density, haze causing proteins are removed (*hot break*), and light caramelization occurs that deepens the flavour and colour of the beer. One of the most important things to occur during the boil is the addition of hops. Hops for bittering are added early in the boil, while those for flavour and aroma are added later.

Cooling: Following the boil, the *wort* is pumped through a heat exchanger to cool it as quickly as possible to fermentation temperature. Rapid cooling minimizes the danger of bacterial contamination and causes more haze causing proteins to precipitate out of the *wort* (*cold break*).

Fermentation: Yeast is added once the *wort* has reached the desired temperature for fermentation. Brewers call the addition of yeast *pitching*. Once the yeast has been *pitched* the *wort* can properly be called *beer*. Fermentation can last a few days or a few weeks depending of the strain of yeast and the strength of the beer. During the process the yeast reproduce and then metabolize the sugars, making CO_2, alcohol, and a host of other flavourful and aromatic compounds that add complexity to the beer. During the height of fermentation the beer is capped by a thick creamy foam called *kreusen*. Once the available sugars have been consumed the yeast cells clump together or *floc* and fall to the bottom of the fermenter.

Conditioning: When fermentation is complete, the beer is removed from the yeast and pumped to a conditioning or *bright tank* where it is stored at near freezing temperatures that cause most of the remaining yeast to drop out of suspension. Hops can be added at this point as well, a step known as *dry hopping* that lends the beer additional hop flavour and aroma. Once the beer is clear it is ready to filter and package.

Beer defects and diseases: Beer defects are certain undesirable characteristics such as turbidity due to unstable proteins, tannins and resins and development of off-flavours due to poor quality ingredients or contact with metals. Beer infections or diseases are primarily due to growth and activity of undesirable microorganisms during fermentation. Butyric acid fermentation by *Clostridium* sp. or lactic acid fermentation by lactics may result in off flavours. Yeasts grown in beer can cause turbidity or cloudiness and bitterness, e.g. *Saccharomyces pastorianus*. Bacteria of genera *Pediococcus, Lactobacillus, Flavobacterium* and *Acetobacter* cause beer diseases resulting in sourness, turbidity, ropiness and bad odour akin to hydrogen sulphide. However, most of the yeasts and bacteria are killed by boiling the wort and hops and maintaining asceptic conditions during subsequent stages of brewing.

Other beer types and related beverages: Variations in beer types are manufactured by varying the concentration of malt and hops, ageing time, initial concentration of fermentable solids and temperature of fermentation to yield beers differing in alcohol content, colour and residual fermentable carbohydrate content. For example, *malt liquor* has a higher alcohol content while *bock beer* is a dark beer with higher alcohol content than the normal beer. 'Low calorie' beers are light in colour and contain no carbohydrates manufactured from prehydrolyzed wort using fungal glucoamylases and amylases, e.g. *Pilsener beer. Ale* is made using a top yeast, a strain of *Saccharomyces cerevisiae. Sake* is a beer of Japanese origin with an alcohol content of 14–17%. *Aspergillus oryzae* is grown on soaked and steamed rice mash to yield a koji, a starter culture, rich in amylases and proteases. The koji is mixed with more rice mash and fermented for about 10 days. During fermentation several yeast species ferment sugars to alcohol. The mixture is filtered to yield the clear liquor, sake. *Sonti* is a rice beer' of Indian origin. The mold *Rhizopus sonti* and yeasts are active in the alcoholic fermentation. *Ginger beer* is an acidic and mildly alcoholic beverage made by the fermentation of sugar solution flavoured with ginger.

Transgenic barley

Transgenic barley lines with immunity to a virus, resistance to fungal root rot, and improved brewing properties have been developed over the past several years. Now, field trials at the University of Giessen and the Friedrich-

Alexander University Erlangen-Nuremberg in Germany are testing some of these lines for potential unintended effects.

GM barley with extra defense against root rotting fungi

The plough may be almost as old as agriculture itself, but progressive-minded farmers in many parts of the world are finding success without it. Tilling damages the soil's delicate crumb-like structure, disturbs beneficial fungi, reduces soil organic matter, compacts deep soil layers, and causes erosion. Like most things, however, no-till agriculture also has its drawbacks. Although it leaves beneficial fungi undisturbed, it also makes a better environment for root rotting fungi. The pathogenic fungi *Rhizoctonia solani* and *Rhizoctonia oryzae* have been becoming more and more widespread in areas where no-till systems are in place, especially the northwest United States and southern Australia. One of the major problems with fungal root rot is that it can significantly reduce yield without causing any conspicuous above-ground damage.

Farmers have spent decades trying to come up with effective ways of managing Rhizoctonia. Fungicides are an option, but treating soil with fungicides is much more complicated than simply spraying the leaves, which also takes a toll on the environment. For all practical purposes, managing Rhizoctonia with fungicides is prohibitively expensive.

Crop rotation can sometimes help by depriving the fungus of its host and starving out the pathogen, but in this case, Rhizoctonia infects a wide range of plants. Most rotation schemes would do nothing to cut back the occurrence of root rot. Rotating barley with an appropriate non-crop plant could help, but it would mean that farmers would have to go a year without income.

Plant breeders have tried for years to improve barley's disease resistance with conventional breeding. To date, searches for genetic material for breeding programmes have been inconclusive and disappointing. Barley has a relatively small genetic base of possible breeding partners, and among these, there are no varieties with good resistance to Rhizoctonia.

One interesting way of managing Rhizoctonia is using a biological approach; in essence, using fungi to fight fungi. Trichoderma harzianum attacks other types of fungi, including Rhizoctonia, and feeds on their cells. The cocktail of substances Trichoderma uses to digest its host includes an enzyme called chitinase, which breaks down a substance known as chitin. Chitin is an essential component in fungal cell walls. There are different forms of chitin, and correspondingly, there are many different forms of chitinase. Barley itself already contains several chitinases, but none are effective against Rhizoctonia.

Field trials in Germany

The field trials now underway in Germany are taking Rhizoctonia resistant barley and barley with heat-stable glucanase and are looking closely for potential unexpected effects. Researchers will be studying the barley lines' gene expression profiles to see if there are any unexpected shifts in gene expression. They will also check to make sure there are no unexpected changes in grain quality. The anti-fungal activity of chitinase and the incidental anti-fungal activity of glucanase will be checked to see if they also block interactions with beneficial fungi.

19.4.2   Wine

Wine (Fig. 19.2) is principally fermented grape juice. It may also be produced by fermentation of juices of fruits such as apple (cider), peaches, apricots, plums, pears, cherries and berries and honey (mead or honey wine). Grape wine is either red or white depending on whether the skins of purple or red varieties of grapes have been fermented to impart the red colour of the pigment, or white grapes or expressed juices of other grapes without the skin have been fermented. The steps involved in the manufacture of grape wine include: (i) juice extraction, (ii) fermentation and (iii) ageing.

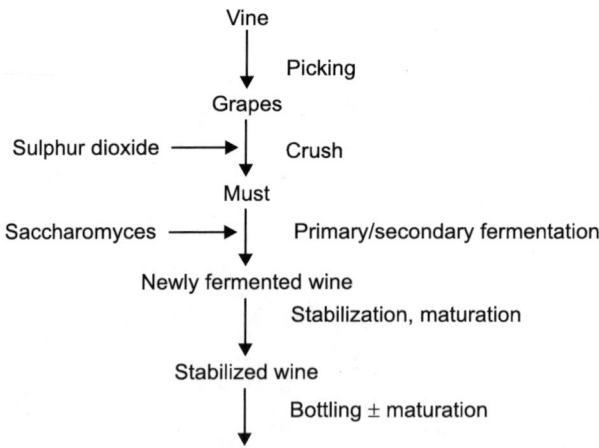

Figure 19.2 An overview of wine making.

Wine manufacturing process

Step 1 – *Harvesting:* As the grapes ripen, the concentration of sugars and aroma compounds rises and the concentration of acids falls. The aim at harvest is to pick the grapes at their optimum composition. This depends on the type of wine to be produced. For example, sparkling wine requires a higher acidity than still table wine. The development of the grapes is followed by taking samples of the grapes at regular intervals from a few weeks before the expected optimum levels will be reached. The samples are analysed for pH (using a pH meter), acid (by titration with sodium hydroxide), sugar (by refractive index or chemical reduction of copper salts) and flavour compounds (by tasting). When optimum levels are reached, the grapes are harvested.

Step 2 – *Crushing and destemming:* Sulphur dioxide (5–10% solution of metabisulphite) is usually added to the grape bunches as they are fed into the crusher/destemmer. The stems are removed as bunches pass through a perforated rotating cylinder in which the grapes fall through the perforations while the stems are separated out by beathers. The berries are then passed through rollers and crushed. The SO_2 inhibits the growth of wild microorganisms and prevents oxidative browning of the juice. Molecular SO_2 is the active biocide, but in solution this is in equilibrium with inactive HSO_3^-. At wine pH only 2–8% of the SO_2 exists in the molecular form, but this is usually sufficient to give the required protection. Wherever possible during the manufacturing process the juice is kept under a blanket of CO_2 to exclude air, and if necessary more SO_2 is added to maintain the level of molecular SO_2 at a minimum of 80 ppm.

Step 3 – *Juice preparation:* The free-run juice is separated from the crushed berries, which are pressed by gentle squeezing to obtain a high quality juice. The juice is allowed to settle overnight or is centrifuged to clarify it. If necessary, pectolytic enzymes are added to remove haze. Finally, the pulp is then squeezed almost dry. This final juice is of low quality and is used for cask wine or fermented for distillation into alcohol for sherry or port production.

Step 4 – *Fermentation:* Fermentation is begun by innoculating the juice with the chosen wine yeast. This yeast catalyses a series of reactions that result in the conversion of glucose and fructose to ethanol:

$$C_6H_{12}O_6 \rightarrow 2C_2H_5OH + 2CO_2$$

The driving force behind this reaction is the release of energy stored in the sugars to make it available to other biological processes. In aerobic conditions, the reaction can proceed further and convert the ethanol to H_2O and CO_2, releasing all of the energy present in the original sugars. This process is undesirable in wine production, so fermentation is usually carried out under a blanket of CO_2 to exclude oxygen and hence maximize alcohol production. Various intermediates in the fermentation process can be converted into other products.

The main products of alcoholic fermentations are listed in Table 19.1.

Table 19.1 Products of alcoholic fermentations

Product	Theoretical % w/w	Industrial fermentation % w/w	Wine yeast % w/w
Ethanol	51.1	48.4	47.86
CO_2	48.9	46.5	47.02
Acetaldehyde	–	0.08	0.01
Acetic acid	–	0.25	0.35
Glycerol	–	3.6	2.99
Lactic acid	–	0.2	0.20
Succinic acid	–	0.7	0.045
Higher alcohols	–	0.33	0.1
Yeast mass (dry weight)	–	1.2	0.55

The yield of ethanol is affected by such factors as temperature, extent of agitation, sugar concentration, acidity, strain of yeast and yeast activity. The lower the temperature, the higher the alcohol yield due to a more complete fermentation (better sugar utilization) and less loss of alcohol entrained with CO_2. While it may be important to maximize the yield of ethanol, it is equally important that this is never achieved with complete efficiency. The various by-products of yeast metabolism formed by this inefficiency contribute to wine's distinctive flavour and aroma and prevent it from simply being alcoholic grape juice.

The juice used to be fermented in wax-lined concrete or plastic vats, but now stainless steel is used for all wines except for certain high-quality ones that are fermented in wood. Wooden barrels are the container of choice for chardonnay, sauvignon blanc and pinot noir as the wood is smoked during processing, forming additional flavour compounds (particularly tannins) which are leached into the wine, giving it further complexity.

Step 5 – *Purification:* In former times, after fermentation is complete, the wine is heavily treated to alter the pH, composition, etc., to give it a desirable flavour, appearance, etc. Very few such measures are used today, but those that are retained are outlined briefly below. Proteins and tannins that are suspended in colloidal form in the wine are precipitated out with substances such as gelatin or adsorbed to the surface of substances such as bentonite. This process is called *fining*. The wine is often also clarified in a process called *racking*. This is the drawing off of the wine from the lees (sediment formed). Wine is often also *cold stabilised* (left at 0 to –3°C for 10–14 days) to crystallise out any potassium bitartrate (KHTa). These treatments are only usually necessary in white wine, as red wine fines and clarifies itself by forming deposits of proteins, tannins and tartrates during the ageing process, although sometimes proteinaceous fining agents are added to modify tannin levels and structures. The wine is continually racked off this precipitate, such that by the end of the ageing process all it needs is simple filtration before bottling and sale.

Wine defects and spoilage: Wine defects include turbidity, cloudiness, precipitation and loss of colour or colouration due to metals such as iron, tin and copper and their salts. Micro-organisms, both aerobic and facultative, cause wine' spoilage by imparting cloudiness, bitter characteristics, ropiness, undesirable flavours, high volatile acidity and low alcohol content. Wild yeasts, molds and bacteria of the genera *Acetobacter, Lactobacillus, Leuconostoc* and *Pediococcus* cause spoilage of wine.

19.4.3   Whisky

The principals for the distillation of whisky have changed little over the last 200 years. Just three basic ingredients are needed: water, barley and yeast. Technology now aids production, but traditionally there are five stages to the process-malting, mashing, fermentation, distillation and maturation. The various important stages of manufacturing whisky are discussed below:

Step 1 – *Malting:* Barley contains starch, and it is this starch which needs to be converted into soluble sugars to make alcohol. For this to occur, the barley must undergo germination and this first part of the process is called 'malting'. Each distiller has their own preference about the type of barley they buy, but they need a type that produces high yields of soluble sugar. The barley is soaked for 2–3 days in warm water and then traditionally spread on the floor of a building called a malting house. It is turned regularly to maintain a constant temperature. This is also carried out on a commercial scale in large drums which rotate.

When the barley has started to shoot, the germination has to be stopped by drying it in a kiln. Traditionally peat is used to power the kiln and it is at this point where the type of peat used and length of drying in the peat smoke can influence the flavour of the final spirit. The barley is now called 'malt' and this is ground down in a mill, with any husks and other debris being removed.

Step 2 – *Mashing:* The ground down malt, which is called 'grist', is now added to warm water to begin the extraction of the soluble sugars. The water is normally from a pure, reliable, local source-this is why most distilleries around the world are next to a river or lake. The character of this water can influence the final spirit as it can contain minerals from passing over or through granite, peat or other rock. The liquid combination of malt and water is called the 'mash'. It is put into a large vessel called a mash tun and stirred for several hours.

During this process, the sugars in the malt dissolve and these are drawn off through the bottom of the mash tun. The resulting liquid is called 'wort'. This process is normally carried out three times with the water temperature being increased each time to extract the maximum amount of sugar. Only wort from the first two times is used. The third lot is put back into the next batch of new grist. Any residue, such as husks, is called 'draff'. This is collected and used in the production of farm feed.

Step 3 – *Fermentation:* The wort is cooled and passed into large tanks called wash backs. These are traditionally made of wood, but now a number of distilleries use stainless steel. Here the yeast is added and the fermentation begins. The yeast turns the sugars that are present into alcohol. As with the barley and water, the distiller will carefully select the strain of yeast that they use and it can also have a small effect on the final flavour of the spirit. The fermentation normally takes around 48 hours to run its natural course, although some distilleries will let it go for longer so as to create further characteristics that they require. The liquid at this stage is called 'wash' and is low in alcohol strength (between 5–10% ABV), like beer or ale. We can make beer from the liquid at this point, but the difference with whisky is that the liquid is now distilled rather than brewed.

Step 4 – *Distillation:* In Scotland, the wash is traditionally distilled twice. In Ireland, it is distilled three times although there are exceptions in both countries. Here is a brief explanation of the double distillation process. The stills are made from copper, which has been found to be the best material for extracting impurities from the spirit as it is being distilled, and consist of a bowl shape at the bottom that rises up to the neck at the top. All are the same in principal, but a different shape will give a different flavour and character to the final spirit. Taller stills with longer necks will give finer, lighter spirits while shorter, fatter stills will produce a fuller, richer spirit.

The stills tend to work in pairs. Firstly, the wash enters the larger wash still and is heated (this was traditionally by coal, but is now largely by gas or steam). The liquid vaporizes and rises up the still until it reaches the neck, where it condenses.

This liquid is called 'low wines' and is unusable as it is. The low wines are passed to the second smaller still, called the spirit still. Any residue from the wash still is collected and used to manufacture farm feed.

Alcohols from the beginning of the distillation (called 'foreshots') are very high in alcohol level and very pungent. Alcohols from the end (called 'feints') are weak but also pungent. It is only the alcohol from the middle or 'heart' of the distillation that is used and this is skillfully removed by a stillman and collected through the spirit safe. The foreshots and feints are then mixed with the next batch of low wines and re-distilled. The heart is the spirit that is then taken to be matured and that will become whisky. This 'heart' has an alcoholic strength of 65–70% ABV.

Step 5 – *Maturation:* The spirit is put into oak casks and stored. The most common types of oak casks are those that have previously been used in the American bourbon and Spanish sherry industries. The spirit must mature in casks for a minimum of three years before it is legally allowed to be called whisky in Scotland. During maturation, the flavours of the spirit combine with natural compounds in the wood cask and this gives the whisky its own characteristic flavour and aroma.

Wood is porous, so over time it will breathe in air from the surrounding environment in which it is stored. This will also give the whisky some unique characteristics. If the distillery storage facilities are next to the sea, on an island or in the middle of the Highlands then the air quality, temperature and humidity will be different and influence the end product. During each year of maturation about 2% of the spirit is lost through natural evaporation. This is called the 'angel's share' and explains why older whiskies are less readily available and more expensive to buy. There is simply less whisky in the cask to bottle.

19.4.4   Distilled liquors

Distilled spirits (or, simply, spirits or liquors) are the alcohol-containing fluids (ethanol, also called ethyl alcohol) obtained via distillation of fermented juices from plants. These juices include wines, distillates of which are termed brandies. The most commonly used plants are sugarcane, potatoes, sugar beets, corn, rye, rice, and barley; various fruits such as grapes, peaches, and apples are also used. Flavours may be added to provide distinctive character. All distilled spirits begin as a colourless liquid, pure ethyl alcohol (as it was called by 1869)—C_2H_6O. This had been called aqua vitae (Latin, water of life) by medieval alchemists; today it is often called grain alcohol, and the amount contained in distilled spirits ranges from 30 to 100 per cent (60 to 200 proof)—the rest being mainly water.

Examples of distilled spirits include brandy, whiskey, rum, gin, and vodka. Brandy was called brandewijn by the Dutch of the 1600s—burned, or distilled, wine. It was originally produced as a means of saving space on trade ships, to increase the value of a cargo. The intent was to add water to the condensate to turn it back into wine, but customers soon preferred the strong brandy to the acidic wines it replaced. Cognac is a special brandy produced in the district around the Charente river towns of Cognac and Jarnac, in France, where wine is usually distilled twice, then put into oak barrels to age. The spirits draw colour and flavour (tannins) from the wood during the required five-year ageing process.

Whiskey is distilled off grains—usually corn or rye, but millet, sorghum, and barley are also used. Traditional American whiskeys are bourbons (named after Bourbon county in Kentucky), which are made from a sour mash of rye and corn. Bourbons typically contain 40–50 per cent ethyl alcohol (called 80–100 proof, doubled by the liquor industry). Canadian whiskey is very similar to bourbon and to rye whiskey, while Irish whiskey is dry (has less sugars), with a distinctive austere flavour gained by filtration. All these whiskeys lack the smoky taste of Scotch whiskeys, which get their unique flavour by using malt that had been heated over peat fires. By using less malt and by ageing for only a few years in used sherry casks (traditionally), a light flavour is produced; by using more malt and long ageing, heavy peaty smoky flavours are produced. Today, some scotches and other whiskeys are blended to achieve uniform taste from batch to batch.

The distillation of fermented sugarcane (Saccarum officinarum) results in rum. Of all distilled spirits, rum best retains the natural taste of its base, because: (i) the step of turning starch into sugar is unnecessary, (ii) it can be distilled at a lower proof, (iii) chemical treatment is minimized and (iv) maturing can be done with used casks.

The amount of added (sugar-based) caramel gives rum its distinctive flavour and colour—which can vary from clear to amber to mahogany. The New England colonists made rum from molasses, which is the thick syrup separated from raw sugar during crystal sugar manufacture. Caribbean colonists grew sugarcane and shipped barrels of molasses to New England. New Englanders shipped back barrels of rum.

Both substances were originally ballast for the barrels, which were made from New England's local forests to hold the sugar shipped from the Caribbean to the mother country—England.

Gin is a clear distillate of a grain (or beer) base that is then reprocessed; juniper berries and other herbs are added to give it its traditional taste. Vodka is also clear liquor, often the same as gin without the juniper flavour. Traditional vodkas, made in Russia, Ukraine, Poland, and other Eastern European countries, are made from grain or potatoes at a very high proof;

typical ranges are 65 to 95 proof, or about 33 to 43 per cent ethyl alcohol. Vodka has no special taste or aroma, although some are slightly flavoured with immersed grasses, herbs, flowers, or fruits. The Scandinavian aquavit is clear, like vodka, distilled from either grain or potatoes, and flavoured with caraway seed; it is similar to Germany's kümmelvasser (kümmel means caraway in German). When any clear liquor is added to fruit syrups, the product is called a cordial or a liqueur. Swiss kirschwasser is, however, a clear high-proof cherry-based brandy (Kirsche means cherry in German); and slivovitz is a clear high-proof Slavic plum-based brandy.

The raw grain alcohol distilled in the American South and in Appalachia has been called white lightning since the early 1900s; this is also known as moonshine, corn whiskey, or corn liquor.

19.4.5   Vinegar

Vinegar is an alcoholic liquid that has been allowed to sour. It is primarily used to flavour and preserve foods and as an ingredient in salad dressings and marinades. Vinegar is also used as a cleaning agent.

The use of vinegar to flavour food is centuries old. It has also been used as a medicine, a corrosive agent, and as a preservative. In the middle ages, alchemists poured vinegar onto lead in order to create lead acetate. Called 'sugar of lead,' it was added to sour cider until it became clear that ingesting the sweetened cider proved deadly.

The transformation of wine or fruit juice to vinegar is a chemical process in which ethyl alcohol undergoes partial oxidation that results in the formation of acetaldehyde. In the third stage, the acetaldehyde is converted into acetic acid. The chemical reaction is as follows:

$$CH_3 CH_2 OH = 2HCH_3 CHO = CH_3 COOH$$

Historically, several processes have been employed to make vinegar. In the slow, or natural, process, vats of cider are allowed to sit open at room temperature. During a period of several months, the fruit juices ferment into alcohol and then oxidize into acetic acid.

The French Orleans process is also called the continuous method. Fruit juice is periodically added to small batches of vinegar and stored in wooden barrels. As the fresh juice sours, it is skimmed off the top.

Both the slow and continuous methods require several months to produce vinegar. In the modern commercial production of vinegar, the generator method and the submerged fermentation method are employed. These methods are based on the goal of infusing as much oxygen as possible into the alcohol product.

Raw materials

Vinegar is made from a variety of diluted alcohol products, the most common being wine, beer, and rice. Balsamic vinegar is made from the Trebbiano and Lambrusco grapes of Italy's Emilia-Romagna region. Some distilled vinegars are made from wood products such as beech. Acetobacters are microscopic bacteria that live on oxygen bubbles. Whereas the fermentation of grapes or hops to make wine or beer occurs in the absence of oxygen, the process of making vinegars relies on its presence. In the natural processes, the acetobacters are allowed to grow over time. In the vinegar factory, this process is induced by feeding acetozym nutrients into the tanks of alcohol.

Mother of vinegar is the gooey film that appears on the surface of the alcohol product as it is converted to vinegar. It is a natural carbohydrate called cellulose. This film holds the highest concentration of acetobacters. It is skimmed off the top and added to subsequent batches of alcohol to speed the formation of vinegar. Acetozym nutrients are manmade mother of vinegar in a powdered form. Herbs and fruits are often used to flavour vinegar. Commonly used herbs include tarragon, garlic, and basil. Popular fruits include raspberries, cherries and lemons.

Design: The design step of making vinegar is essentially a recipe. Depending on the type of vinegar to be bottled at the production plant—wine vinegar, cider vinegar, or distilled vinegar—food scientists in the test kitchens and laboratories create recipes for the various vinegars. Specifications include the amount of mother of vinegar and/or acetozym nutrients added per gallon of alcohol product. For flavoured vinegars, ingredients such as herbs and fruits are macerated in vinegar for varying periods to determine the best taste results.

Manufacturing process

(a) Orleans method

1. Wooden barrels are laid on their sides. Bungholes are drilled into the top side and plugged with stoppers. Holes are also drilled into the ends of the barrels (Fig. 19.3).

2. The alcohol is poured into the barrel via long-necked funnels inserted into the bungholes. Mother of vinegar is added at this point. The barrel is filled to a level just below the holes on the ends. Netting or screens are placed over the holes to prevent insects from getting into the barrels.

3. The filled barrels are allowed to sit for several months. The room temperature is kept at approximately 85°F (29°C). Samples are taken periodically by inserting a spigot into the side holes and drawing

liquid off. When the alcohol has converted to vinegar, it is drawn off through the spigot. About 15% of the liquid is left in the barrel to blend with the next batch.

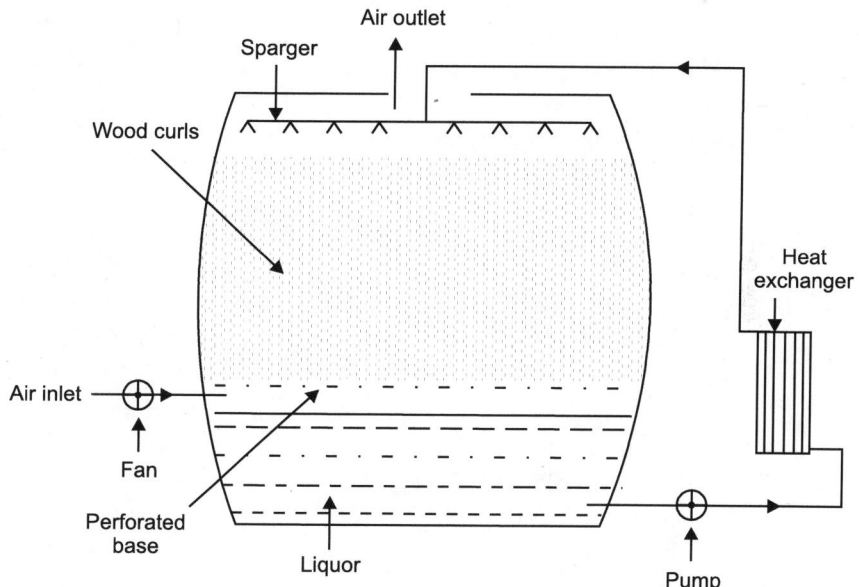

Figure 19.3 Vinegar generator

(b) Submerged fermentation method

1. The submerged fermentation method is commonly used in the production of wine vinegars. Production plants are filled with large stainless steel tanks called acetators. The acetators are fitted with centrifugal pumps in the bottom that pump air bubbles into the tank in much the same way that an aquarium pump does.

2. As the pump stirs the alcohol, acetozym nutrients are piped into the tank. The nutrients spur the growth of acetobacters on the oxygen bubbles. A heater in the tank keeps the temperature between 80°F and 100°F (26–38°C).

3. Within a matter of hours, the alcohol product has been converted into vinegar. The vinegar is piped from the acetators to a plate-and-frame filtering machine. The stainless steel plates press the alcohol through paper filters to remove any sediment, usually about 3% of the total product. The sediment is flushed into a drain while the filtered vinegar moves to the dilution station.

(c) Generator method

1. Distilled and industrial vinegars are often produced via the generator method. Tall oak vats are filled with vinegar-moistened beech wood shavings, charcoal, or grape pulp. The alcohol product is poured into the top of the vat and slowly drips down through the fillings.

2. Oxygen is allowed into the vats in two ways. One is through bungholes that have been punched into the sides of the vats. The second is through the perforated bottoms of the vats. An air compressor blows air through the holes.

3. When the alcohol product reaches the bottom of the vat, usually within in a span of several days to several weeks, it has converted to vinegar. It is poured off from the bottom of the vat into storage tanks. The vinegar produced in this method has very high acetic acid content, often as high as 14%, and must be diluted with water to bring its acetic acid content to a range of 5–6%.

4. To produce distilled vinegar, the diluted liquid is poured into a boiler and vinegar brought to its boiling point. A vapour rises from the liquid and is collected in a condenser. It then cools and becomes liquid again. This liquid is then bottled as distilled vinegar.

(d) Balsamic vinegar

1. The production of balsamic vinegar most closely resembles the production of fine wine. In order to bear the name balsamic, the vinegar must be made from the juices of the Trebbiano and Lambrusco grapes. The juice is blended and boiled over a fire. It is then poured into barrels of oak, chestnut, cherry, mulberry, and ash.

2. The juice is allowed to age, ferment, and condense for 5 years. At the beginning of each year, the ageing liquid is mixed with younger vinegars and placed in a series of smaller barrels. The finished product absorbs aroma from the oak and colour from the chestnut.

Quality control

The growing of acetobacters, the bacteria that creates vinegar, requires vigilance. In the orleans method, bungholes must be checked routinely to ensure that insects have not penetrated the netting. In the generator method, great care is taken to keep the temperature inside the tanks in the 80–100°F (26–38°C) range. Workers routinely check the thermostats on the tanks. Because a loss of electricity could kill the acetobacters within seconds, many vinegar plants have backup systems to produce electrical power in the event of a blackout.

By-products/waste

Vinegar production results in very little by-products or waste. In fact, the alcohol product is often the by-product of other processes such as wine making and baker's yeast. Some sediment will result from the submerged fermentation method. This sediment is biodegradable and can be flushed down a drain for disposal.

Biotechnology of brewer's yeast

20.1　Introduction

Yeasts are eukaryotic microorganisms classified as members of the fungus kingdom with 1,500 species currently identified and are estimated to constitute 1% of all described fungal species. Yeasts are unicellular, although some species may also develop multicellular characteristics by forming strings of connected budding cells known as *pseudohyphae* or false hyphae. Yeast sizes vary greatly, depending on species and environment, typically measuring 3–4 μm in diameter, although some yeasts can grow to 40 μm in size. Most yeasts reproduce asexually by mitosis, and many do so by the asymmetric division process known as budding. Yeasts do not form a single taxonomic or phylogenetic grouping. The term 'yeast' is often taken as a synonym for *Saccharomyces cerevisiae*, but the phylogenetic diversity of yeasts is shown by their placement in two separate phyla: the Ascomycota and the Basidiomycota. The budding yeasts ('true yeasts') are classified in the order *Saccharomycetales*. By fermentation, the yeast species *Saccharomyces cerevisiae* converts carbohydrates to carbon dioxide and alcohols–for thousands of years the carbon dioxide has been used in baking and the alcohol in alcoholic beverages. Other species of yeasts, such as Candida albicans, are opportunistic pathogens and can cause infections in humans.

20.2　Biotechnology of brewer's yeast

The progress of chemistry, physiology and microbiology during the 19th Century, allowed a scientific approach to brewing that caused a tremendous advancement on the production of beer. Louis Pasteur demonstrated that alcoholic fermentation is a process caused by living yeast cells. His conclusion was that fermentation is a physiological phenomenon by which sugars are converted in ethanol as a consequence of yeast metabolism.

20.3　Genetic constitution of brewer's yeast

Saccharomyces cerevisiae is one of the best genetically characterised yeast as its genome is fully sequenced and analyzed exhaustively. Being a eukaryotic, the key of its success lies in the selection of a model strain with a perfect heterothallic life cycle. In contrast, brewer's yeast is refractory to the genetic procedures used with laboratory strains. The main reason is its low sexual fertility. Like most other industrial yeast, brewing strains do not sporulate or do so with low efficiency. Even in those cases that they show a suitable sporulation frequency, most spores are not viable.

20.3.1   Strain types

There are basically two kinds of yeast used in brewing that correspond to the ale and lager types of beer. Ale beer is produced by top-fermenting yeast that works at room temperature, ferments quickly, and produces beer with a characteristic fruity aroma. The bottom-fermenting lager yeast works at lower temperatures, about 10–14°C, ferments more slowly and produces beer with a distinct taste. The vast majority of beer production worldwide is lager. It is difficult to make generalizations concerning the yeast strains used for the industrial production of beer, since they are generally ill characterized and very few comparative studies have been reported. Bottom fermenting, lager strains are usually labelled *Saccharomyces carlsbergensis*. Although strains from different sources show differences regarding cell size, morphology and frequency of spore formation, it is unlikely that these differences reflect a significant genetic divergence. Only one strain, Carlsberg production strain 244, has been extensively analyzed and most of the studies described in the following sections have been conducted with this strain.

20.3.2   Genetic crosses

Early attempts to carry out conventional genetic analysis with brewer's yeast faced the problems of poor sporulation and low viability. To overcome this difficulty, several researchers hybridized brewing strains with laboratory strains of *S. cerevisiae*. Notwithstanding the poor performance of brewing strains, viable spores were recovered from them. Some of the spores had mating capability and could be crossed with *S. cerevisiae* to generate hybrids easier to manipulate. The meiotic offspring of the hybrids was repeatedly backcrossed with laboratory strains of *S. cerevisiae* to bring particular traits of the brewing strain into an organism amenable to analysis. This procedure was followed to study flocculence, an important character in brewing. Gjermansen and Sigsgaard carried out a detailed analysis of the meiotic offspring of *S. carlsbergensis* strain 244. They obtained viable spore clones of both mating types. Celllines with opposite mating type were crossed pairwise to generate a number of hybrids that were tested for brewing performance. One of them was as good as the original strain. Additionally, the clones derived from strain 244 with mating capability served as starting material for further genetic analysis which are described in the following section. The kar mutations have been particularly useful tools to investigate cytoplasmic inheritance. Additionally, the kar mutations supplied new genetic techniques. For instance, the chromosome number of virtually any *Saccharomyces* strain can be duplicated upon mating with a *kar2* partner. These new tools and techniques opened a new way for the characterization of the brewer's yeast. Since the brewing strain does not mate normally, the strain used in *kar* crosses was a meiotic derivative of strain 244 with mating capability. When disomic strains for chromosome III (also referred to as chromosome addition strains) were crossed to haploid *S. cerevisiae* strains, normal spore viability was obtained, allowing tetrad analysis. In this process, one of the two copies of chromosome III can be lost. If the original *S. cerevisiae* copy is lost, the result is a 'chromosome substitution strain' carrying a complete *S. cerevisiae* chromosome set, except chromosome III, which comes from *S. carlsbergensis*. Meiotic analysis of crosses between chromosome III addition strains and laboratory strains of *S. cerevisiae* revealed two important facts: (i) the functional equivalence of chromosome III for the brewing strain and *S. cerevisiae*, since ascospore viability and chromosome segregation were normal, and (ii) in grite of the functional equivalence, the two copies of chromosome III were different since the overall frequency of recombination between them was much lower than that expected for perfect homologues. The new procedure allowed the analysis of entire chromosomes from the brewing strain, placed into a laboratory yeast that could easily be manipulated genetically.

Molecular analysis

A clear picture of the genetic composition of *S. carlsbergensis* emerged from Southem hybridization experiments and from the first gene sequences from this yeast. Nilsson-Tillgren reported the transfer of a chromosome III from the brewing strain to *S. cerevisiae*. Determination of the nucleotide sequence of a number of *S. carlsbergensis* genes provided a precise characterization of the difference between the two types of homologous alleles present in the brewing yeast.

Ploidy

Finding a sound answer for the long-standing question of how many chromosomes are contained in brewer's yeast has taken a long time. The relative DNA content of *S. carlsbergensis* 244 has been recently determined by flow cytometry. Results obtained show that the genetic constitution of this strain must be close to tetraploidy.

Origin of brewing strains

The hybrid nature of the brewing yeast explains its poor sexual performance. Divergence between homologous sequences impairs chromosome pairing and recombination, which are requisites for a proper meiotic function. Sexual reproduction appears in Evolution as a mechanism that recombines the genetic material of organisms to generate variability. It offers adaptive advantages to a changing environment through the random generation of new genotypes. On the contrary, abolition of sex is advantageous when the purpose is to keep unchanged a given property.

The maintenance over the centuries of a brewing procedure to produce beer with particular organoleptic properties likely caused the selection of a particular type of yeast. The hybrid, vegetative vigor of this yeast assured a good fermentative capability, whereas its sexual infertility would keep fixed the genetic constitution responsible for the 'good beer' phenotype.

Sequence analysis shows that one of the two parental species that generated *S. carisbergensis* was *S. cerevisiae*, but the precise identification of the other contributor is less clear.

Thus, lager strains of different origin, labelled *S. carlsbergensis*, could be independently generated hybrids of slightly different genetic constitution.

20.3.3   Genetic manipulation

Yeast and barley play an active, primary role in the brewing process. The other two beer ingredients, water and hops, have secondary roles. Yeast is the fermenting agent, which transforms the carbohydrates stored in the grain of barley into ethanol. It produces a battery of compounds that ultimately result in the aroma and flavour of the beer. Barley is not solely a source of fermentable sugars. During the process of malting, cells in the germinating barley seeds secrete enzymes that are required to digest the starch into simpler sugars, mainly maltose and glucose, which can be assimilated by the yeast. Many properties of barley, in particular those affecting its carbohydrate content and composition, but also other characteristics, are very important for the quality of beer. Genetic engineering can be used to modify the properties of yeast and barley in ways that improve their performance in brewing. Different experimental approaches directed to the modification of the brewer's yeast, to produce beer with better properties or new characteristics. In most cases, technical advances allow the construction of new strains of yeast with the desired properties. Currently however, public concern about the use of genetically modified food poses a barrier to the industrial use of these strains.

Accelerated maturation of beer

The production of lager beer comprises two separate fermentation stages. The main fermentation, in which the fermentable sugars are converted in ethanol, is followed by a secondary fermentation, referred to as maturation or lagering. The most important function of maturation is the removal of diacetyl, a compound that causes an unwanted buttery flavour in beer. Diacetyl is formed by the spontaneous (non-enzymatic) oxidative decarboxylation of α-acetolactate, an intermediate in the biosynthesis of valine. In yeast, as in other organisms, the two branched-chain amino acids, isoleucine and valine, are synthesized in an unusual pathway in which a set of enzymes, acting in parallel reactions, lead to the formation of different end products. Like diacetyl is formed as a by-product of valine biosynthesis, a related compound, 2-3-pentanedione, is formed by decarboxylation of α-aceto-α-hydroxybutirate in the isoleucine biosynthesis. Both compounds, diacetyl and α-aceto-α-hydroxybutirate, produce a similar undesirable effect in beer, although much more pronounced in the case of diacetyl. Together, they are referred to as vicinal diketones. Diacetyl is converted to acetoin by the action of diacetyl reductase, an enzyme from the yeast. The maturation period, which lasts several weeks, assures the conversion of the available α-acetolactate into diacetyl and the subsequent transformation of diacetyl into acetoin. The amounts formed of this last compound do not have a significant influence on beer flavour. Preventing diacetyl formation would reduce or even make unnecessary the lagering period. This would represent a considerable benefit for the brewing industry. Different approaches have been devised to eliminate diacetyl. A first one requires the manipulation of the isoleucine-valine biosynthetic pathway, either by blocking the formation of the diacetyl precursor α-acetolactate, or by increasing the flux of the pathway at a later stage, channelling the available α-acetolactate into valine before it is converted into diacetyl. Masschelein and collaborators were first to suggest that a deleterious mutation of the brewer's yeast *ILV2* gene would solve the diacetyl problem. This gene encodes the enzyme acetohydroxyacid synthase, which catalyses the synthesis of α-acetolactate, from which diacetyl is formed. This or any alternative action on the valine pathway requires the manipulation of specific genes encoding enzymes of the pathway. These genes have been cloned from *S. cerevisiae* and characterized. *S. carlsbergensis*- specific alleles of the *ILV* genes from the brewer's strain have also been cloned. Because of the genetic complexity of the brewing strain (a hybrid with about four copies of each gene, two from each parent), the abolition of the *ILV2* function requires the very laborious task of eliminating each of the four copies of the gene present in the yeast. An alternative could be to boost the activity of the enzymes that direct the following steps in the conversion of α-acetolactate into valine: the reductoisomerase, encoded by *ILV5* and possibly the dehydrase, encoded by *ILV3*. To achieve the desired effect, it could be sufficient to manipulate only one of the four copies of the *ILV* genes present in the brewer's yeast. A clever procedure to inhibit the *ILV2* function, by using an antisense RNA of the gene, has been reported.

However, a later note from the same laboratory stated that the reported results were incorrect. Another approach makes use of an enzyme, α-aceto-lactate decarboxylase, which catalyses the direct conversion of acetolactate into acetoin, by-passing the formation of dyacetyl. This enzyme is produced by different microorganisms.

20.3.4   Beer attenuation and the production of light beer

Conversion of barley into wort that can be fermented requires two previous processes: malting and mashing. During malting, the barley grain is subjected to partial germination, achieved by moistening, and subsequent drying. Germination induces the synthesis of amylase and other enzymes that allow the seed to mobilize its reserves. The dried malt is milled and the resulting powder is mixed with water and allowed to steep at warm temperatures. During mashing, amylases digest the seed's starch, liberating simpler sugars, chiefly maltose.

This process is critical, since the brewing yeast is unable to hydrolyze starch. The enzymatic action of barley's amylases on starch yields fermentable sugars, but also oligosaccharides (dextrins) which remain unfermented during brewing. Dextrins represent an important fraction of the caloric content of beer. In current brewing practice, it is quite common to add exogenous enzymes. Thus glucoamylase can be added to the mash to improve the digestion of the starch. If the enzymatic treatment is carried out exhaustively, the dextrins are completely hydrolyzed, and the result is a light beer with substantially lower caloric content, for which there is a significant market demand in some parts of the world. A convenient alternative to the addition of exogenous glucoamylase is to endow the brewer's yeast with the genetic capability of synthesizing ibis enzyme. A variety of *S. cerevisiae*, formerly classified as a separate species (*S. diastaticus*), produces glucoamylase. Because of its close phylogenetic relationship with the brewing yeast, *S. diastaticus* is an obvious source of the glucoamylase gene.

The percentage of the sugar in the wort that is converted into ethanol and CO_2 by the yeast is called attenuation. Microbial contamination of beer is often associated with a pronounced increase in the attenuation value, which is known as superattenuation. This effect is due to the fermentation of dextrins, which are hydrolysed by amylases produced by the contaminant microorganisms. *S. diastaticus* was characterized as a wild yeast that caused superattenuation. Similarly to the synthesis of invertase or maltase by Saccharomyces, the synthesis of glucoamylase is controlled by a set of at least three polymeric genes, designated *STA1*, *STA2* and *STA3*. This genetic system is complicated by the existence in normal *S. cerevisiae* strains of a gene, designated *STA10*, which inhibits the expression of the other *STA* genes. Recently, the *STA10* gene has been identified with the absence of *FLO8p*, a transcriptional regulator of both glucoamylase and flocculation genes. The sequence of the *STA1* gene was first determined by Yamashita and others. Different species of filamentous fungi, in particular some of the genus *Aspergillus*, produce powerful glucoamylases. The gene that encodes the enzyme of *A. awamori* has been expressed in *S. cerevisiae*. Available information about the genetic control of glucoamylase production by *Saccharomyces* and current technology makes the construction of brewing strains with this capability relatively easy.

Beer filterability and the action of β-glucanases

Brewing with certain types or batches of barley, or using certain malting or brewing practices, can yield wort and beer with high viscosity, very difficult to filtrate. When this problem arises, the beer may also present hazes and gelatinous precipitates. Scott pointed out that this problem was caused by a deficiency in β-glucanase activity. The substrate of this enzyme, β-glucan, is a major component of the endosperm cell walls of barley and other cereals. During the germination of the grain, β-glucanase degrades the endosperm cell walls, allowing the access of other hydrolytic enzymes to the starch and protein reserves of the seed. Insufficient β-glucanse activity during malting gives rise to an excess of β-glucan in the wort, which causes the problems. The addition of bacterial or fungal β-glucanases to the mash, or directly to the beer during the fermentation is a common remedy. The construction of brewing yeast with appropriate β-glucanase activity would make unnecessary the treatment with exogenous enzymes. Suitable organisms to be used as sources of the β-glucanase gene are *Bacillus subtilis* and *Thricoderma reesei*, from which the commercial enzyme preparations used in brewing are prepared. The genes from both have been characterized and brewer's yeast expressing β-glucanase activity has been constructed. An alternative is to make use of the gene encoding barley β-glucanase, the enzyme that naturally acts in malting. This gene has been characterized and expressed in *S. cerevisiae*. However, the barley enzyme has lower thermal resistance than the microbial enzymes, which is a limitation for its use against the β-glucans present in wort. Consequently, the enzyme has been engineered to increase its thermal stability.

Control of sulphite production in brewer's yeast

Sulphite has an important, dual function in beer. It acts as an antioxidant and a stabilizing agent of flavour. Sulphite is formed by the yeast in the assimilation of inorganic sulphate, as an intermediate of the biosynthesis of sulphur-containing amino acids, but its physiological concentration is low. Hansen and Kielland-Brandt have engineered a brewing strain to enhance sulphite level to a concentration that increases flavour stability. The formation of sulphite from sulphate is carried out in three consecutive enzymatic steps catalysed by ATP sulphurylase, adenylsulphate kinase and phosphoadenyl sulphate reductase. In *S. cerevisiae*, these enzymes are encoded by *MET3*, *MET14* and *MET16*. In turn, sulphite is converted firstly into sulphide, by sulphite reductase, and then into homocysteine by homocysteine synthetase. This last compound leads to the synthesis of cysteine, methionine and *S-adenosylmethionine*. It has been proposed that *S-adenosylmethionine* plays a key regulatory role by repressing the genes of the pathway. However, more recent evidence assigns this function to cysteine. Anyhow, because of the regulation of the pathway, yeast growing in the presence of methionine contains very little sulphite. To increase its production in the brewing yeast, Hansen and Kielland-Brandt planned to abolish sulphite reductase activity. This would increase sulphite concentration, as it cannot be reduced. At the same time, the disruption of the methionine pathway prevents the formation of cysteine and keeps free from repression the genes involved in sulphite formation. Sulphite reductase is a tetramer with an a_2 b_2 structure. The α and β subunits are encoded by the *MET10* and *MET5* genes, respectively. Hansen and Kielland-Brandt undertook the construction of a brewing strain without *MET10* gene function.

The allotetraploid constitution of *S. carlsbergensis* made it extremely difficult to perform the disruption of the four functional copies of the yeast. Therefore, they used allodiploid strains, obtained as meiotic derivatives of the brewer's yeast. These allodiploids contains two homologous alleles of the *MET10* gene, one similar to the version normally found in *S. cerevisiae* and another which is *S. carlsbergensis*-specific. It is known that some allodiploids can be mated to each other to regenerate tetraploid strains with good brewing performance. The functional *MET10* alleles present in the allodiploids were replaced by deletion-harbouring, non-functional copies, by two successive steps of homologous recombination. New allotetraploid strains with reduced or abolished *MET10* activity were then generated by crossing the manipulated allodiploids. The brewing performance of one of these strains, in which the *MET10* function was totally abolished, met the expectations. Hansen and Kielland-Brandt have used another strategy to increase the production of sulphite which relies in the inactivation of the *MET2* gene function. The *MET2* gene encodes *O*-acetyl transferase. This enzyme catalyses the biosynthesis of *O*-acetyl homoserine, which binds hydrogen sulphide to form homocysteine. Similarly to the inactivation of *MET10*, inactivation of *MET2* impedes the formation of cysteine, depressing the genes required for sulphite biosynthesis.

Yeast flocculation

As beer fermentation proceeds, yeast cells start to flocculate. The flocs grow in size, and when they reach a certain mass start to settle. Eventually, the great majority of the yeast biomass sediments. This phenomenon is of great importance to the brewing process because it allows separation of the yeast biomass from the beer, once the primary fermentation is over the small fraction of the yeast that is left in the green beer is sufficient to carry out the subsequent step, the lagering.

Flocculation is a cell adhesion process mediated by the interaction between a lectin protein and mannose. Stratford and Assinder carried out an analysis of 42 flocculent strains of *Saccharomyces* and defined two different phenotypes. One was the known pattern observed in laboratory strains that carried the *FLO1* gene. They found, in some ale brewing strains, a new flocculation pattern characterized by being inhibited by the presence in the medium of a variety of sugars, including mannose, maltose, sucrose and glucose, whereas the *FLO1* type was sensitive only to mannose. The genetic analysis of flocculation has revealed the existence of a polymeric gene family analogous to the *SUC*, *MAL*, *STA* and *MEL* families.

The *FLO1* gene has been extensively characterized, which encodes a large, cell wall protein of 1,537 amino acids. The protein is highly glycosylated. It has a central domain harbouring direct repeats rich in serine and threonine (putative sites for glycosylation). Kobayashi and others have isolated a flocculation gene homolog to *FLO1* that corresponds to the new pattern described by Stratford and Assinder.

This result is consistent with the hybrid nature of the brewing yeast. In addition to the structural genes encoding flocculins, other *FLO* genes play a regulatory role. For instance, the *FLO8* gene (alias *STA10*) encodes a transcriptional activator that in addition to flocculation regulates glucoamylase production, filamentous growth and mating.

Beer spoilage caused by microorganisms

Microbial contamination of beer, caused by bacteria or wild yeast is a serious problem in brewing. To overcome the contamination, commonly sulphur dioxide and other chemicals are added, but this practice faces restrictive legal regulation and consumer rejection. An attractive alternative is to endow the brewing yeast with the capability of producing anti-microbial compounds. A specific example is the expression in *S. cerevisiae* of the genes required for the biosynthesis of pediocin, an antibacterial peptide from *Pediococcus acidilactici*. Another example is the transfer to brewing strains of the killer character, conferred by the production of a toxin active against other yeasts.

Enhanced synthesis of organoleptic compounds

The yeast metabolism during beer fermentation gives rise to the formation of higher alcohol, esters and other compounds which make an important contribution to the aroma and taste of beer. A first group of compounds important to beer flavour are isoamyl and isobutyl alcohol and their acetate esters. These compounds derive from the metabolism of valine and leucine. Two genes, *ATF1* and *LEU4*, encoding enzymes involved in the formation of these compounds, have been successfully manipulated to increase their synthesis. *ATF1* encodes alcohol acetyl transferase. It has been shown that its over-expression causes increased production of isoamyl acetate. *LEU4* encodes α-isopropylmalate synthase, an enzyme that controls a key step in the formation of isoamyl alcohol from leucine. This enzyme is inhibited by leucine. Mutant strains resistant to a toxic analog of leucine are insensitive to leucine inhibition. Mutants of this type, obtained from a lager strain, produce increased amounts of isoamyl alcohol and its ester.

Genetically modified organisms in the wine industry

21.1 Introduction

The process used in creating the transgenic grape was originally developed by McKersie to genetically engineer cold tolerance into alfalfa. The grape was created by inserting naturally occurring genes for cold tolerance into single grape plant cells. The gene introduced into the grape was isolated from a wild cousin of broccoli called *Arabidopsis thaliana*. This gene produces the enzyme superoxide dismutase, which detoxifies the toxic metabolites of oxygen called oxygen-free radicals, which cause many of the damaging reactions associated with stress. During freezing and drought, these molecules attack plant cells and cause decay. Cold tolerance depends on how well a plant's proteins can detoxify the oxygen molecules.

The gene was inserted into single grape plant cells via common soil bacteria, *Agrobacterium tumefaciens*, which was genetically programmed to deliver the cold tolerance genes. The transgenic grape cells grew in an incubator, a growth chamber and the Guelph Transgenic Plant Research Complex before being planted at Château des Charmes. This new transgenic technique could eventually be applied to almost all perennial fruit crops and winter annual plants growing in a cold climate.

Wine is arguably the oldest biotechnological endeavor, with humans having been involved in wine production for at least 7000 years. Despite the artisan nature of its production, work by pioneering scientists such as Antoine-Laurent de Lavoisier and Louis Pasteur placed wine research in a prominent position for the application of cutting-edge biological and chemical sciences, a position it still holds to this day. Technologies such as whole-genome sequencing and systems biology are now revolutionizing winemaking by combining the ability to engineer phenotypes rationally, with a precise understanding of the genetic makeup and key phenotypic drivers of the key organisms that contribute to this age-old industry.

Vitis vinifera is thought to have originated in Europe and consists of approximately 5000 cultivars used in wine, table and dried grape industries around the world. Initial improvement of grapewines was reliant on random selection of natural mutations which led to an improvement in cultivation and/or some aspect of the fruit quality. Grapewine improvement has remained relatively untouched however by classical breeding programs, and few new cultivars, e.g. Pinotage, have become commercial successes. In comparison, classical breeding has played a major role in the development of rootstock varieties which are resistant to soil-borne pests and pathogens. Plant genetic engineering has been proposed to have potential for grapewine/ wine improvement in the wine industry.

21.2   Genetically modified grapewines

Vitis vinifera consists of 38–40 chromosomes, the entire DNA sequence of which has recently been determined. Worldwide, all major viticultural research centres are carrying out some form of grapewine biotechnology, including Australia, Chile, France, Germany, Italy, South Africa, Spain and the USA. Initial studies of plant genetic engineering proved difficult since *Vitis vinifera* is a woody perennial, and hence difficult to genetically manipulate (recalcitrant). Only a few limited successes have been reported and there is still no universal protocol for genetically modifying all grapewine cultivars. The first significant progress was made in 1989 using embryonic cell suspensions derived from the plant's anthers as the target tissue for grapewine transformations. Both the techniques of *Agrobacterium*-mediated transformation and particle bombardment have been used for such experiments. Trials are also being carried out in other countries such as Canada, South Africa and Australia. In South Africa, genetically modified grapewines have recently been planted at Welgevallen, an experimental farm at the University of Stellenbosch.

In South Africa, Winetech (Wine Industry Network of Expertise and Technology) has also been established, which is an organization performing research into various aspects of genetically improving organisms for the wine industry. Their program involves 'Improving grapewine, wine yeast and bacteria for a quality focused, market directed wine industry'. The mission of Winetech is 'To provide the South African wine industry with a sustainable basis of forefront technology and human resources in order to strengthen both local and international competitiveness and profitability'.

21.2.1   Fungal resistance

Since 1999 there have been field trials with gene-modified fungal-resistant grape wines in two areas in Germany, the Pfalz and Franken. The trials were planned to last for at least 10 years, and examined mainly the varieties Riesling and Chardonnay, for which it had not been possible previously to breed fungus-resistant wines (e.g. against grey mould, powdery and downy mildew). The trials however, were suspended at the beginning of 2005.

Recently however, new trials have been initiated in Germany where the wines have been modified with genes from barley, to protect the grapes from fungal infections. The Institute of Wine Cultivations said there would be no noticeable difference in the wine's taste. The first wine made from these altered grapes will most likely only be available in several decades.

In South Africa, a field trial for fungal resistance is being carried out on the Welgevallen Experimental Farm in Stellenbosch. One hundred non-transgenic and virus-free 238 US Vit8-7 rootstocks have been planted and onto these, transgenic Chardonnay or Sultana has been grafted. The grape wines have been genetically modified with a grape gene (Vvpgip) which protects against fungal pathogens, and a grape gene (VvNCED) which protects against water stress. The wines will be monitored for at least 5 seasons as to their performance. The Institute of Wine Biotechnology (IWBT) is largely focusing on fungal resistance to grey rot (*Botrytis cinerea*), as well as powdery and downy mildew. Researchers believe the production of fungal disease resistant plants using transgenic technology is an attractive alternative to chemical treatments and should encourage environmentally friendly practices in the wineyard. The IWBT is focusing on the use of chitinases, polygalacturonase-inhibiting proteins.

21.2.2   Virus resistance

Grapewines with genetically engineered resistance to Fanleaf degeneration caused by a virus are being tested in field trials in Colmar (Alsace, France). The virus is transmitted by nematodes and is the most widespread nepovirus involved in grapewine degeneration.

In South Africa, Winetech established a 'Grapewine Virus research program' in 2000 which 'aims to alleviate the serious virus disease problems in the South African wine industry by thorough characterization of grapewine viruses, to develop strategies to manage the diseases they cause, to prevent further spread of these diseases, and to use the latest technologies for the establishment of genetic virus resistance in wine grape cultivars in the longer term'.

Resistance/tolerance to abiotic stress

Apart from biotic threats such as viruses, fungi, bacteria and insects, grape-wines are also threatened by abiotic stresses such as drought, heat, cold, water logging and salt. Grapewines can be affected significantly by water loss in the form of factors such as canopy growth, bunch quality and photosynthetic efficiency. The IWBT is researching into the development of grapewines with enhanced capabilities to grow under adverse conditions including water stress, and are studying the carotenoid biosynthetic pathway of grapewines in this regard. This pathway has been found to produce compounds involved in environmental stress responses. Examples of genes which have been isolated in the IWBT biotechnology program include:

- Carotenoid biosynthetic pathway: *VvPSY, VvPDS, VvZDS, VvCiso, VvLECY, VvECH, VvLBCY, VvBCH, VvZEP, VvVDE, VvNSY.*
- Abscisic acid biosynthetic pathway: *VvNCED, VvCCD.*

These genes have been used in:

1. Isolation and characterization of carotenoid pathway genes and promoters from *Vitis vinifera* as resources towards stress-tolerant grapes with superior quality.

2. Functional analysis of central metabolic pathways with regards to roles in stress-tolerance, colour development or sugar metabolism.

3. Metabolic engineering of grapewine towards enhanced abiotic stress resistance and improved quality parameters.

Quality traits and plant development

Australia currently has 4 GM field trials underway. They have field-tested grapes which have had their composition modified of components such as sugar content, colour and grape size. It is believed that modifying flower and fruit development may allow an increase in the harvest.

In South Africa, IWBT are using their research into the carotenoid biosynthetic pathway to also investigate quality parameters of grapes. These metabolic pathways have been found to also be involved in quality aspects and are thus being investigated as to their ability to improve flavour and aroma compounds in grapes. Carotenoids serve as important precursors for apocarotenoids, which are involved in a wide range of functions in plants including pigments, flavours and aromas.

21.3   Genetically modified yeasts

Yeasts are single-celled organisms that are classified as fungi. According to American Tartaric Products, the first genetically modified wine yeast, ML01, was released in 2005 by Springer Oenologie (a division of Lesaffre Yeast Corporation). This yeast was first available only in North America where GMOs were unregulated, but has since spread to other countries. The yeast was developed by Dr. HJJ van Vuuren from the Wine Research Centre, University of British Columbia, Vancouver, Canada, and was modified by inserting two foreign genes, one from the pombe yeast, a yeast found in Africa and used to make beer, and one from the bacteria *O. oeni*, so that the alcoholic and malolactic fermentations, could occur at the same time. This may thus be a convenience to winemakers, especially those producing large quantities of wine. In 2003, the FDA designated the yeast as Generally Recognized As Safe (GRAS), however, Professor Joseph Cummins, genetics professor at the University of Western Ontario, has stated that wine yeasts are unstable, and that by genetically altering them, it could lead to unexpected toxicity in the final product. There has also been concern that should certain wineries choose to use ML01, the GM wine yeast could contaminate native and traditional wine yeasts through the air, surface waste and water runoff. Many wineries in the Napa Valley are very particular about their choice of wine yeast, and contamination of these other yeast strains would be undesirable to them.

There are now two commercially available GM yeasts. One such yeast has been genetically manipulated to better degrade urea during the wine making process (*Saccharomyces cerevisiae* strain ECMo01). The benefit of such a characteristic is that the wine contains less ethyl carbamate, a chemical considered by some regulatory bodies to be a human health risk. In January 2006, this yeast strain was declared GRAS by the FDA. The second GM yeast (ML01, been designed to allow malolactic fermentation to proceed more efficiently, thereby producing fewer biogenic amines, such as histamines, which cause headaches and asthmatic-type reactions in some people. Other experiments to genetically modify yeasts include improvement in culture maintenance and the viability of cells, improved yield of fermentable sugars and an improved assimilation of nitrogen. They have also involved an improved tolerance to anti-microbial substances, a reduction in the formation of foam, improved flocculation ability, and an improvement in wine processing with the use of polysaccharide degrading yeast strains which are capable of degrading natural wine polymers (glucans, xylans, pectins). In addition, some yeasts have been genetically modified in order to correct wine acidity.

21.4   Genetically modified wine bacteria

Lactic acid bacteria have historically been associated with food and beverage fermentations as they occur naturally in the starting materials used. Lactic acid bacteria occur in must and wine and perform the secondary fermentation known as malolactic fermentation. They are also considered beneficial to the wine's sensory qualities due to flavour modification, and have been found to control spoilage bacteria. *Oenococcus oeni* is a lactic acid bacteria used commercially in the wine industry. Internationally, rapid progress has been made in the last 20 years in the development of tools for the genetic modification of lactic acid bacteria. Countries involved in the improvement of lactic acid bacteria strains for wine making include Australia, France, Germany, Italy, South Africa and the USA. Major targets of the IWBT are to select for strains that are better adapted as starter cultures for malolactic fermentation. They have also been investigating specific enzymes that are involved in the production of wine aroma compounds, and bacteriocins that can be used as an alternative to chemical preservatives. The genome sequences of several lactic acid bacteria have recently become available contributing to the study of genes of these bacteria.

Section VII
Special topics

Carbon footprint of food

22.1   Introduction

The carbon footprint is a measure of the amount of greenhouse gases (GHG) produced by our activities in relation to carbon dioxide (CO_2) or carbon. All activities caused by mankind from building our homes, using our cars to flying on holiday can be the subject of carbon footprinting. The carbon footprint on food is an estimate of all the emissions caused by the production (e.g. farming), manufacture and delivery to the consumer and the disposal of packaging. For example a 1.4 liter petrol car emits about 160g CO_2 per kilometer. So a carbon footprint of 80g CO_2 for a standard packet of crisps is about the same as driving a typical petrol powered car half of a kilometer.

Thus, the carbon footprint is a measure of the exclusive total amount of carbon dioxide emissions that is directly and indirectly caused by an activity or is accumulated over the life stages of a product. This includes activities of individuals, populations, governments, companies, organizations, processes, industry sectors, etc. Products include goods and services. In any case, all direct (on-site, internal) and indirect emissions (off-site, external, embodied, upstream, downstream) need to be taken into account.

22.2   Ecological footprint of the global food system

After air and water, food is the most essential resource people require to sustain themselves. These resources are provided by the layer of inter-connected life that covers our planet: the biosphere. Yet the way the food system provides food often severely damages the health of the biosphere through soil and aquifer depletion, deforestation, aggressive use of agro-chemicals, fishery collapses, and the loss of biodiversity in crops, livestock, and wild species.

The global food system has become such a dominant force shaping the surface of this planet and its ecosystems that we can no longer achieve sustainability without revamping the food system. At the same time sustainable food systems provide great hope for building a sustainable future—a future in which *all can lead satisfying lives within the means of the biosphere.*

This section discusses the ecological footprint analysis to document the current food system's demand on the biosphere. Ecological footprint accounts track the area of biologically productive land and water needed to produce the resources consumed by a given population and to absorb its waste. In order to create a sustainable food system, we must break down the food footprint into its primary components: the cropland footprint, the pasture footprint, the fisheries footprint, and the energy footprint. By understanding consumption patterns by sector, it becomes easier to target specific areas of consumption.

22.2.1   Cropland footprint

The world's cropland produces human food, animal feed, fiber, and other non-food crops, and makes up 53 per cent of the global food footprint. Footprint accounts analyze the consumption of 75 primary crop products and 15 secondary products.

The cropland footprint has steadily increased with global population. The intensification of farming using agrochemicals, irrigation, and monoculture cropping has slowed the expansion of cropland: the cropland footprint grew by less than 10 per cent over the last 40 years, while the world population doubled. But these gains have ecological costs: a swollen energy footprint and increased demands on neighbouring ecosystems to cope with nutrient loading, soil erosion, toxicity, and water shortages.

The concentration of global food production under the control of a few transnational corporations, bolstered by free trade agreements, structural adjustment policies, and subsidies for the overproduction of crop commodities, has created North-South food trade imbalances and import dependencies that underlie a growing food insecurity in many countries. Production of cash crop exports in exchange for food imports can undermine food self-sufficiency and threaten local ecosystems, adding to the global footprint.

Agribusiness consolidation and large-scale, monoculture cash cropping also leads to the loss of crop and livestock diversity. Wheat, rice, and corn are now the three most abundant plants on earth, providing 60 per cent of human food. At the same time, industrial agriculture threatens crop diversity through the replacement of native varieties with hybrid strains and the contamination of crop and wild species from the introduction of genetically modified organisms. As the global food supply relies on a diminishing variety of crops, it becomes vulnerable to pest outbreaks, the breeding of superbugs, and climate disruptions, all of which could further expand the human footprint even as it must shrink.

22.2.2   Pasture footprint

The world's grazing lands provide us with meat, milk, wool, and hides and represent 13 per cent of the global food footprint. Footprint accounts analyze eight pasture dependent categories and show a growing pasture footprint as the world consumes more animal products. While the pasture and grassland footprint has not grown as rapidly as the consumption of animal products, this is due to the increased use of fertilized pastures for grazing, breeding and managing livestock to boost production efficiency, and feeding livestock from cropland production. In many countries, livestock are at least partially, sometimes exclusively, fed from corn, soyabeans, other crops and crop residues, and fishmeal.

22.2.3   Fisheries footprint

Industrial scale fisheries are rapidly changing the ecology of ocean eco-systems, while an increasing number of studies document the damaging effects of aquaculture and farmed fish. Footprint accounts analyze 22 fish and aquaculture categories, incorporating 40 species groups. The global fisheries footprint has risen more dramatically than other food categories, as the world craves more and bigger fish—that is, fish higher on the food chain. Overall, the world's fisheries are losing productivity. There are fewer and smaller fish. While the catch tonnage remains constant, the quality of fish is declining, as measured by their average trophic level—their status on the food chain. If trends continue, and fish populations from higher trophic levels continue to be overfished or collapse, we may be moving toward oceans of jellyfish or other sea life low on the food chain and with little economic value.

22.2.4   Energy footprint

The global food system is responsible for a sizeable portion of the world's fossil fuel consumption and corresponding carbon dioxide emissions. Estimates vary depending on how the food system is defined or bounded— we use 10 per cent as a conservative placeholder for calculating the global food energy. This 10 per cent includes the energy used in food production, for inputs like fertilizers, pesticides, and irrigation, and in post-production.

Post-production, which accounts for 80–90 per cent of the food system's fossil fuel use, includes processing, packaging, transportation, storage, and retail. An increasingly globalized food supply means a hefty transport Footprint. The food system's thirst for fossil fuel energy leads to stunning imbalances: the energy required to produce, process, package, and distribute a can of corn is six times the food energy contained in that corn. The packaging alone uses more than twice the energy of production; driving the corn home from the store and preparing it also uses more energy than production.

22.2.5   Other food system impacts: Beyond the ecological footprint

As we have seen, the global food Footprint represents a significant portion of the Earth's total biomass production, yet even this is a conservative underestimate of the true area required for food production. Several other factors could be included, as described below.

Unsustainable yields: Footprint accounts currently do not reflect the environmental damage associated with industrial yields, such as soil degradation from intensive agricultural practices, water eutrophication, salinization from irrigation, or pesticide toxicity.

Climate change: Besides the CO_2 from its fossil fuel use, agricultural production adds to the atmospheric carbon stock through forest clearing and the release of soil carbon through cultivation. Food production also contributes to global warming through the release of methane from livestock, rice cultivation, and the burning of agricultural residues. Yet agriculture has the potential to act beneficially as a carbon sink, through farming practices like conservation tillage that build up organic carbon in soil rather than release it to the atmosphere.

Fresh water: The shortage of fresh water is one of the most immediate and potentially devastating environmental challenges facing humanity. Agriculture depletes water stocks and compromises water quality through increasing loads of organic and inorganic pollutants. Despite its significance, current ecological footprint accounts leave out the consumption of water due to a lack of adequate data documenting the impact of a given unit of water, which varies widely depending on soil composition, watershed hydrology, seasonal availability, withdrawal methods, and water quality.

Footprint accounts also do not incorporate human activities that cause irreversible damage to the environment, such as aquifer depletion or the bioaccumulation of persistent toxins from pesticides.

22.3   Eating less beef will reduce carbon footprint more than cars

Beef's environmental impact dwarfs that of other meat including chicken and pork. New research reveals that eating less red meat would be a better way for people to cut carbon emissions than giving up their cars.

The heavy impact on the environment of meat production was known but the research shows a new scale and scope of damage, particularly for beef. The popular red meat requires 28 times more land to produce than pork or chicken, 11 times more water and results in five times more climate-warming emissions. When compared to staples like potatoes, wheat, and rice, the impact of beef per calorie is even more extreme, requiring 160 times more land and producing 11 times more greenhouse gases.

Agriculture is a significant driver of global warming and causes 15% of all emissions, half of which are from livestock. Furthermore, the huge amount of grain and water needed to raise cattle is a concern to experts worried about feeding an extra 2 billion people by 2050. But previous calls for people to eat less meat in order to help the environment, or preserve grain stocks, have been highly controversial.

Researchers analyzed how much land, water and nitrogen fertilizer was needed to raise beef and compared this with poultry, pork, eggs and dairy produce. Beef had a far greater impact than all the others because as ruminants, cattle make far less efficient use of their feed. 'Only a minute fraction of the food consumed by cattle goes into the bloodstream, so the bulk of the energy is lost.' Feeding cattle on grain rather than grass exacerbates this inefficiency, although Eshel noted that even grass-fed cattle still have greater environmental footprints than other animal produce.

'The biggest intervention people could make towards reducing their carbon footprints would not be to abandon cars, but to eat significantly less red meat.' 'Another recent study implies the single biggest intervention to free up calories that could be used to feed people would be not to use grains for beef production.'

The study of British people's diets was conducted by University of Oxford scientists and found that meat-rich diets defined as more than 100 g per day resulted in 7.2 kg of carbon dioxide emissions. In contrast, both vegetarian and fish-eating diets caused about 3.8 kg of CO_2 per day, while vegan diets produced only 2.9 kg. The research analyzed the food eaten by 30,000 meat eaters, 16,000 vegetarians, 8,000 fish eaters and 2,000 vegans.

22.4   Shrink your food footprint

A person's food footprint (foodprint) is all the emissions that result from the production, transportation and storage of the food supplied to meet their consumption needs. We chose to focus on food supply, rather than only food consumption, because a large proportion of food is lost at retail and consumer level. Although emissions also occur when people transport, store and cook food, these emissions are omitted from our calculations as they are captured in travel and housing footprints.

The biggest sources of emissions are from the beef (0.8 T), dairy (0.5 T) and chicken (0.3 T) food groups. This isn't because these food groups dominate intake in the US diet, but rather that they are relatively carbon intensive to produce, in particular beef and dairy.

The remaining food footprint is reasonably well distributed across other groups like cereals, vegetables, fruit, oils, sugars and drinks. In terms of shares, the footprint breaks down like this.

22.4.1   Food waste

The simplest and most cost-effective way to reduce your food footprint is to minimize food waste. Although not all food waste is within your control, your purchasing and cooking habits can play a large part in reducing food losses.

For each of our food groups, we divide total food supply into three groups: retail losses, consumer losses and consumption. Each is expressed as a share of total food supplied. Retail losses are food that is supplied to stores but never sold, due to spoilage and processing. Consumer losses are food purchased but not eaten due to a combination of spoilage, the non-edible share of food, cooking waste and plate waste. Consumption is the portion of total food supplied that is actually eaten. Even though not all food supplied can be eaten (e.g. bones, cores, skins), the scale of food loss is still surprising. While small amount of fresh food can be lost during repackaging within stores, the vast majority of retail losses occur because shops are unable to sell food before it goes out of date. In fruit, vegetables, chicken and beef, this is more than a quarter of all food supplied.

Consumer losses are between 10% and 25% of total food supplied for each group, meaning that as a share of purchased food they are actually much higher. Take the beef example, 23% of total supply is consumer loss, which corresponds to a loss of 33% of purchased food. Consumer losses are made up of food that is non-edible, allowed to go out of date, wasted during cooking or wasted from the plate.

With the exception of the non-edible share of food, you can generally make large reductions in your food footprint by simply ensuring that you eat everything you buy. Doing so just takes some common sense and is action we would naturally take if food was more expensive or scarce. By being careful about the portions you cook, eating what you cook or storing leftovers, you can easily reduce plate waste. By being careful about what you buy, eating perishable things on time and being less squeamish about used by dates where sensible, you can also minimize food spoilage in your home. A large proportion of retail losses is food that is perfectly fine to eat but is discarded due to supermarket caution and food aesthetics. Although retail losses are largely out of our hands we can still affect them to a degree. Buying food which is close to its 'best before' date often avoids such waste. This makes sense if you know you are going to eat it soon after purchase and such food will often be subject to discounts.

By reducing food waste you may be able to shrink your food footprint by as much as a quarter. Although it can be hard to quantify this improvement in your calculations, limiting food waste should result both in a reduction in your kitchen waste volume and food costs. As such reducing food waste is the natural place to start shrinking your food footprint.

22.4.2   Carbon intensity of food

The great variation in how foods are produced, processed and transported means their footprints are very different. The vast majority of emissions, typically around 80%, occur during food production. This means how your food is produced is the most important factor in your food footprint.

Unlike in other sectors of your personal footprint, which are typically dominated by carbon dioxide emissions, both nitrous oxide and methane play an important role in food production. Nitrous oxide emissions are significant in most food groups due to the widespread use of nitrogen-based fertilizers in agriculture. In the US, nitrous oxide emissions make up around a third of total food emissions. Methane emissions occur mostly due to enteric fermentation in animals like cows, sheep and goats so they are largely limited to the beef and dairy food groups. Despite this, methane emissions account for around a quarter of total food emissions in the United States.

By using weighted averages for the production of foods within each group, we can calculate average carbon intensity for each. Although such averages aren't accurate for specific foods, they are very good for assessing complete diets. Having converted our intensities to food energy, we also want to account for the emissions from retail and consumer losses. By dividing our intensities by one minus the percentage loss for both consumer and retail losses, we produce intensities in terms of grams of carbon dioxide equivalents per kilocalorie (g CO_2e/kcal) of food eaten. This captures all the emissions from the food typically supplied for each kilocalorie consumed.

After being adjusted for energy content and losses, our results look similar but there are subtle changes. Because of their high energy content, snacks, oils and cereals are the least carbon intensive ways to supply food energy. Beef at 14 g CO_2e/kcal remains the most intensive while fruit, dairy and chicken are also relatively high. Again, these are weighted averages so within each group there may be wide variation depending on production methods, food losses and energy content.

Using food energy rather than food weight to assess footprints is very useful, because it gives us an idea of the most and least carbon intensive ways we can supply our energy needs. Although daily kilocalorie intake is not the only important factor in diet, it is a simple and clear way to think about different diets while ensuring estimated consumption is in a reasonable range. And the better you understand the carbon intensity of the food you eat, the more effectively you will be able to reduce your footprint.

22.4.3   Different diets

Because the majority of food supply emissions occur during production, changing our diet is the most effective way to shrink our footprint. Reducing our consumption of beef, lamb and dairy products will have the largest effect, due to their high carbon intensity. Additional reductions may also be found by limiting consumption of other meats and certain types of fruits. Choosing to switch food consumption from high carbon intensity foods (e.g. beef, cheese) to low-carbon intensity foods (bread, potatoes, grains) can reduce emissions by as much as 90% for that food consumption.

22.4.4   Food miles

Despite often being highlighted in the media the significant distance that much of our food travels is not the major source of food emissions. This is confirmed by bottom up studies of individual foods and well as top down studies of entire food supply markets. Although this varies from country to country, a rough guide of food supply emissions is something like 80% production, 10% transport and 10% storage at wholesale and retail level.

Transport emissions are in fact dominated by the upstream emissions of moving foodstuffs to production facilities, with the remainder of transport emissions being from delivering final products to wholesalers and retailers. Reducing these transport emissions can help reduce total emissions, but only if it makes sense in the context of the foods entire footprint. Foods for which transport is a significant factor are often those which are shipped by air due to their perishing quickly and having high value per unit weight.

22.4.5   Cooking and storage

After its sale, food is generally transported home, stored and often cooked before being eaten. Because emissions from these processes are accounted for in our travel and housing footprints, we do not calculate them as part of our food footprint. This is not to say they are unimportant, it merely avoids double counting.

Because fridges and freezers are on all day every day they are often a major use of electricity. In the US refrigeration accounts for 12% of domestic electricity use, which has a footprint of around 400 kg CO_2e a year per person. To a limited degree, you reduce the electricity your fridge and freezer consume by not setting them too cold, insuring they have proper seals, are well defrosted and located in the coolest area possible. Much more important however is the choice you make when replacing an old fridge and freezer.

In most countries there are efficiency standards, like the A-G energy labelling in Europe and Energy Star in the United States, which you can use to inform your choice. While these tell you how efficient the fridge or freezer is for its class they are only so helpful, because they don't account for size. If you buy the highest class possible, and also keep your fridge-freezer as small as possible given you needs, you will limit how many kilowatt-hours a year your fridge uses. And thus reduce its carbon footprint. Cooking can also be a significant source of emissions. In the US, it is around 200 kg CO_2e a year per person per year. In general cooking with gas creates fewer emissions than cooking with electricity, unless you have access to low-carbon electricity.

22.4.6   Land use emissions

Emissions from land use, land-use change and forestry (LULUCF) are a major part of global emissions. Although these emissions are generally not calculated as part of the food supply footprint, the expansion of agriculture is a major driver of deforestation, which dominates LULUCF emissions. More than 60% of these land use emissions occurred in Brazil and Indonesia, largely as the result of deforestation. The major driver of this deforestation was conversion to agricultural land. In Brazil as much as 80% of deforestation occurs due to the reclaiming of land for cattle rearing or growing feeds like soya for cattle. In Indonesia deforestation in recent years has been largely driven by clearing land for palm oil plantations, which as one of the cheapest vegetable oils is used widely in food production. By helping to drive deforestation, the production of beef, soya and palm oil, along with other crops in tropical regions, has a very large indirect footprint. Though these emissions are generally not quantified as part of the food supply footprint, they are hugely important and an additional reason to consider switching away from particular foods.

22.5   Food transportation issues and reducing carbon footprint

Transportation is the largest end-use contributor toward global warming in the United States and many other developed countries. Transportation has a significant impact within the food and beverage sector because food is often shipped long distances and not infrequently via air.

Although the impact of transportation is important, full life cycle analyses indicate that for most foods transportation does not have the largest environmental impact.

Some analysts, such as Weber and Matthews, estimate that given the typical household food basket, aggregate transportation accounts for just 11% of total carbon emissions associated with food production.

Therefore, it is still worthwhile to consider improving the food distribution system. There are often many options for delivering food to consumers, and these supply chain configurations can result in vastly differing energy and emissions profiles. In this section provides the background and tools for analyzing the energy intensity and resultant emissions of a food distribution system, evaluating tradeoffs and identifying opportunities for significant improvement.

22.5.1   Supply chain basics

Before we can further investigate transportation impacts, we must first introduce the concept of the supply chain: the sequenced network of facilities and activities that support the production and delivery of a good or service. A supply chain starts with basic suppliers and extends all the way to consumers via stages. These stages may include such facilities as suppliers, factories, warehouses and other storage facilities, distribution centers, and retail outlets.

Figure 22.1 shows a sample supply chain, where the arrows denote the flow of a product toward the consumer. This figure depicts both inbound logistics (the delivery of raw materials and packaging to the manufacturer) as well as outbound logistics (the transportation and storage of the finished good to the end consumer). This section focuses on outbound logistics, colloquially known as 'gate-to-kitchen' and 'farm-to-fork' in the food and beverage industry. The emissions associated with outbound logistics vary by origin and type of food.

Weber and Matthews estimate that food transportation may account for 50% of total carbon emissions for many fruits and vegetables, but less than 10% for red meat products. Although inbound logistics can require substantial energy use, it is considered part of the production process.

Supplier   Manufacturer   Distributor   Retailer   Consumer

Figure 22.1 A simple supply chain

Although the interrelationships between supply chain stages may be quite complex, all supply chains have one aspect in common—they end with a consumer. Supply chains for different products may be interlinked; one supply chain's end consumer may represent an intermediate node for another supply chain. Examples include a firm that buys components and assembles them into consumer items, and a soft drink producer that buys cylinders of compressed CO_2 to carbonate its products.

Much supply chain complexity results from the fact that few supply chains are completely controlled by one firm or vertically integrated. For example, producers and retailers are not typically owned by the same firm. Companies may outsource supply chain activities, especially transport and storage activities, which are handled more effectively by third party logistic (3PL) providers. Outside firms that form a part of a company's supply chain are channel partners. These partnerships require collaboration across organizations. The supply chain management (SCM) can be defined as the coordination of business functions within an organization and between the organization and its channel partners. SCM strives to provide goods and services that fulfill customer demand responsively, efficiently, and sustainably. SCM includes such functions as demand forecasting, purchasing (also known as sourcing), customer relationship management (CRM), and logistics. Logistics concerns the movement and storage of goods, services, and information. It is an umbrella term for such important functions as transportation, inventory management, packaging and returns/reverse logistics.

Some terminology will be helpful to understand who is doing what. The shipper initiates the movement of the product forward into the supply chain, the carrier is the party that does the actual moving of the product, and the consignee receives the product.

22.5.2   Transport modes

Within the developed world there are four basic transport modes for shipping large quantities of packaged products: water, rail, truck, and air. Trucking dominates, comprising more than 75% of the total US freight transit bill. Trucking variables include truck type, ownership model (such as 3PL or company-owned fleet), and loading option (less-than-truckload or full-truckload). The dominant transport mode has shifted over time.

Short sea shipping, using ocean-going vessels for delivering cargo domestically, is popular in Europe and also holds promise for replacing many truck deliveries in the United States.

To compare transport modes with regard to energy usage and resultant emissions, we define a ton-km as the movement of 1 metric ton of cargo over 1 km. Table 22.1 shows that these modes have very different energy and emissions profiles.

Table 22.1 Energy and emissions per T-km.

	Megajoules per T-km	kg CO_2e per T-km
International water-container	0.2	0.14
Inland water	0.3	0.21
Rail[a]	0.3	0.18a
Truck[b]	2.7	1.8
Air[c]	10	6.8

Note that utilization and backhaul rates will affect all figures.
[a]May depend on whether diesel or electric power is used.
[b]Depends on size and type of truck, power source.
[c]Includes effects from radiative forcing.

The water and rail transport modes are contingent upon the availability of navigable water and established railroad tracks. An additional consideration is the potential need for supply chain responsiveness: Air freight may be the only viable option for long-distance transport when customer orders require immediate fulfillment.

Intermodal transport

Before we choose one mode over another, we should consider intermodal transport. Defined as using more than one transportation mode to move a shipment between two points, an intermodal route might involve shipping cargo by water, then by rail, then by truck. Intermodal transport became practical with the advent of containerization, where products stay in the same container throughout their entire journey. Containerization was made possible through global standardization of container size and features, which dramatically reduced intermodal transfer times and significantly increased cost efficiency. From a sustainability viewpoint, the advantage of intermodal transport is that we can utilize more efficient modes for major transport corridors, and then shift to trucks for transport to remote destinations. Shippers can also use a 3PL provider to oversee the entire shipping process. One disadvantage of intermodal transport is its inherent complexity of coordination and the information technology support required to address that complexity. Another issue is the movement and repositioning of empty containers.

Utilization and backhaul

Many carbon analyzers base calculations on only transport mode and shipping distance. In analyzing, we will take into account additional factors, including vehicle utilization (how full the vehicle is) and backhaul (whether or not the vehicle carries freight on its return journey). Although fully laden vehicles use more fuel than nearly empty ones, most of the energy expended during a trip is used to move the vehicle and not its cargo. Underutilized vehicles waste energy, as do vehicles that return empty. Also, weight and volume limits must be respected, and all but the lightest and bulkiest cargo loads tend to 'weigh out' rather than 'cube out.'

It can be difficult to determine utilization fractions and backhaul percentages, as these are likely to vary with each trip. Such information is even more challenging to obtain when transportation functions have been outsourced. However, some assumptions can be made. For example, vehicles chartered by 3PL providers are likely to have higher utilization fractions because they often carry cargo from multiple companies. Third party logistic providers are also likely to have higher backhaul rates, because they have more opportunities for obtaining return freight owing to their broader customer base.

Warehousing

Logistics involves not only the movement of goods, but their storage. Unless a product is custom ordered by an onsite client, it is likely that the product will enter storage at some point in its journey to the consumer. Such storage can occur at any supply chain stage: at the producer, distributor warehouse, and/or retailer stockroom. Intermediate supply chain stages range from pure storage centers to dedicated cross-dock facilities, in which cargo from upstream supplier trucks/railcars is transferred directly to outbound trucks/railcars destined for downstream stages. In addition to storage, warehouses can provide additional services: pick and pack (repackaging palletized products to smaller quantities destined for either retailers or end consumers), customs clearance, or even house product-finishing functions such as customizing goods to the local marketplace.

Packaging

Packaging decisions are inherently linked to the supply chain. Goods are frequently shipped in bulk and broken into consumer-sized quantities at a warehouse or other facility, and individual commodities are sometimes bundled into larger end-items, such as multipacks, and palletized. Packaging materials (pallets, boxes, totes, slipsheets, etc.) for both finished goods and intermediate support functions may be designed to be recyclable, compostable, or reusable. Non-landfilled packaging is highly desirable, but creates other challenges, such as the impact of reusable packaging in the reverse supply chain. Packaging can often be reengineered to reduce package weight or bulk, which can translate into savings in raw materials, landfill impacts, and transport/storage energy use; but extra costs may be incurred elsewhere.

Complexity of food supply chains

The supply chains can be long and complex. Food supply chains are some of the most difficult to manage as they must often address time constraints to avoid spoilage, as well as concerns about contamination, high weight-to-value ratios, fragility, unique packaging requirements, and the potential impact of food being wasted rather than consumed. We show here how these considerations affect outbound logistics.

One challenge relates to food production being inherently dependent on nature. Not only is the cultivation of many foods restricted geographically, but also temporally. Fruits, vegetables, and grains typically have fixed growing cycles with short and specific annual harvest periods. There are three options for supplying fresh produce that is out of season locally: sourcing from distant growing areas, using long-term storage, or cultivating in a protected environment such as a greenhouse. Importing produce often results in lower overall emissions than harvesting and storing local produce for several months. Indeed, energy needed for long-term cold storage can dominate a product's overall emissions profile. Carlsson-Kanyama shows, for example, that storage accounts for 60% of the carbon emissions associated with carrots. Higher emissions can result not only from the energy needed for climate control, but also from the inherent yield losses that occur during storage. Protected cultivation is even more energy intensive. Carlsson-Kanyama and others show that tomatoes produced locally in Swedish greenhouses require ten times the energy as field-grown tomatoes imported from Southern Europe. Thus, long-distance supply chains, even though they are energy intensive, may yield the lowest overall footprint for providing out-of-season product to consumers.

A second challenge is related to situations where similar food commodities are produced locally as well as imported from distant locations, the emissions intensity of the production methods must be considered in any comparison of overall supply chain emissions. For example, Saunders and Barber find that milk solids produced locally in the United Kingdom generate 34% more emissions than the same product imported from New Zealand, even with transport included. This result reflects the more energy-intensive dairy production system in the United Kingdom.

A third challenge is that highly perishable foods require special handling to avoid yield loss and potential health issues. These foods often require cooling, refrigeration, or freezing during transport and/or storage. It may also be necessary to control other conditions, such as humidity, exposure to air, or contact with other items.

These requirements increase energy usage and emissions.

A fourth challenge is that the location of facilities within a food supply chain can also affect emissions. For example, Sim and others find that overall carbon emissions can be significantly reduced by locating processing and storage facilities in countries where more electricity is generated from renewable fuels or cleaner energy.

Fifth, when time is of the essence, as in the transport of highly perishable produce such as berries, airfreight may be the only viable transport option. Air freighting may also be necessary in regions such as Africa where no other viable alternative exists for transporting produce to market. As previously shown, air freighting is highly energy intensive. Scholz and others report that fresh salmon air freighted from overseas has about twice the environmental impact as frozen salmon transported by container ships over the same distance. The difference owing to transport modes is far more significant in this case than production choices such as wild versus farmed or organic versus conventional.

A sixth challenge is that safe food storage not only requires climate controls, but also a high degree of sanitation. In most developed countries, warehouses must be built and maintained to stringent guidelines to be certified as 'food grade.' In the United States, wood pallets may not be reused and may soon be phased out as unsanitary.

The process of packaging food is yet another challenge. Twede and others emphasize that packaging beverage products is a high-speed automated process involving expensive equipment. Such capital investment and the need for a controlled environment favours centralizing packaging at the point of production, even if it might be more energy efficient to ship product in bulk. Food and beverage products typically require extensive packaging, which adds both weight and volume to the product. Additional energy and materials are required to create the packaging and transport it to the production site. Smith performs a life cycle assessment of the Nova Scotia wine industry and finds that the largest contribution to emissions is owing to the production and transport of wine bottles.

Measuring transportation-related carbon emissions

This section presents the basics of performing a carbon audit and concludes with some examples from practice. Although other gases such as methane and nitrous oxide may contribute to global warming, aggregate greenhouse gas measures are typically reported in CO_2 equivalents (CO_2e), which is kilograms of CO_2 emitted per kg of product. Carbon dioxide dominates, comprising 95% of total greenhouse gas emissions by volume. Emissions are colloquially called 'carbon emissions,' which can lead to confusion as some older studies only weigh the carbon component of the gas, which is 30% of the total mass of CO_2. It is also now standard practice to report all significant greenhouse gas emissions as single carbon emissions.

The scope of the analysis depends on the purpose of the study. Scope 1 includes only direct emissions, whereas Scope 2 also includes indirect emissions from any consumption of purchased electricity, heat, or steam. Scope 3 is the broadest, including all other indirect emissions, such as the extraction and production of purchased materials and fuels, all outsourced activities, and waste disposal. Scope 3 can also include the substantial impact of radiative forcing from the contrails in tallying airplane emissions.

Scope must be carefully considered because incomplete framing (inappropriate scope) may lead to incorrect conclusions. For example, food miles are defined as the distance between the production source and the retail store, or 'farm-to-fork.' In addition to providing consumers with information, carbon audits can provide useful insights for companies evaluating their operations. However, the figures from carbon audits should be viewed as guidelines rather than as precise and absolute truths.

To sum up, the transportation-related carbon footprint varies from a few per cent to more than half of the total carbon footprint associated with food production, distribution, and storage. Supply chains are complex and varied, and food supply chains are especially challenging because of seasonality, freshness, spoilage, and sanitary considerations. Measuring transportation-related carbon footprint involves careful choice of the scope of the analysis, and there is much uncertainty in the results. Caution is warranted regarding the absolute numbers from carbon assessments, so it may be best to focus primarily on relative comparisons.

Supply chain planners must carefully consider the trade-off between transportation-related energy cost and carbon footprint and storage-related energy cost and carbon footprint. Also, the frequent small deliveries called for by lean manufacturing practices, although optimizing efficiency within a facility can increase overall carbon footprint. Packaging is another important consideration, and the use of plastics rather than glass tends to lower carbon footprint. Benefits for the environment and health are further accrued by plastic recycling.

To reduce carbon footprint, suppliers are consolidating their operations, increasing their use of rail and water transit, and increasing transport efficiency by filling trucks and considering backhaul opportunities. Food waste is another potentially significant contributor to carbon emissions, which could potentially be reduced via alternative packaging options.

22.6   Carbon footprint ranking of food

Table 22.2 shows the greenhouse gas emissions produced by one kilo of each food. It includes all the emissions produced on the farm, in the factory, on the road, in the shop and in our home. It also shows how many miles we need to drive to produce that many greenhouse gases. For example, we need to drive 63 miles to produce the same emissions as eating one kilogram of beef.

Table 22.2 Greenhouse gas emission produce by 1 kg of each food.

Rank	Food	CO_2 kilos equivalent	Car miles equivalent
1	Lamb	39.2	91
2	Beef	27.0	63
3	Cheese	13.5	31
4	Pork	12.1	28
5	Turkey	10.9	25
6	Chicken	6.9	16
7	Tuna	6.1	14
8	Eggs	4.8	11
9	Potatoes	2.9	7
10	Rice	2.7	6
11	Nuts	2.3	5
12	Beans/tofu	2.0	4.5
13	Vegetables	2.0	4.5
14	Milk	1.9	4
15	Fruit	1.1	2.5
16	Lentils	0.9	2

Meat, cheese and eggs have the highest carbon footprint. Fruit, vegetables, beans and nuts have much lower carbon footprints. If we move towards a mainly vegetarian diet, we can have a large impact on our personal carbon footprint.

22.6.1   Tips for reducing carbon footprints

- Eat vegetarian
- Bring back home-cooking
- Cooking smartly
- Eat organic
- Save water
- Shop wisely
- Shop local
- Reuse and recycle
- Grow your own food

Nanotechnology in agriculture and food industry

23.1   Introduction

Nanoscience and nanotechnology are concerned with the understanding and rational manipulation of materials at the atomic and molecular levels, generally with structures of less than 100 nm in size. Scientifically, nanoscience is defined as the study of phenomena and the manipulation of materials at the atomic, molecular, and macromolecular scales, where the properties differ from those at a larger scale and have unique novel functional applications.

Nanotechnology has the potential to revolutionize the agricultural and food industry with new tools for the molecular treatment of diseases, rapid disease detection, enhancing the ability of plants to absorb nutrients, etc. Smart sensors and smart delivery systems will help the agricultural industry combat viruses and other crop pathogens. Nanostructured catalysts will increase the efficiency of pesticides and herbicides, allowing lower doses to be used. Nanotechnology will also protect the environment indirectly through the use of alternative (renewable) energy supplies, and filters or catalysts to reduce pollution and clean-up existing pollutants. Agriculture is the backbone of most developing countries, with more than 60% of the population reliant on it for their livelihood. As well as developing improved systems for monitoring environmental conditions and delivering nutrients or pesticides as appropriate, nanotechnology can improve our understanding of the biology of different crops and thus potentially enhance yields or nutritional values. In addition, it can offer routes to added value crops or environmental remediation.

23.2   Applications of nanotechnology

23.2.1   Nanotechnology in seed science

Seed is most important input determining productivity of any crop. Conventionally, seeds are tested for germination and distributed to farmers for sowing. In spite of the fact that seed testing is done in well-equipped laboratories, it is hardly reproduced in the field due to the inadequate moisture under rainfed conditions. A group of research workers are currently working on metal oxide nano-particles and carbon nanotube to improve the germination of rainfed crops. Khodakovskaya and others have reported the use of carbon nanotube for improving the germination of tomato seeds through better permeation of moisture. Their data show that carbon nanotubes (CNTs) serve as new pores for water permeation by penetration of seed coat and act as a passage to channelize the water from the substrate into the seeds. These processes facilitate germination which can be exploited in rainfed agricultural system.

23.2.2   Nano-fertilizers for balanced crop nutrition

Nano-fertilizer technology is very innovative; however, some of the reports and patents strongly suggest that there is a vast scope for the formulation of nano-fertilizers. Significant increase in yields has been observed due to foliar application of nano particles as fertilizer. It was shown that 640 mg ha^{-1} foliar application (40 ppm concentration) of nanophosphorus gave 80 kg ha^{-1} P equivalent yield of cluster bean and pearl millet under arid environment.

23.2.3   Nano-herbicide for effective weed control

Weeds are menace in agriculture. Herbicides are designed to control or kill the above ground part of the weed plants. None of the herbicides inhibits activity of viable below ground plant parts like rhizomes or tubers, which act as a source for new weeds in the ensuing season. Soils infested with weeds and weed seeds are likely to produce lower yields than soils where weeds are controlled. Improvements in the efficacy of herbicides through the use of nanotechnology could result in greater production of crops.

The encapsulated nano-herbicides are relevant, keeping in view the need to design and produce a nano-herbicide that is protected under natural environment and acts only when there is a spell of rainfall, which truly mimics the rainfed system.

23.2.4   Nano-pesticide

In order to protect the active ingredient from the adverse environmental conditions and to promote persistence, a nanotechnology approach, namely 'nano-encapsulation' can be used to improve the insecticidal value. Nano-encapsulation comprises nano-sized particles of the active ingredients being sealed by a thin-walled sac or shell (protective coating). Nano-encapsulation of insecticides, fungicides or nematicides will help in producing a formulation which offers effective control of pests while preventing accumulation of residues in soil. In order to protect the active ingredient from degradation and to increase persistence, a nanotechnology approach of 'controlled release of the active ingredient' may be used to improve effectiveness of the formulation that may greatly decrease amount of pesticide input and associated environmental hazards. Nano-pesticides will reduce the rate of application because the quantity of product actually being effective is at least 10–15 times smaller than that applied with classical formulations.

23.2.5   Nanotechnology in water management

Nanotechnology offers the potential of novel nanomaterials for the treatment of surface water, groundwater and wastewater contaminated by toxic metal ions, organic and inorganic solutes and microorganisms. Due to their unique activity towards recalcitrant contaminants, many nanomaterials are under research and development for use for water purification.

To maintain public health, pathogens in water need to be identified rapidly and reliably. Unfortunately, traditional laboratory tests are time consuming. Faster methods involving enzymes, immunological or genetic tests are under development.

Water filtration may be improved with the use of nanofiber membranes and the use of nanobiocides, which appear promisingly effective. Biofilms contaminating potable water are mats of bacteria wrapped in natural polymers which are difficult to treat with antimicrobials or other chemicals. They can be cleaned up only mechanically, which cost substantial down-time and labour. Work is in progress to develop enzyme treatments that may be able to break down such biofilms.

23.2.6   Nano-scale carriers

Nanoscale carriers can be utilized for the efficient delivery of fertilizers, pesticides, herbicides, plant growth regulators, etc. The mechanisms involved is the efficient delivery, better storage and controlled release include: encapsulation and entrapment, polymers and dendrimers, surface ionic and weak bond attachments among others.

These help to improve stability against degradation in the environment and ultimately reduce the amount to be applied, which reduces chemical runoff and alleviates environmental problems. These carriers can be designed in such a way that they can anchor plant roots to the surrounding soil constituents and organic matter.

This can only be possible if we unravel the molecular and conformational mechanisms between the nanoscale delivery and targeted structures, and soil fractions. Such advances as and when they happen will help in slowing the uptake of active ingredients, thereby reducing the amount of inputs to be used and also the waste produced.

23.3   Biosensors to detect nutrients and contaminants

Protection of the soil health and the environment requires the rapid, sensitive detection of pollutants and pathogens with molecular precision. Soil fertility evaluation is being carried out for the past 60 years with the same set of protocols which may be obsolete for the current production systems and in the context of precision farming approaches.

Biosensors provide high performance capabilities for use in detecting contaminants in food or environmental media. They offer high specificity and sensitivity, rapid response, user-friendly operation, and compact size at a low cost. While the direct enzyme inhibition sensors currently lack the analytical ability to discriminate between multiple toxic substances in a sample (such as simultaneous presence of heavy metal and pesticide), they may prove useful as a screening tool to determine when a sample contains one or more contaminants. These methods are amenable to deployment in single-use test strips (making them useful to those in the field).

According to Campbell and others, detection of multiple residues of organo-phosphorus pesticides has been accomplished using a nanomagnetic particle in an enzyme-linked immunosorbent assay (ELISA) test.

23.3.1   Agricultural engineering issues

Nanotechnology has many applications in the field of agricultural machinery. These cover: application in machine structure and agricultural tools to increase their resistance against wear and corrosion and ultraviolet rays; producing strong mechanical components with use of nano-coating and use of bio-sensors in smart machines for mechanical–chemical weed control; production of nano-cover for bearings to reduce friction. The use of nano-technology in production of alternative fuels and reduction of environmental pollution are also worth mentioning.

23.3.2   Application of nanotechnology in animal sciences

Nanotechnology has the ability to provide appropriate solutions for addressing the issues of food items, veterinary care and prescription medicines as well as vaccines for domesticated animals.

Taking certain medications such as antibiotics, vaccines, and probiotics, would be effective in treating the infections, nutrition and metabolic disorders, when used at the nano level. Medicines used at the nano level have multilateral properties to remove biological barriers for increased efficiency of the applied medicine. Appropriate timing for the release of drug and self-regulatory capabilities are the main advantages of the use of nanotechnology in the application of drugs.

The C-60 carbon particle (bucky ball) is spherical molecule having nearly 1 nm diameter. It is non-toxic to the live cells and biocompatible in nature. It can be used as a carrier to deliver the water-soluble peptides and drugs. The nanotechnology can help to understand certain drug behaviour in an animal body. The nano particles can penetrate the skin through minor abrasions; these are reported to be used as sensor to detect the altered cell behaviour. The dendrimers are synthetic three-dimensional macromolecules having a core particle surrounded by branches like a tree. They can be conjugated with the target molecule like drug as they are biocompatible and are easily cleared from blood through the kidney. It was observed that *in vivo* delivery of dendrimer-methotrexate reduce the tumour size ten times more than the free methotrexate.

The nano-magnets can be used as drug delivery system specially to treat the cancerous growth without any harm to the surrounding tissues. Different types of proteins like albumin, gelatin, gliadin and legumin can be used to prepare nanoparticle-based drug delivery system. Inert nanobeads were used to neutralize the antigen causing osteoarthritis in racing horses. Use of nano-based antibiotics in treatment of animal diseases requires less amount of antibiotics leaving less antibiotic residues.

Nanoparticle-based chromium supplementation has beneficial effects on growth performance and body composition and it increases tissue chromium concentration in the muscles. Iron deficiency is a common problem in animals, especially during the early stage of life, gestation and parasitic infestation due to less bioavailability. The bioavailability can be increased with the supplementation of ferric phospholic nano-particles.

Nanotechnology is used to produce the chicken/goat meat in the laboratory in large quantities maintaining the same nutritive value, taste, texture without any hazard (*vegetarian meat*). Use of nanotechnology in *designer egg* production is well known. It can produce the eggs with low cholesterol, less yolk content, more nutrients, and desired antibodies. In addition, nano-based sensors can help in early detection of egg-borne pathogens.

23.3.3　Nanotechnology application in fisheries and aquaculture

Nanotechnology has tremendous potential to revolutionize fisheries and aquaculture sector. Nanotechnology tools like nano-materials, nano-sensor, DNA nano-vaccines, gene delivery and smart drug delivery have the potential of solving many puzzles related to fisheries nutrition and health production, reproduction, prevention and treatment of disease. Nanotechnology will help fish processing industry for producing quality products by detecting bacteria in packaging, producing strong flavour, colour quality and safety.

23.4　Smart delivery systems

Nanoscale devices are envisioned that would have the capability to detect and treat diseases, nutrient deficiencies or any other maladies in crops long before symptoms were visually exhibited. 'Smart Delivery Systems' for agriculture can possess timely controlled, spatially targeted, self-regulated, remotely regulated, pre-programmed, or multi-functional characteristics to avoid biological barriers to successful targeting. Smart delivery systems can monitor the effects of delivery of nutrients or bioactive molecules or any pesticide molecules. This is widely used in health sciences wherein nano-particles are exploited to deliver required quantities of medicine to the place of need in human system. In the smart delivery system, a small sealed package carries the drug which opens up only when the desirable location or infection site of the human or animal system is reached. This would allow judicious use of antibiotics than otherwise would be possible.

23.4.1　Nanodevices for identity preservation (IP) and tracking

One of the major constraints in Indian agriculture is the quality maintenance of agricultural produce. Proper monitoring of production system through nanotechnology will be appropriate to promote quality and make a clear distinction with organic products. Identity Preservation (IP) is a system that creates increased value by providing customers with information about practices and activities used to produce a particular crop or other agricultural products. Certifying inspectors can take advantage of IP as a better way of recording, verifying, and certifying agricultural practices. Through IP, it is possible to provide stakeholders and consumers with access to information, records and supplier protocols. Quality assurance of agricultural products safety and security could be significantly improved through IP at the nano-scale. Nano-scale IP holds the possibility of continuous tracking and recording of the history which a particular agricultural product experiences. The nano-scale monitors may be linked to the recording and the tracking devices to improve identity preservation of food and agricultural products. The IP system is highly useful to discriminate organic versus conventional agricultural products.

23.4.2   Nanolignocellulosic materials

Recently, nano-sized lignocellulosic materials have been obtained from crops and trees which had opened up a new market for innovative and value-added nanosized materials and products, e.g. nano-sized cellulosic crystals have been used as lightweight reinforcement in polymeric matrix. These can be applied in food and other packaging, construction, and transportation vehicle body structures. Cellulosic nano-whisker production technology from wheat straw has been developed by the Michigan Biotechnology Incorporate (MBI) International, and is expected to make biocomposites that could substitute for fiberglass and plastics in many applications, including automative parts.

23.4.3   Photocatalysis

One of the processes using nanoparticles is photocatalysis. The mechanism of this reaction is that when nanoparticles of specific compounds are subjected to UV light, the electrons in the outermost shell (valence electrons) are excited resulting in the formation of electron hole pairs, i.e. negative electrons and positive holes. Analogy with n- and p-type semiconductors, i.e. the Group IV elements, e.g. Germenium of the Periodic Table, dopted with, respectively, minute quantities of the Group V and Group III impurities, is worth noting. Due to their large surface-to-volume ratio, these have very efficient rates of degradation and disinfection. As the size of the particles decrease, surface atoms are increased, which results in tremendous increase in chemical reactivity and other physico-chemical properties related to some specific conditions such as photocatalysis, photoluminescence, etc. So this process can be used for the decomposition of many toxic compounds such as pesticides, which take a long time to degrade under normal conditions. Nanoparticles can be used for the bioremediation of resistant or slowly degradable compounds like pesticides.

The removal of toxins from wastewater is an emerging issue due to its effects on living organisms. Many strategies have been applied for wastewater treatment with little success. Photocatalysis can be used for purification, decontamination and deodorization of air. It has been found that semiconductor sensitized photosynthetic and photocatalytic processes can be used for the removal of organics, destruction of cancer cells, bacteria and viruses. Application of photocatalytic degradation has gained popularity in the area of wastewater treatment.

23.4.4   Nanobarcode technology

In our daily life, identification tags have been applied in wholesale agriculture and livestock products. Due to their small size, nanoparticles have been applied in many fields ranging from advanced biotechnology to agricultural encoding. Nanobarcodes (>1 million) have been applied in multiplexed bioassays and general encoding because of their possibility of formation of a large number of combinations that render them attractive for this purpose. The UV lamp and optical microscope are used for the identification of micrometer-sized glass barcodes which are formed by doping with rare earth containing a specific type of pattern of different fluorescent materials. The particles to be utilized in nanobarcodes should be easily encodeable, machinereadable, durable, sub-micron-sized taggant particles. For the manufacture of these nanobarcode particles, the process is semi-automated and highly scalable, involving the electroplating of inert metals (gold, silver, etc.) into templates defining particle diameter, and then the resulting striped nanorods from the templates are released.

Nanobarcodes have been used as ID tags for multiplexed analysis of gene expression and intracellular histopathology. In the near future, more effective identification and utilization of plant gene trait resources is expected to introduce rapid and cost effective capability through advances in nanotechnology based gene sequencing. Nanobarcodes serve as uniquely identifiable nanoscale tags and have been applied for non-biological applications such as for authentication or tracking in agricultural food and husbandry products. Such nanobarcode technology will enable one to develop new auto-ID technologies for the tagging of items previously not practical to tag with conventional barcodes. With the enhanced importance of traceability in food trade, such technologies will be helpful in promoting biosafe international food trade.

23.4.5   Quantum dots (QDs) for staining bacteria

Bacteria, the most primitive life forms present almost everywhere, are useful as well as harmful for life. There are numerous bacteria which are responsible for many diseases in human-like tetanus, typhoid fever, diphtheria, syphilis, cholera, food-borne illness, leprosy and tuberculosis caused by different species. As a remedial process, we need to detect bacteria and for this, dye staining method is used. To stain bacteria, the most commonly used biolabels are organic dyes, but these are expensive and their fluorescence degrades with time. Fluorescent labelling by quantum dots (QDs) with bio-recognition molecules has been discovered through the recent developments in the field of luminescent nanocrystals. QDs are better than conventional organic fluorophores (dyes) due to their more efficient luminescence compared to the organic dyes, narrow emission spectra, excellent photostability, symmetry and tunability according to the particles sizes and material composition. By a single excitation light source, they can be excited to all colours of the QDs due to their broad absorption spectra. Bio-labelled *Bacillus* bacteria with nanoparticle consisting of ZnS and Mn^{2+} capped with biocompatible 'chitosan' gave an organ glow when viewed under a fluorescence microscope. For the detection of *E. coli* O157: H7, QDs were used as a fluorescence marker coupled with immune magnetic separation. For this purpose, magnetic beads were coated with anti-*E. coli* O157 antibodies to selectively attach target bacteria, and biotin-conjugated anti-*E. coli* antibodies to form sandwich immune complexes. QDs were labelled with the immune complexes via biotin-streptavidin conjugation after magnetic separation.

23.4.6   Nanobiotechnology

Nanobiotechnology has the potential to increase the efficiency and quality of agricultural production and food storage, to enhance the safety of food supplies for the protection of consumers and producers and to introduce new functionality (value-added products) for food, fiber and agricultural commodities. Nanobiotechnology will pave the ways for new researchable areas and applications such as DNA chip, protein identification and manipulation, novel nucleic acid engineering based films, smart delivery of DNA using gold nanoparticles. Biological tests measuring the presence or activity of selected substances become quicker, more sensitive and more flexible when nano-particles are put to work as tags or labels. Magnetic nanoparticles, bound to suitable antibody, are used to label-specific molecules, structures or microorganisms.

23.4.7   Nano-food industry

Nanostructures in food

Understanding the nature of nanostructures in foods allows for better selection of raw materials and enhanced food quality through processing. Techniques such as electron microscopy and the newer probe microscopies, such as atomic force microscopy (AFM), have begun to reveal the nature of these structures, allowing rational selection, modification and processing of raw materials. For example, the creation of foams (e.g. the head on a glass of beer) or emulsions (e.g. sauces, creams, yoghurts, butter and margarine) involves generating gas bubbles, or droplets of fat or oil, in a liquid medium. This requires the production of an air-water or oil-water interface and the molecules present at this interface determine its stability.

Nanoparticles in foods

Nanotechnology has applications in food safety (e.g. detecting pesticides and microorganisms), in environmental protection (e.g. water purification), and in delivery of nutrients.

Food contact materials

Various additives or ingredients are approved for use in food contact materials. The types of materials approved and the regulations on their use will vary for different countries. If the materials do not partition into foods, then there should be little concern over the safety of their use in food contact applications. Should some degree of partitioning occur, then the limitations on their use are based on an acceptable daily uptake (ADI) for these additives or ingredients. The inclusion of nanoparticles in food contact materials can be used to generate novel types of packaging materials and containers. Nanoparticles of pigments such as TiO_2 become transparent but retain their ultraviolet (UV) absorption characteristics.

This suggests applications in transparent wraps, films, or plastic containers where absorption of UV radiation needs to be avoided.

Foods

Worldwide commercial foods and food supplements containing added nano-particles are becoming available. A major growth area appears to be the development of 'nanoceuticals' and food supplements. The general approach is to develop nanosized carriers or nanosized materials, in order to improve the absorption and, hence, potentially the bioavailability of added materials such as vitamins, phytochemicals, nutrients, or minerals. The materials can be incorporated into solid foods, delivered as liquids in drinks, or even sprayed directly on to mucosal surfaces.

Food packaging

Novel food packaging technology is by far the most promising benefit of nanotechnology in the food industry in the near future. Companies are already producing packaging materials based on nanotechnology that are extending the life of food and drinks and improving food safety. While the nanofood industry struggles with public concerns over safety, the food packaging industry is moving full-speed ahead with nanotechnology products. Leading the way is active or 'smart' packaging that promises to improve food safety and quality and optimizes product shelf life. Numerous companies and universities are developing packaging that would be able to alert if the packaged food becomes contaminated, respond to a change in environmental conditions, and self-repair holes and tears. Some examples of the use of nanotechnology in food products are cooking oils that contain nutraceuticals within nanocapsules, nanoencapsulated flavour enhancers, and nanoparticles that have the ability to selectively bind and remove chemicals from food. The main reasons for the late incorporation of food into the nanotechnology sector are issues associated with the possible labelling of the food products and consumer-health aspects.

Smart packaging and active packaging: One of the most promising innovations in smart packaging is the use of nanotechnology to develop antimicrobial packaging. Packaging that incorporates nanomaterials can be 'smart,' which means that it can respond to environmental conditions or repair itself or alert a consumer to contamination and/or the presence of pathogens. These smart packages, will also detect public health pathogens such as *Salmonella* and *E. coli.*

Scientists in the Netherlands are taking smart packaging a step further with nanopackaging that will not only be able to sense when food is beginning to spoil, but will release a preservative to extend the life of that food. Because of their ability to improve safety and extend the life of food, these nanopackaging solutions are some of the most exciting innovations in the food industry today.

Clay nanoparticles to improve plastic packaging for food products: The nanoclay also makes the plastic lighter, stronger, and more heat-resistant. Clay nanocomposites are being used in plastic bottles to extend the shelf life of beer and make plastic bottles nearly shatter proof. Embedded nanocrystals in plastic create a molecular barrier that helps prevent the escape of oxygen. The technology currently keeps beer fresh for 6 months, but developers at several companies are already working on a bottle that will extend shelf life to 18 months. Several large beer makers, including South Korea's Hite Brewery and Miller Brewing Company, are already using the technology.

23.4.8   Nanoparticles for filtration

Nano-enabled water treatment techniques rely on membranes and filters made of carbon nanotubes, nanoporous ceramics, and magnetic nano-particles rather than the use of chemicals and ultraviolet light used in conventional water treatment. Carbon nanotube filters can be used to remove impurities from drinking water. A fused carbon nanotube mesh that can filter out water-borne pathogens, lead, uranium, and arsenic has been suggested as a useful nanotechnology. The application of the NanoCeram filter, which uses a positively charged filter to trap negatively charged bacteria and viruses. This filtering device removes endotoxins, DNA, viruses, and micron-sized particles. A simple handheld magnet can be used to remove the nanocrystals and the arsenic from water. Such a treatment could be used as a point-of-use water filtration process. However, the problem remains as how to dispose of the enriched product generated during the filtration process in an eco-friendly manner.

23.4.9   Wireless nanosensors for precision agriculture

Crop growth and field conditions like moisture level, soil fertility, temperature, crop nutrient status, insects, plant diseases, weeds, etc. can be monitored through advancement in nanotechnology. Such real-time monitoring is done by employing networks of wireless nano-sensors across the cultivated fields, providing essential data for agronomic processes like optimal time of planting and harvesting of the crops. It is also helpful for monitoring the time and amount of water application, fertilizers, pesticides, herbicides and other treatments. This has moved precision agriculture to a much higher level of control, for instance, in water usage, leading eventually to conservation of water. More precise water delivery systems are likely to be developed in the near future. The factors critical for such development include water storage, *in situ* water holding capacity, water distribution near roots, water absorption efficiency of plants, encapsulated water released on demand, and interaction with field intelligence through nano-sensor systems.

23.4.10   Future for nanofood

Nanotechnology can confer unique advantages on processed foods in many ways. Programmable foods, considered the ultimate dream of the consumer, will have designer food features built into it and a consumer can make a product of desired colour, flavour, and nutrition using specially programmed microwave ovens. The trick is to formulate the food at the manufacturer's end with millions of nanoparticles of different colours, flavours, and nutrients and under the program in the oven set by the consumer based on his preferences, only selective particles are activated while others stay inert, giving the desired product profile. Nano-based polymers with silica-based nanoparticles sandwiched can enhance the properties of pressure-sensitive adhesive labels and create biodegradable properties in them. Enhanced solubility, improved bioavailability, facilitating controlled release and protecting the stability of micronutrients in food products, are other virtues of nanotechnology.

The impact of nanotechnology is huge, ranging from basic food-to-food processing, from nutrition delivery to intelligent packaging.

Nanotechnology is becoming increasingly important for the food sector. Promising results and applications are already being developed in the areas of food packaging and food safety. Nanotechnology has begun to find potential applications in the area of functional food by engineering biological molecules toward functions very different from those they have in nature, opening up a whole new area of research and development. Of course, there seems to be no limit to what food technologists are prepared to do to our food and nanotechnology will give them a whole new set of tools to go to new extremes. Many have taken critical view of food nanotechnology in the past. Today, though, we look at the potentially beneficial effects nanotechnology enabled innovations could have on our foods and, subsequently, on our health.

Nanotechnology also has the potential to improve food processes that use enzymes to confer nutrition and health benefits. For example, enzymes are often added to food to hydrolyze antinutritive components and hence increase the bioavailability of essential nutrients such as minerals and vitamins. To make these enzymes highly active, long-lived, and cost-effective, nanomaterials can be used to provide superior enzyme-support systems due to their large surface-to-volume ratios compared to traditional macroscale support materials.

The incorporation of nanomaterials into food packaging is expected to improve the barrier properties of packaging materials and should thereby help to reduce the use of valuable raw materials and the generation of waste. Edible nanolaminates could find applications in fresh fruits and vegetables, bakery products, and confectionery, where they might protect the food from moisture, lipids, gases, off-flavours, and odours. Natural biopolymers of nanosize scale, such as polysaccharides, can be used for the encapsulation of vitamins, prebiotics, and probiotics and for delivery systems of drugs or nutraceuticals. In the food sector, one of the most important problems is the time-consuming and laborious process of food quality-control analysis. Innovative devices and techniques are being developed that can facilitate the preparation of food samples and their precise and inexpensive analysis. From this point of view, the development of nanosensors to detect microorganisms and contaminants is a particularly promising application of food nanotechnology. Table 23.1 summaries the application of nanotechnology in agriculture and food industry.

Table 23.1 Nanotechnology applications in agriculture and food.

Agriculture packaging	Food processing	Food	Supplements
Single molecule detection to determine enzyme substrate interactions	Nanocapsules to improve bioavailability of nutraceuticals in standard ingredients such as cooking oils	Antibodies attached to fluorescent nanoparticles to detect chemicals or foodborne pathogens	Nanosize powders to increase absorption of nutrients
Nanocapsules for delivery of pesticides, fertilizer and other agrichemicals more efficiently	Nanoencapsulated flavour enhancers	Biodegradable nanosensors for temperature, moisture and time monitoring	Cellulose nanocrystal composite as drug carrier
Delivery of growth hormones in a controlled fashion	Nanotubes and nanoparticles as gelation and viscosifying agents	Nanoclays and nanofilms as barrier materials to prevent spoilage and prevent oxygen absorption	Nanoencapsulation of nutraceuticals for better absorption, better stability of targeted delivery
Nanosensors for monitoring soil conditions and crop growth	Nanocapsule infusion of plant based steroids to replaces meat's cholesterol	Electrochemical nanosensors to detect ethylene	Nanocochleates (coiled nanoparticles) to deliver nutrients more efficiently to cells without affecting colour or taste of food
Nanochips for identity preservation and tracking	Nanoparticles to selectively bind and remove chemicals or pathogens from food	Antimicrobial and antifungal surface coatings with nanoparticles (silver, magnesium, zinc)	Vitamin sprays dispersing active molecules into nano-droplets for belter absorption
Nanosensors for detection of animal and plant pathogens	Nanoemulsions and particles for better availability and dispersion of nutrients	Lighter, stronger and more heat-resistant films with silicate nanoparticles	
Nanocapsules to deliver vaccines		Modified permeation behaviour of foils	
Nanoparticles to deliver DNA to plants (targeted genetic engineering)			

To sum up as developments in nanotechnology continue to emerge, its applicability to the food industry is sure to increase. Most aspects of incremental nanotechnology are likely to enhance product quality and choice and will be perceived as progressive changes in standard and accepted technology. There are a few issues, particularly regarding the accidental or deliberate use of nanoparticles in food, or food-contact materials, which may provoke consumer concern. It is particularly important to ensure that consumers are able to exercise choice in the use of the products of nanotechnology and that they have the information to assess the benefits and risks of such products. The success of these advancements will be dependent on consumer acceptance and the exploration of regulatory issues.

Food producers and manufacturers could make great strides in food safety by using nanotechnology, and consumers would reap benefits as well. Many companies are conducting research in nanotechnology and its application to food products and as more of its functionalities become evident, the level of interest is certain to increase. Nanotechnology has already made inroads into the food industry and it is claimed that more than 300 foods have already been developed with this technology. To maintain leadership in food and food-processing industry, one must work with nanotechnology and nanobio-info in the future. The future belongs to new products and new processes with the goal to customize and personalize the products. Improving the safety and quality of food will be the first step. Designing and producing food by shaping molecules and atoms is the future of the food industry worldwide.

23.5   Way forward: Policy options and actions

Nanotechnology has emerged as a cutting edge technology with profound realized and potential outcomes and impacts. It is a powerful tool for food and nutritional security, management of abiotic and biotic stresses, enhanced input use efficiency, elevated yield potential and superior quality traits. However, the process and products if not handled properly, may pose human health and environmental risks. Hence, strategic research and policy options and actions, as informed by rigorous science, are needed to judiciously develop this new emerging technology to be used in agriculture and food industry for congruent enhanced productivity, profitability, social equity, biosafety and environmental sustainability.

Bioethics and biotechnology

24.1   Introduction

Biotechnology is at the intersection of science and ethics. Technological developments are shaped by an ethical vision, which in turn is shaped by available technology. Much in biotechnology can be celebrated for how it benefits humanity. But technology can have a darker side. Biotechnology can produce unanticipated consequences that cause harm or dehumanize people. The ethical implications of proposed developments must be carefully examined. The ethical assessment of new technologies, including biotechnology, requires a different approach to ethics. Changes are necessary because new technology can have a more profound impact on the world; because of limitations with a rights-based approach to ethics; because of the importance and difficulty of predicting consequences; and because biotechnology now manipulates humans themselves.

The ethical questions raised by biotechnology are of a very different nature. Given the potential to profoundly change the future course of humanity, such questions require careful consideration. Rather than focusing on rights and freedoms, wisdom is needed to articulate our responsibilities towards nature and others, including future generations. The power and potential of biotechnology demands caution to ensure ethical progress.

Biotechnology, at its core, is about understanding life and using this knowledge to benefit people. Many see biotechnology as a significant force in improving the quality of people's lives in the 21st century. Obviously, biotechnology is intimately tied to science and scientific knowledge. At the very least, biotechnology promotes a certain vision of life, one in which some things are viewed as good and to be encouraged or pursued, and other things are bad and should be avoided or eliminated. That vision influences people's choices and what is viewed as ethically appropriate. A two-way flow exists in which ethics influences biotechnology even while the science impacts ethics. At times, the relationship between biotechnology and ethics is portrayed as one of conflict. Sometimes the impression is conveyed that ethics is needed only when someone wants to tell others that what they are doing is wrong. To a degree, this is understandable since controversy, debate and argument are usually integral to ethics discussions.

But ethics is just as important when there is consensus that a direction is good and right. The role of ethics is often invisible at this stage. There wasn't an ethical debate over whether to search for a cure for cancer. But the decision to pursue such research was motivated by a common vision that curing cancer was the ethical thing to do. Ethical examination of issues is important not only as a form of critique but also to identify and celebrate the right things people do.

The effort, resources and creativity focused on developing better treatments are ethically laudable. As such, there is much to celebrate about biotechnology. Society and individuals have benefited in many ways from technology. Many technological developments protect people from illnesses and natural disasters, giving some people 'liberation from the tyranny of nature'. In some parts of the world, people have higher living standards. Travel and communication have developed in unprecedented ways. Many of these changes can be welcomed as ethical developments.

Yet at the same time, other ethical considerations must be considered. At what price are some of these developments realized? Some developments seem motivated by a desire to find treatment at any price. Assisted human reproduction is a particularly controversial area where biotechnological treatment of infertility leads to many ethical dilemmas. Even with less controversial conditions like heart disease or cancer, developments have left people with high expectations that cures should exist. Some are concerned that technological developments lead to dehumanization or in healthcare lead to less emphasis on caring. Ethical concerns exist about justice, and how fairly these technological benefits are distributed—both within society and around the world. With all the options now available for some, concerns are raised about whether too much choice is bad for us.

Overall, though, technology has a strong ethical foundation. The appropriate response to misgivings and concerns is not to reject technology. 'By turning our backs on technological change, we would be expressing our satisfaction with current levels of hunger, disease, and privation.'

The benefits of technology, realized and potential, point to a technological mandate: biotechnology should strive to benefit people's lives. Many of the concerns about technology can be traced to the technological imperative the idea that something should be developed because we can, or we think we can.

24.2   Goals of biotechnology

Ethics includes assessment of the rights and wrongs of specific technologies and applications (like cloning or genetic diagnosis). Another important pursuit within ethics is examining the broader goals and aims of enterprises like biotechnology. The relief of sickness is one goal, but there are others that can be more ethically controversial.

Developing the necessary biotechnology for engineered negligible senescence assumes that indefinite life extension is good for humanity. Even if accepted as an ethical goal, it would be one goal among many.

Taking the time to reflect on these aspects of scientific developments can be difficult, especially with the pace and focus within biotechnology. The pressures of competing for funding, making breakthroughs, securing intellectual property, and obtaining market share all push against calls for caution or time-consuming reflection. Technological development can seem like a motorway, everyone on the fast track to success. Ethics, even when well intentioned, can seem like a diversion or a roadblock that prevents biotechnology reaching its destination, or delays it inexcusably.

However, there is a growing realization that ethics must be a part of the planning process within biotechnology. In many areas of research, ethics does impact the design of scientific experiments. Any research involving human or animal participants will be scrutinized by ethics committees. The methodology must conform to ethical codes and guidelines. An argument can be made that publicly funded research should be conducted in ways that conform to society's values. 'When the nation decides an activity is worth its public money, it declares that the activity is valued, desired, and favoured'. Therefore it is important to ensure that what is publicly funded is ethically acceptable in society. The goal of relieving suffering is widely accepted, yet it must be balanced against other societal goals. The ethics of proposed biotechnological developments must be scrutinized carefully.

24.2.1   The darker side

Even such a laudable goal as relieving human suffering cannot be taken as condoning any and all biotechnology. Humanity's creativeness and resource-fulness have long been recognized and praised. But human activity can have a darker side. The ancient Greek philosopher Sophocles reflected on these two sides of technological development. On the one hand he noted many human accomplishments in transport, agriculture and medicine. But he also pointed to problems with this same inventiveness.

Biotechnology is a particularly fitting example of technology with such fundamentally different characteristics that it requires a careful re-examination of how its ethical dimensions are evaluated. Biotechnology 'raises moral questions that are not simply difficult in the familiar sense but are of an altogether different kind'.

Challenging characteristics of biotechnology

The capacity for new technology to have global impact shows that ethics needs to broaden its focus. Environmental problems and the existence of nuclear technology demonstrate the importance of ethical examination of more than just human–human interactions.

New technology also highlights the vulnerability of nature. Previous technological developments appeared to assume that natural resources were in endless supply and that nature could rebound from any human impact. Environmental changes show these assumptions were problematic. Ethical evaluations of biotechnology need to take the vulnerability of nature into account. These issues also point to limitations in previous ethical approaches that focused only on humans. At the same time, a concern for these broader issues can lead to new technological challenges and exciting research opportunities, such as has occurred with research into renewal energy sources stemming from ethical concern for the environment.

Limitations with rights

Rights-based approaches to ethics have made important contributions to human welfare. They provide a means by which vulnerable humans can argue for more ethical treatment. However, such approaches have their limitations.

A rights-based approach to ethics must include some method of identifying those who bear rights. Those who have rights place duties on others to uphold those rights. It has proved very difficult to find consensus on how rights are to be ascribed. One approach is that all humans are inherently entitled to all human rights. This raises questions about when a human is given these rights (at fertilization or birth or some other point). It also leaves no guidance on how to treat the non-human world. Biotechnology requires answers to these questions to address ethical concerns about non-human species and nature as a whole. This has led to an approach where rights are granted based on particular abilities and attributes. There is little consensus over what abilities entitle an organism to rights. Philosophically, it is also difficult to justify why any particular attribute should lead to the granting of rights. The whole approach is criticized as being motivated by a desire to treat unethically those not given rights. This is particularly relevant to research on human embryos, especially embryonic stem cell research.

Developments in biotechnology point to serious limitations with a rights-based approach to ethics. Rather than providing insurmountable problems for ethics, these point to the need for a different approach to ethics. Jonas and others point out that rather than focusing exclusively on human rights and entitlements, the new technological era requires a greater focus on human responsibility.

Future consequences: Earlier technology impacted humans and their lives, but did not have the potential to change human nature but biotechnology does. With that comes the potential for broader and long-range consequences. Predictions about these consequences can be difficult and unreliable. This is particularly cogent with genetic technology. The consequences of our ability to manipulate the human genome could impact many, if not all, future generations. The way genes interact with one another means that manipulating one gene could have unintended effects on other genes or their expressed proteins. This is especially important given the recent realization that the human genome contains fewer genes than originally presumed.

The medicalization of patients and the instrumentalization of people are consequences of technology's successes. This can have a dehumanizing effect on human life, which makes it easier to treat some humans as less than fully human. This is a way in which technology can take on a life of its own and have much more profound ethical consequences.

Biotechnology has the added capacity to produce products that literally do take on life. The technology humans developed in the past was inanimate and could be left unused if found to be ethically problematic—as difficult as that might have been.

Impact on human nature and personhood: No area of biotechnology more clearly brings to focus the need for careful ethical reflection than its potential to impact human nature. Previous technology has provided new tools that impacted human activities and society. Humans were the makers of technology. Some aspects of biotechnology now make humans the objects of technology. Humans have turned upon themselves and are ready 'to make over the maker of all the rest'.

Recent developments with stem cell research and cloning have been the lightning rod for debate over human personhood. These discussions point to the gulf between proponents on the different sides. Some have viewed embryos as 'featureless bundles of cells'. From this perspective the human embryo is a human non-person that can be used and destroyed in research. Others disagree and maintain that the human embryo should be treated as a person, making it unethical to treat it merely as a means to others' ends.

Personhood can be viewed as an inherent attribute of all humans. This confers all humans with certain rights and determines how persons should be treated ethically. This approach protects humans, especially the vulnerable, from unethical treatment. The other approach makes personhood conditional on reaching some stage of development or possessing certain abilities. Only humans with those capacities are then entitled to protection. A fundamental problem with this approach is that it always arises to justify killing those declared to be human non-persons. How will it affect us to treat human lives as commodities to be manipulated and destroyed at will? When we justify doing so with embryos, will it become easier to do so at later stages of development?

This points to the difficulty of determining public policy when sections of society have irreconcilable positions on matters of fundamental importance. We must also examine how biotechnology itself impacts our view of human nature. Leon Kass asks how will it affect us 'to look upon nascent human life as a natural resource to be mined, exploited, commodified. The little embryos are merely destroyed, but we—their users—are at risk of corruption'. This is much more than a debate over rights. This is about human dignity, including what it means for humans to act with dignity. This changes the focus from ascribing rights to determining responsibilities.

Central place of responsibility: The enormity of the potential impact of biotechnology on human nature should cause us to proceed cautiously. Biotechnology has the potential to do great good. But it also has the potential to cause much harm. This could arise in the physical realm through unexpected consequences of the technology itself. But other harms could arise through the non-physical impacts of biotechnology. Cars and computers have affected many aspects of human life and society. Biotechnology could change what it means to be human.

A rights approach to ethics makes clear where people have rights. Each right carries a corollary duty or responsibility. If people have a right to healthcare, someone has the responsibility to provide healthcare resources. Much energy has been expended identifying and defending human rights. We now need a similar emphasis on human responsibilities.

Responsibility is also a corollary of power. Biotechnology brings new powers to humanity. These powers should remind us of our responsibility to nature and the environment, to all of life, to the future, and to human nature and personhood. To understand these responsibilities entails the development of wisdom. That wisdom requires ethical reflection before developing specific forms of biotechnology. Taking the time for that reflection can go against the pace of biotechnological developments and hubris over human wisdom.

Researchers warned that new technology was propelling us towards a utopian future. These developments have the potential for much good, but also risk changing, harming or even destroying some species, including ourselves. To make the right ethical decisions 'requires supreme wisdom— an impossible situation for man in general, because he does not possess that wisdom, and in particular for contemporary man, because he denies the very existence of its object, objective value and truth. We need wisdom most when we believe in it least'.

In contrast, ethics searches for better answers to ethical questions. It acknowledges the limitations in current wisdom, and strives to improve our understanding. The way forward is muddied by our inability to accurately predict the consequences of proposed biotechnological developments. Some argue that we should push ahead and deal with problems as they arise. But given the scale of disaster that biotechnological mistakes could trigger, Jonas' guiding principle contains much wisdom. He argued that 'ignorance of the ultimate implications becomes itself a reason for responsible restraint—as the second best to the possession of wisdom itself'.

Time and resources must be committed to examining the ethical implications of proposed biotechnological developments. The potential impact on all aspects of nature must be considered.

The social, emotional and spiritual implications of developments in biotechnology must also be examined. When humans themselves are the objects of biotechnology, great caution is necessary least we promote a view of ourselves and our neighbours as nothing more than living bits of technology.

24.3   Ethical aspects of food and agricultural biotechnology

Many of the expressed concerns about food and agricultural biotechnology are described as ethical. Decision leaders should interpret the expression of ethical concerns as a demand for competing visions of nature and the public good to be expressed in public dialogue about food and agricultural biotechnology, for those who feel that their values have been neglected to have an adequate opportunity to express their concerns in their own words, and for their voices to be heard.

Those who call for attention to ethical issues appeal to many diverse values. Their concerns can be classified into two broad categories. On the one hand, some see the very act of using genetic technology to raise ethical issues that would not apply to other applications of food and agricultural technology. On the other hand, some believe that specific applications of biotechnology raise ethical issues that are not being adequately addressed, even if these issues may be raised in connection to other, more conventional types of agricultural technology, as well.

24.3.1   Special arguments pertaining to the use of rDNA technology

There are several types of concern noted by those who question whether the use of biotechnology may be intrinsically questionable.

- *Genes and essences:* Longstanding religious and cultural traditions associate the idea of a particular essence with different species of living organisms, and specify an obligation for human beings to respect these essences. Some may associate the modern notion of genes with this traditional notion of essence.

- *Species boundaries and natural kinds:* The idea that there is a specified order of nature may involve the belief that the species of plants and animals we find around us represent natural kinds. Some may fear that biotechnology disturbs this order and thereby violates absolute limits on what human beings are ethically permitted to do.

- *Religious arguments:* Many religious traditions prohibit acts that involve transpecies reproduction, or ban the consumption of some species groups for food, and the mixing of foods from different groups. Biotechnology may be interpreted as contrary to some of these religious traditions.

- *Emotional repugnance:* Cultural traditions dictate that some potentially consumable substances (e.g. species such as cat and dog, or particular parts of plants and animals) are not suitable for use as food. Western food systems currently respect the repugnance that people feel toward these substances as a sufficient ground for policies that help people avoid consuming them. Some individuals may feel a similar repugnance toward bioengineered foods.

General technological ethics

There are a number of ethical questions that can be raised with respect to virtually any new food or agricultural technology. As they are raised in connection with biotechnology, these questions suggest the following types of ethical concern:

- *Environmental ethics:* Technology raises environmental issues when there are environmental exposures that pose risk to humans, wildlife or to ecosystem integrity. It has been alleged that agricultural biotechnology may pose risks to wildlife in or near farm fields. There are also issues associated with the question of whether agricultural ecosystems can themselves to exhibit features of ecological integrity.

- *Food safety:* Many of the issues associated with the safety of eating bioengineered foods are technical, but the question of whether regulators should make this decision based on an assessment of the risks, or whether individual consumers should be placed in a position to make the choice themselves is an ethical one.

- *Moral status of animals:* If genetic engineering of livestock would compromise animal welfare, there are ethical questions that can be raised. There are also ethical questions about whether it would be ethical to use biotechnology to make animals more tolerant of production settings that are currently regarded as inimical to animal welfare.

- *Impact on farming communities:* Some critics of agricultural biotechnology have alleged that it will contribute to farm bankruptcies and the depletion of farming population in rural communities. There has been a longstanding ethical debate as to whether technology or policy that has these effects on farming communities can be ethically justified in virtue of offsetting benefits in the form of efficient production and lower food prices. The concern is particularly relevant to the impact of biotechnology in developing regions where many farm at the subsistence level.

- *Shifting power relations:* Related to the concern on farming communities, some have argued that biotechnology will help a few well-capitalized firms control decision making in agriculture (including future research), and limit farmers ability to choose from an array of production possibilities. This concern is related to a general ethical concern with the distribution of economic power and wealth in democratic societies.

Responses to these issues

This section discusses several approaches that have been discussed as a possible response to these various ethical issues:

- *Uncertainty and the precautionary principle:* Many of these ethical issues involve uncertainty about the risks or outcomes associated with biotechnology. The precautionary principle has been suggested as the appropriate decision rule to utilize in response to such situations. It suggests that decision makers should not permit technological innovations to go forward simply because alleged harms have not been proven to exist. However, it is not clear how the precautionary principle should be applied in the case of food and agricultural biotechnology.

- *Consent, labels and consumer choice:* Various proposals for labelling products of biotechnology have been discussed. On the one hand, these proposals are supported by an informed-consent approach to issues in food safety, and may be the most satisfactory response to concerns based on religious values, emotional repugnance and other intrinsic objections to biotechnology. Labels might give individuals who have these concerns an opportunity of exit, to opt out of a food system that causes them anxiety or concern. On the other hand, labels may stigmatize bioengineered foods, and may not provide information that would be useful for consumers trying to make choices on the basis of nutrition and food safety.

- *Methods in applied ethics:* How do methods in ethics suggest a response to these concerns. One approach suggests that common ethical principles can be applied to provide definitive answers to the questions raised above. A more promising approach suggests that only open public discussion of these issues can produce an adequate basis for responding to the questions that critics of biotechnology raise.

Trust and public confidence

As debates over food and agricultural biotechnology become politicized, with activist organizations opposing both industry and governmental spokespersons, there is a growing tendency for public discourse on biotechnology to reflect the strategic interests of industry and activists. There is a grave risk that as science becomes deployed in these debates, scientists themselves will become so tainted by the strategic character of debate that the public will begin to lose confidence in the objectivity and judgment of scientists. Scientific spokes persons thus have an ethical responsibility to develop a capacity to participate in ethically charged public discussions of biotechnology without either denigrating the values of others by characterizing them as irrational or presuming uncritically that their science-based perspectives are the ethically proper approach to take.

In the light of above mentioned facts, the ethical issues associated with food and agricultural biotechnology must be regarded as open-ended and in great need of more structured and serious dialogue. Both specialists and members of the public should be encouraged to articulate their concerns, and to respond to the views of others in a considered and respectful manner.

24.4   Variety of changes shaping bioethics today

The following development reflects a variety of changes that are shaping bioethics today:

- The rapid pace of development in science and technology: Ethical guidelines cannot be developed quickly enough to keep up.
- The lack of public trust in scientists, corporations and regulators, partly due to the failure to safeguard the public against infected blood and, more recently, against 'mad cow' disease in Europe.
- The poor ability of institutions to educate the public on complex scientific issues without obfuscating or patronizing the public.
- Increasing democratization and demands for accountability and transparency.

- Weakening of political control.
- Rapid increase in availability of information, especially through the Internet.
- Development of technology for which there is little precedent in its own terms and in terms of its ethical use.
- The lack of coherence within the bioethical establishment itself in terms of having strong foundational values accepted by all.
- The technical complexity of the issues.
- Globalization of economies, in which policy decisions in one country have implications for many other countries.
- Vocal, well-organized, protest groups.
- An increasing public recognition of the impact of our actions on the environment.

These factors have resulted in two primary outcomes a public that has become empowered to demand consultation and policymakers who are reluctant to make risky decisions—including those that may have been acceptable in the past—without public consultation.

For bureaucrats, seeking to involve the public has been a way of coping with their diminished credibility.

In addition to public consultation, there is a need for dialogue among different constituencies. Academics, industries, the public, the media and applied ethicists must collectively identify issues, plan an agenda for vigorous in-depth studies of those issues, engage in transparent discussion of the results of the research in a truth-seeking, non-confrontational way, and reach consensus in the development of guidelines.

If this is done well, the guidelines will be clear, understood by all and, more importantly, supported by all stakeholders. The current confrontational methods have not served the public well. Attempts by industry public relations officers to provide more and more information to a distrusting public have also been ineffective. In this context, bioethicists can have an important function: to bring their specialized knowledge and analytical skills to clarify and facilitate, rather than simplistically to preach or propound on what is right or wrong in their opinion.

An interesting recent phenomenon, likely to become more common as biotechnology issues become more complex, is that bioethicists are being held legally responsible for their advice. The parents of a teenager who died as a result of a gene therapy experiment have recently sued a prominent bioethicist from a well-known US university. The bioethicist had advised the scientist-clinicians who performed the procedure.

Greenhouse gases released by human activities (including feeding 6 billion people) are building up in the atmosphere, trapping heat that would otherwise escape into space and altering the fundamental climate processes that drive global weather patterns. Runaway climate change poses serious threats to humanity and the natural world, but with deliberate and concerted efforts to reduce our emissions, these consequences can be largely avoided.

Whether we live to eat or eat to live, one thing is sure: staying fed represents a substantial portion of our total impact on the climate. The greenhouse gas impacts of food are complex and far-reaching, as every bite of food we eat takes energy to grow, process, store, transport, sell, cook, and discard. But by understanding how our eating habits affect global warming, we gain the power to reduce those impacts through conscious daily living. (The social and environmental impacts of food extend far beyond climate change; our food system also affects biodiversity, water quality, ecosystem functions, human health, and human rights, to name a few).

Our 'carbon foodprint' is the sum of all the greenhouse gases our meals produce as they wind their way through the food system. Three main gases comprise the vast majority of food-related emissions: carbon dioxide, methane, and nitrous oxide.

- Carbon dioxide (CO_2) is released whenever fossil fuels like coal, gasoline, or natural gas are burned to generate energy. CO_2 accounts for about 71% of our total food impact.
- Methane (CH_4) is released when food scraps and packaging decompose in landfills, and during livestock digestion and manure treatment. While methane is released in relatively low volumes, it is 25 times as potent as carbon dioxide. In all, it accounts for about 13% of our total food emissions.
- Nitrous oxide (N_2O) comes predominantly from chemical fertilizers used on crops. Although little nitrous oxide is released, each pound has a global warming impact equivalent to 300 pounds of carbon dioxide. N_2O makes up about 15% of the average American's foodprint.
- The remaining 1% of food's global warming impact comes from a number of gases that are released in very small quantities, primarily SF_6 from electricity production and HFCs from refrigeration systems.

- To simplify things, we combine all of these gases and their relative potencies into a single comprehensive measure of the climate impact of a given activity, called carbon dioxide equivalent (CO_2e). When we talk about 'CO_2e,' 'carbon emissions,' 'climate impact,' or 'foodprints,' we're referring to the combined impact of the various greenhouse gases.

Estimating the climate impact of food is a tricky process, and estimates of that total number vary.

The average American has a carbon foodprint of over 12,000 pounds CO_2e each year. That includes emissions from growing, processing, distributing, and selling food, emissions from getting it home and cooking it, and emissions from discarding or recycling the leftover waste products. In all, food represents 21% of the typical American's total annual carbon footprint of 28.6 tons CO_2e. Of course, that's just the average—our personal foodprint depends on how much and what kinds of food we eat, where and how that food is produced, how it's prepared, and what we do with the leftovers.

The good news is there's lot we can do to reduce our carbon foodprint, and many of the steps we take will have substantial non-climate benefits as well. Shrinking our foodprint will help improve environmental and social conditions in near and distant places touched by our food system. It can also increase our quality of life by saving our time and money, improving the healthiness and tastiness of our diet, and building community connections.

A1.   American carbon foodprint at a glance

A1.1   Foodprint as fraction of total footprint

Food-related emissions comprise 21% of total emissions, or 6.1 tons out of 28.6 tons per person per year. 15% of personal transportation relates to food, as does 20% of housing energy use, while 23% of the emissions from all other activity are food-related.

Breakdown by life cycle phase

54% of food-related emissions are released in the supply chain upstream of the consumption point. A further 14% come from personal transport to grocery stores and restaurants, while 29% are released during the cooking and serving process in restaurants and home kitchens. The final 3% released downstream via decomposition in the landfill.

Breakdown by food group

A quarter of the average American's foodprint derives from red meat, with another quarter coming from other animal products. Plant foods make up the remaining half, of which the majority result from grains, fats, and sugars. This emissions breakdown differs markedly from caloric breakdown, however—for example, red meat makes up only 11% of calories despite causing 25% of emissions.

Breakdown by greenhouse gas

The majority of food's global warming impact comes from carbon dioxide, released during fossil fuel combustion. The remainder comes mostly in the form of methane and nitrous oxide, coming largely from livestock and chemical fertilizers, respectively. Other gases, including SF_6 from electricity production and HFCs from refrigeration systems, comprise the remaining 1%. All values are in CO_2e.

A1.2   Ins and outs of food emissions

Not all foods are created equal

'Eat less meat' is one of the most common suggestions for reducing our diet's climate impact—and it's a sage advice. Food groups vary widely in greenhouse gas emissions per calorie or serving, and red meat tops the chart. By understanding the relative carbon impacts of the various food groups, we can make informed decisions about which foods we choose to purchase.

More than a third of the greenhouse gas emissions generated by the US agricultural system come from the livestock industry. And while red meat comprises just 11% of the typical American's caloric intake, it accounts for a full 25% of their carbon foodprint. There are two main reasons that red meat is so much more carbon-intensive than other foods.

- Eating meat is less efficient than eating vegetables, because animals only convert a small portion of the energy in plants into meat. When we eat meat, we again only capture a small portion of the energy in the animal. And some animals (e.g. cows) are less efficient at converting plant energy into meat than others (e.g. chickens).
- Raising livestock creates emissions from 'enteric fermentation'—essentially, burping and farting. Ruminants like cattle and sheep possess digestive systems that release large quantities of methane (a phenomenon that is exacerbated by feeding them corn, which is not their natural diet). Substantial quantities of methane are also released from the decomposition of livestock manure.

By comparison, every other food group averages substantially fewer greenhouse gases per calorie than red meat. Fish and dairy are the second most carbon-intensive food groups, producing 68% and 58%, respectively, the emissions per calorie of red meat. Poultry, eggs, and vegetables emit approximately half as much carbon per calorie as red meat, while cereals and grains, oils and sugars, and nuts produce roughly a quarter as much.

To illustrate the carbon effects of making choices among the various food categories, it's helpful to consider the emissions profiles of three common diet types, each with the same number of total calories: a standard omnivorous diet representing the average American; a vegetarian diet that excludes red meat, poultry, and fish; and a vegan diet that excludes all animal products including eggs and dairy. While the average diet results in 4.3 grams of embodied CO_2e emissions per calorie consumed, a vegetarian diet emits only 79% of that average, and a vegan diet releases just 68%. Effectively, the typical vegetarian would thus emit about 1,300 pounds less CO_2e per year than the average American–and that's assuming they eat the same number of calories, when many vegetarians in fact eat fewer.

The simplest thing we can do to reduce our foodprint is to eat types of food that are less emissions intensive.

A huge portion of food's emissions come from the industrial supply chain –indeed, 54% of the average foodprint is comprised of 'embodied' emissions from producing, distributing, and selling food, with the other 46% coming from cooking, household transport, and waste. Just making the packaging that encases so many food accounts for 10% of all food production emissions, or about 655 pounds of CO_2e annually per American. Cartons and aluminium packaging are the biggest contributors, primarily because they emit more per pound of packaging than other materials. While it's difficult to avoid packaging and processing entirely, we can decrease our carbon foodprint by shopping for fresh, whole, bulk foods with little or no packaging.

Taking it a step further and harvesting our own food directly from the land can completely eliminate the energy used to transport and sell food. Done properly, gardening, hunting, and gathering are as close as it comes to carbon neutral dining. Of course, these activities all take equipment, supplies, and some personal transportation, so the carbon benefits of gardening or hunting have to be weighed against the energy that goes into making them possible. But if we minimize new equipment, motorized travel associated with these activities, and the energy and supplies that go into preparing and storing these foods, living off the land is a major boon for the climate.

A2.   Organic agriculture

Generally speaking, organic agriculture is significantly less carbon-intensive per acre, but conventional agriculture can be significantly more productive per acre. From a global warming standpoint, the starkest differences between conventional and organic agriculture relate to fertilizers. Conventional farmers typically use substantial quantities of nitrogen-based chemical fertilizers, where as organic farmers instead rely on manure and other natural supplements to increase soil fertility. Chemical fertilizer is produced in an energy-intensive process that converts inert nitrogen (N_2) from the air into ammonia (NH_3) by combining it with hydrogen extracted from water and methane. When this nitrogen fertilizer is applied to fields, much of the ammonia eventually degrades and is converted into nitrous oxide (N_2O), a greenhouse gas 300 times as potent as CO_2 that escapes into the atmosphere.

Globally, N_2O from agriculture is responsible for approximately 6.3% of anthropogenic climate change. There's also evidence that chemical fertilizers and other conventional farming practices can cause the release of soil carbon that has accumulated through centuries of plant growth. Still, from a greenhouse gas standpoint organic agriculture may or may not be preferable. Depending on the situation, nitrogen fertilization, chemical pesticides, and other methods employed in conventional farming can permit substantial increases in productivity, with conventionally managed cropland sometimes yielding up to twice the harvest per acre of organically managed land.

This means that producing a given quantity of food organically would require more acreage than producing it conventionally. But studies also indicate that organic farming uses less than half the energy per acre of conventional farming, and that proper organic management can substantially increase the rate at which atmospheric carbon is removed from the atmosphere into the soil.

No overarching conclusion can yet be drawn as to which farming school is preferable from a carbon emissions standpoint, since their relative advantages depend on many variables including the crop, the local climate and soil, and the specific farming techniques employed. Still, while the climate benefits of organic agriculture may be open to debate, it bears remembering that carbon emissions are just one issue in the larger picture of organics, which also relates to human and environmental health beyond the issue of climate change.

A2.1   Food miles in perspective

The average meal travels 4,200 miles en route to the store, and this journey is powered almost entirely by fossil fuels. But of those food miles, only a quarter are 'final delivery' from farm or factory to retailer, while three quarters are delivery of inputs like fertilizer or raw ingredients. Food groups vary in their average travel distances from a low of fruit (at 1,265 miles) to a high of red meat (at 13,273 miles).

Working to reduce the distance our food travels before we purchase it can certainly have a positive climate impact. But we shouldn't assume that sourcing food closer to home necessarily emits less carbon. That's because of what economists call 'comparative advantage,' the idea that goods can be more efficient to produce in one location than another. Buying local foods that could have been more efficiently produced elsewhere can be counter-productive if producing them locally ends up emitting more carbon than transporting them from afar. Buying a local Vermont tomato might make sense in August, but in February that local tomato was either grown in a heated greenhouse under artificial light or stored for months in a refrigerator, making it more carbon-intensive than a sun-ripened tomato shipped from southern California. In short, in-season local specialties are perfect to source locally, and in general should be prioritized in our diet over imported foods and inefficiently produced local novelties. Despite the impressive distance that most food travels, transportation accounts for a modest 11% of the 'embodied' emissions of the typical cartload of groceries (that's all the emissions that went into producing, transporting, and retailing the food), or 700 pounds CO_2e per person annually.

The portion of a meal's footprint that comes from transportation also varies depending on the kind of food. For example, while transportation accounts for only 6% of the embodied emissions of red meat, it accounts for three times that amount in vegetables, or a full 18%. From a carbon standpoint, food groups in which a greater fraction of emissions come from transport should be a higher priority for local sourcing. But remember that final delivery only accounts for a fraction of the total transportation. Even for vegetables, the food group with the highest portion of miles from final delivery, two-third of the total miles come from delivery of inputs. In other words, our steak's carbon emissions may depend less on how locally you source your beef than on how locally our rancher sources his cattle feed. But a singular focus on supply chain food miles belies the reality that all the miles a meal travels en route to your local vendor represent only 30% of total food transport emissions—the remaining 70% of that meal's transportation footprint comes from our own personal food-related transportation (travel to grocery stores and restaurants). While food travels much farther from farm to store than from store to pantry, travelling in a densely packed container ship or tractor trailer is orders of magnitude more efficient than travelling in the back seat of a station wagon. Personal food-related errands comprise 14% of our total carbon foodprint, while supply chain transport accounts for just 6%. Indeed, going grocery shopping and going out to eat account for 15% of the typical American's total personal transportation, emitting 1,750 pounds of CO_2e every year.

The carbon benefits of eating local food are nuanced, and the emissions we reduce this way represent a minority of the total transportation emissions associated with our diet. Driving to grocery stores and restaurants produces a much larger quantity of greenhouse gas emissions, and the benefits of avoiding these errands are clear-cut. From a global warming perspective, we should focus at least as much on minimizing our own food-related travel as on pursuing a locavore diet.

Dining in, eating out

Producing and transporting ingredients accounts for the majority of food-related emissions, but the energy to transform those ingredients into a steaming meal is an important contributor as well. Taking into account storage, cooking, and cleanup, kitchen energy use amounts to 15% of the average American's total food-related emissions. Home kitchens account for 1,850 pounds of CO_2e per person annually, or 15% of the average food footprint, while food service (including restaurants and cafeterias) accounts for another 14%, or 1,740 pounds. In all, the energy to cook and serve a meal accounts for nearly a third of its entire life cycle emissions.

The average American eats out at a fast food or full service restaurant about 4 times a week, or roughly one in five meals. And yet food service operations consume roughly the same amount of energy as home kitchens—and that's before we consider the carbon impact of travelling to the restaurant. A simple way to reduce our carbon foodprint is to cook more and eat out less. On the home front, kitchens consume 21% of total household energy while occupying just 13% of household area, making them one of the most energy-intensive rooms in the house. Of this kitchen energy, a third comes from heating, cooling, and lighting the kitchen itself, with the remainder more centrally related to the cooking process. Refrigeration is the biggest energy hog in the kitchen, consuming 30% of all kitchen energy, and emitting 1,440 pounds of CO_2e annually per household while adding $350 a year to the utility bill. One should get rid of old refrigerator which probably uses at least twice as much energy as the one in your kitchen and buy energy efficient fridge. The actual act of cooking comes in second to refrigeration in terms of appliance energy use, with stoves, ovens, and microwaves consuming a combined 14% of kitchen energy, or 685 pounds of carbon emissions annually per household. But there are important differences between them. Microwaves are the most energy-efficient at cooking food, as they generate very little waste heat. Stoves come in second place, while ovens (which transfer a relatively small portion of the energy they consume into the food) are the least efficient cooking method.

The aftermath of the meal is also an important part of our foodprint, with hot water and dishwasher energy use consuming a combined 14% of kitchen energy. Of the 9% consumed by dishwashers, one quarter is hot water, meaning that a total of 7% of kitchen energy goes toward powering the dishwasher, while another 7% goes toward heating water for the sink and dishwasher. Whether it's more energy efficient to wash dishes in the sink or the dishwasher depends on your specific model of dishwasher and your hand washing techniques, but either method presents opportunities for energy savings. Unless we are a super-efficient handwasher, using a new, efficient dishwasher and only running it when fully loaded is the best bet for minimizing dish-washing energy.

A2.2   Garbage

Even once a meal has been grown, transported, cooked, and eaten, all is not quite said and done. Americans discard over 620 billion pounds of garbage (not including recycling and compost) every year, 29% of which is food-related. This waste doesn't just disappear—it's trucked to the landfill, where it decomposes underground, producing methane that seeps into the atmosphere. This final pulse of greenhouse gas from the ground accounts for the last 3% of the average meal's total carbon impact, or about 315 pounds CO_2e per person per year. Food-related waste is responsible for 28% of all US landfill gas emissions.

Food scraps themselves constitute 66% of food-related garbage by weight, while glass, metal, plastic, and paper packaging from food and drink products make up the rest. But food scraps contribute a much larger share of landfill gas emissions (93%), because inorganic food packaging doesn't decompose to release methane. Composting food scraps rather than land filling them can drastically reduce their climate impact. That's because in the oxygenated environment of a compost pile decomposition occurs aerobically and produces carbon dioxide, whereas in the oxygen-poor depths of a landfill food decomposes anaerobically and generates methane, which is 25 times as potent in its global warming potential. Unfortunately, just 3% of food scraps are composted. Better still than composting, get creative with leftovers before they head south, it will reduce both waste in the landfill and the amount of food we buy in the first place.

Recycling also plays an important role in reducing the carbon emissions from the food industry. About 35% of food-related waste that can be recycled actually is, while the rest is bound for the landfill. Recycling reduces the energy needed to mine and refine virgin materials to produce new food and drink packaging. Indeed, the average American saves 655 pounds of carbon emissions every year by recycling food and drink packages—although this number could be tripled if recycling occurred with higher frequency.

To sum up for an average American, the greenhouse gas impact of his diet is greater than the combined impact of his driving and flying. This may come as a surprise because public discussions of carbon emissions focus heavily on transportation, while discussions about the impacts of food are typically centered around non-climate issues. But what it means is that individually and collectively, we have a huge opportunity to reduce our impact on the climate by changing how we feed ourselves.

Eating to fight climate change is within reach for all of us, but it requires a carefully revised approach, even for those already accustomed to thinking about the social and environmental impacts of their diets. Local and organic foods may or may not be good indicators of low climate impact, although they do support the health of our communities, our planet, and our bodies. When it comes to climate change, a stronger focus on our meal's life cycle energy use is the key to a smaller footprint.

Whatever our current carbon foodprint, following these seven basic rules will get us on the fast track to a climate-friendly diet.

- Eat fewer animals and more plants
- Buy unprocessed foods with less packaging
- Grow and harvest your own food
- Minimize car trips to restaurants and stores
- Cook at home more and eat out less
- Cook with efficient appliances and techniques
- Compost, recycle, and relish leftovers

References

Aiba, S., Humphrey, A.E., and Millis, N.F. (1973). Biochemical Engineering, Academic, New York, pp. 182–183.

Batt, C.A. and Sinskey, A.J. (1984). Use of Biotechnology in the production of single–cell protein. Food Technology, 38:108–111.

Belitz, H.D. and Grosch, W. (1987). Food Chemistry, Springer Verlag.

Bender, A.E. (1978). Food Processing and Nutrition, Academic, London.

Beuchat, L.R. (ed.) (1987). Food and Beverage Mycology. 2nd Ed. New York: Van Nostrand Reinhold.

Bjurstrom, E. (1985). Biotechnology, Fermentation and Downstream Processing, Chem. Eng. 92:126–158.

Brennan, J.G., Butters, J.R., Cowell, N.D. and Lilly, A.E.V. (1979). Food Engineering Operations, 2nd ed., Applied Science Pub. Ltd., London.

Brown, M.H., and Emberger, O. (1980). Oxidation–reduction potential. In Microbial Ecology of Foods, 112–115. New York: Academic Press.

Considine, D.M. (ed.) (1982). Foods and Food Production Encyclopedia, Van Nostrand Reinhold, New York, pp. 27–32.

Food Technology (1984). Overview: Genetic Engineering in Food Production, Food Technology, 38:69–127.

Glazer, A.N. and Nikaidi, H. (1995). Microbial Biotechnology: Fundamentals of Applied Microbiology, W.H. Freeman, New York.

Hobbs, B.C. (1968). Food Poisoning and Food Hygiene. 2nd Ed. London: Edward Arnold.

Hudson, B.J.F. (ed.) (1982). Development in Food Proteins, Vol. I, Applied Science Pub., London.

Jarvis, B. and Holmes, A.W. (1982). Biotechnology in relation to the food industry. J. Chem. Tech. Biotechnol. 32:224–232.

Jay, J.M. (1987). (Fungi in) Meats, poultry and seafoods. In Food and Beverage Mycology, 2nd Ed., Ed. L.R. Beuchat, 155–173. New York: Van Nostrand Reinhold.

Johnson, J.C. (1977). Industrial Enzymes. Recent Advances, Noyes Data Corp., Park Ridge, N.J.

Khachatourians, G.G. and Haffie, T.L. (1985). Biotechnology: Applications to genetics. In Biotechnology Handbook (P. Cheremisinoff and R.P. Ouellette, Eds.), Technomic, Lancaster, Mich., pp. 242–249.

Kirsop, B.H. (1981). Biotechnology in the food processing industry. Chem. Ind. 7:218–222.

Knorr, D. and Sinskey, A.J. (1985). Biotechnology in food production and processing. Science 229:1224–1229.

Kon, S.K. (1959). Milk and Milk Products in Human Nutrition, FAO Nutritional Series #17, Food Agric, Org. United Nations Rome.

Kunkee, R.E. (1975). A second enzymatic activity for decomposition of malic acid by malo-lactic bacteria in Lactic Acid Bacteria in Beverages and Food. Ed. J.G. Carr, C.V. Cutting and G.C. Whitting, 29–42. New York: Academic Press.

Lampert, L.M. (1970). Modern Dairy Products, Chemical Publishing Co. New York.

Pederson, C.S. (1971). Microbiology of Food Fermentations, Westport, CT: AVI.

Primrose, S.B. (1991). Molecular Biotechnology, 2nd Ed. Blackwell Scientific Publishers, Oxford.

Robinson, D.S. (1987). Food—Biochemistry and Nutritional Value, Longman Scientific and Technical, London.

Spencer, J.F.T. and Spencer, D.M. (1983). Genetic Improvement of Industrial Yeasts. Annu. Rev. Microbial. 37:121–142.

Steinkraus, K.H. (1983). Handbook of Indigenous Fermented Foods. Marcel Dekker, New York.

Tanner, F.W. (1994). The Microbiology of Foods, 2nd Ed. Champaign, IL: Garrard Press.

Walker, J.M. and Gingold, E.B. (1993). Molecular Biology and Biotechnology. Panima Educational Book Agency, New Delhi.

Wang, D. Cooney, C., Demain, A. Dunnill, P. Humphrey, A. and Lilly, M. (1979). Fermentation and Enzyme Technology, Wiley, New York. p. 185.

Williams, R. (ed.) (1981). Genetic Engineering, Vols. 1–3, Academic, New York.

Index